T0340098

PERFORMANCE VEHICLE DYNAMICS

PERFORMANCE VEHICLE DYNAMICS

Engineering and Applications

JAMES BALKWILL

Department of Mechanical Engineering and Mathematical Sciences, Oxford Brookes University, Oxford, United Kingdom

Butterworth-Heinemann is an imprint of Elsevier
The Boulevard, Langford Lane, Kidlington, Oxford OX5 1GB, United Kingdom
50 Hampshire Street, 5th Floor, Cambridge, MA 02139, United States

Library of Congress Cataloging-in-Publication Data
A catalog record for this book is available from the Library of Congress

British Library Cataloguing-in-Publication Data
A catalogue record for this book is available from the British Library

ISBN: 978-0-12-812693-6

For information on all Butterworth-Heinemann publications
visit our website at https://www.elsevier.com/books-and-journals

Publisher: Matthew Deans
Acquisition Editor: Carrie Bolger
Editorial Project Manager: Carrie Bolger
Production Project Manager: Anusha Sambamoorthy
Cover Designer: Miles Hitchen

Typeset by SPi Global, India

CONTENTS

FOREWORD

'Everything should be as simple as possible but not simpler'. I make no apology for starting the foreword to this book with an aphorism, particularly as this one is one attributed to that great thinker, Albert Einstein. While Einstein was no doubt referring to the complexities of theoretical physics, his wisdom is equally applicable in other areas.

One such area is teaching, something James Balkwill excels at as is witnessed by the awards he has won for his teaching modules. It is no coincidence that the latest of these awards was for his Advanced Vehicle Dynamics module. The skill of teaching is to transfer the requisite amount of knowledge to the student. That transfer should be no more than is necessary at the time lest it confuses and certainly no less than is appropriate as a gap in knowledge negates the whole purpose of learning.

It is my belief that Einstein's wisdom is equally applicable in the field of vehicle modelling. When designing a vehicle, for example, it is remarkable how close one can get to an acceptable weight distribution using a simple two-degree-of-freedom model with linear tyre characteristics. Of course, such an approach will not tease out the nuances of limit handling, but nevertheless, it has its place in the design process. Even in the mid-1970s, when, straight out of university, I began my career designing racing cars, I was able to use calculation to guide my designs. The results spoke for themselves, and while today it is expected of race car designers to have knowledge of vehicle dynamics, in those days, it was rare. It was this knowledge that helped me gain the competitive edge.

In this book, James disseminates that knowledge by taking us through every significant aspect of vehicle dynamics giving, in each chapter, just the right amount of information to allow the reader to become a competent vehicle dynamicist while also sowing the seeds necessary to give further understanding. Those seeds can be grown using the Internet research suggestions, directed reading and learning projects that are detailed at the end of each chapter. If this is not enough, the reader is prompted with keywords for each section of the book to allow further information to be trawled from the depths of the Internet.

Chapter 1 gives some fascinating insights into the history of racing from recorded origins dating back 3000 years through to the early twentieth century and in doing so emphasises that the spirit of competition applies to everything that has wheels and in fact has always done so. I am also pleased to see that it pays homage to some of the great names that played such an important part in the early development of the scientific approach to vehicle dynamics: names like Maurice Olley, Leonard Segel and Bill Milliken. I am a proud owner of a copy of the original papers published by Segel, Milliken and others and was honoured some years ago to meet Bill Milliken. It is fitting that a book such as this recognises the important role that these men played in developing the science that we exploit on a daily basis.

Chapter 2 takes us into the world of tyres. These are one of the most fundamental and yet one of the most difficult aspects of vehicle performance to understand. James introduces us to the mechanisms of hysteretic and chemical grip and gives us a brief overview of the thermal aspects of tyres before discussing the important concept of the friction circle and the forces and moments reacted through the tyre. He also introduces us to tyre modelling with particular reference to the Pacejka model, still one of the most important and useful tyre models whose empiricism is both its strength and its weakness.

In Chapter 3, we look into the important aspects of weight transfer. After the generation of tyre force, there is no singular subject that has a greater effect on the handling of a vehicle. Over the many years that I have spent in Formula One, I have always encouraged drivers to understand the basics of weight transfer. To many, it is something that comes naturally without any detailed understanding, but even amongst the great drivers, including many world champions that I have had the privilege of working with, I have found that a brief seminar on this subject has improved their lap times and enhanced their ability to utilise the tyre to its fullest potential.

In Chapter 4, we have an easy introduction to the basic equations of motion that govern straight-line performance. It is an ideal way to bring the reader into the physics and mathematics of vehicle handling before Chapter 5 that concerns itself with, to us racers at least, the far more important subject of cornering. Of course, the equations get slightly more complex, but again, James is able to introduce the complexities in an orderly fashion that gives the reader confidence to progress to the next stage.

I am particularly pleased to see in this chapter an introduction to derivatives. They were the analysis technique of choice of those early heroes of vehicle handling who had come from the world of aircraft dynamic analysis, an area where derivative notation was in common use. It is a practice I was taught by the great Professor John Ellis at Cranfield University as a student. It was eminently suited to the days before we had the computing power that we take for granted today, but even now, it is an often overlooked and yet fundamentally sound analysis technique.

In Chapter 6, we return to the straight line to examine the physics of braking, reminding ourselves that Newton's laws are the fundamentals of all equations of motion.

To many people, suspension design is the basis of obtaining satisfactory vehicle handling, and of course, in the broadest sense, there is truth in this. Chapters 7 and 8 recognise this dealing in turn with the kinematics and then the dynamics of those vital elements that connect the sprung and unsprung masses.

Thus far in the book, we have investigated subsystems, all of which are important to vehicle performance, but none of which, on their own, give us the knowledge of lap time or understanding of the vehicle characteristics that we require. This is remedied in the final chapter where the author amalgamates the knowledge meted out thus far to determine the true performance of the vehicle. In racing terms, this amounts to minimum lap time manoeuvres, something that exists in the time domain, but even in racing, I continually encourage performance engineers to think in the frequency domain and combine those elements that we learnt in Chapter 5 to aid our path to the minimum time we seek in Chapter 9.

Overall, this excellent book shows us that engineering is the understanding of physics, underpinned by the manipulation of mathematics, to optimise physical systems. In this case, it is vehicle performance that is the object of our attentions and that is something that has been dear to my heart over the forty years that I have been involved in professional motorsport. I hope and believe that this work will guide another generation along the lines that have served me so well.

Pat Symonds
Monaco

PREFACE

Vehicles with wheels aren't going to go out of fashion any time soon. Quite the reverse, with the emerging trends in automotive engineering for the replacement of internal combustion engine vehicles with hybrids and purely electrical vehicles, there is even more engineering needed on vehicles now than ever before. My aim in this text is to give you the knowledge about vehicles that will not pass away, will always be with us and is the foundation upon which the rational vehicle design approach stands. I hope you enjoy it and, more importantly, that you learn from it.

The text is drawn heavily from notes I have prepared for my students studying Vehicle Dynamics at Oxford Brookes University. The University has a strong presence in Motorsport and Performance Vehicle design and attracts very able students who almost all of whom go on to work in the industry, be that road or racing. Students of mine have gone on to work in every F1 team, and I often joke this now means that, whenever I watch a Grand Prix, I always win! Other students have gone on to work in a wide range of racing organisations and performance vehicle manufacturers and in a full range of rolls. It has been an absolute joy teaching vehicle dynamics all these years, and I am much indebted to all my students. It might sound a funny thing to say, but I have learnt so much from teaching them!

James Balkwill, Principal Lecturer
Vehicle Dynamics, Oxford Brookes University, Oxford, United Kingdom

WEBSITE

There is a website especially for readers of this text. You are asked to respect copyright issues when using material from the site. It is for purchasers of the text and not anyone else. The site contains more material, videos to further explain points, questions, etc. There is also a forum where readers can exchange questions and points of view. Before you can access the website, you need to send an email to the following email address:

AdvVehDynamics@gmail.com

You will then receive instructions for enrollment, and once completed, you can visit the website.

ACKNOWLEDGEMENTS

I would like to take a moment to thank those involved in helping me with this text. Elsevier, of course, for taking it on and my editor Carrie Bolger for all the help and guidance. A big thank you to Elizabeth for proofreading, which really was quite an undertaking. A special thanks to Pat Symonds for doing the foreword. Pat is an extraordinary engineer with two particular gifts, first for connecting measured performance trackside into design improvements on the car and second for managing people. He has been tremendously helpful in the development of the courses at the University and has hired its students right back to our very first cohorts in the subject.

A big thanks also to those far too numerous to mention in the motorsport and automotive community in general who have always been so helpful and supportive. Formula student deserves a mention too. For many years, I built up and ran the University team and am indebted to all who helped; it became and still is a very successful team with learning very much a core value.

George, Leo, Harry, Helen and Freya also get thanks, not because they lifted a finger to help with the book, but because they are my children and grandchildren of whom I am understandably very proud. Thanks also of course to the wider circle of family and friends for all the encouragement along the way. Lastly, another extra thanks to Elizabeth for being such an excellent mother, wife and codriver in life's lengthy Grand Tour.

CHAPTER 1

Introduction: Man and Cars

Cars are so vastly much more to us than simply an engineering artefact. For some, they are objects that engender passion, objects to be revered, perhaps because of a fascination for the rich and varied range of different marques, each with their own personality. Or perhaps as racing thoroughbreds in which rivals bitterly contest each race with their hearts set on the chequered flag. For others, still they are merely a utilitarian means to the end, that of transport. Everyone seems to have a view of some kind.

In response to developing technology, cars have been transformed beyond all recognition. In racing, performance is now hundreds of times better than when vehicles were first raced. In response to environmental pressures, vehicle emissions have been greatly reduced to the extent that some now believe the global cow population to be more problematic than cars from this point of view. Within a few years, we shall see most of the global fleet swapped for electric vehicles, and there is more going on now with the evolution and development of vehicles than there has ever been in the past.

Beyond all this, a car symbolises freedom and liberty; we express ourselves through their design. They are a part of our culture; they feature in our literature and our music. In *The Great Gatsby*, Scott Fitzgerald's story pivoted on exactly who was driving Gatsby's car when it crashed. When Janis Joplin asked the Almighty for a gift in her most famous song, she didn't mention climate change or want money or even ask for peace on earth and an end to war—what she wanted was a Mercedes-Benz!

Cars are ubiquitous and all pervading, central to economies, essential for transport; we live our lives through with them. Some of us have even been born in a car, but thanks to modern engineering, fewer of us than ever will come to die in one. Love them or loathe them, wheeled vehicles have a history and a future as long as our own. They are here to stay.

1.1 THE FIRST WHEELS

The earliest known wheel was made in Mesopotamia, now Iraq, and dates from 5000 BC. People often think that the wheel would have simply been

Performance Vehicle Dynamics
http://dx.doi.org/10.1016/B978-0-12-812693-6.00001-8

Fig. 1.1 The Bronocice pot with the first image of a wheeled vehicle.

obvious once a society reached a certain level of sophistication, but there is no evidence for this. The Inca and Mayan people, for example, never used wheels at all. It took time for the invention to spread. The first depiction of a wheeled waggon is on the *Bronocice pot* found in southern Poland and which dates from around 3400 BC. It shows a highly stylised image of a waggon; its wheels splayed out, viewed from above (Fig. 1.1).

Wheels were used before this time for making pottery, and despite the Bronocice pot being the first depiction of a wheeled vehicle, there were clearly vehicles in existence before that time. Wheeled vehicles have been found in Mesopotamia and Central Europe dating to around 4000 BC. In China, the wheel is thought to date to around 2000 BC. In the United Kingdom, it is around 1000 BC, and one was found with a horse skeleton nearby, no doubt because it was drawn by a horse (perhaps that particular one). However, there is no clear recorded origin for the invention of the wheel, and perhaps, there just isn't one earliest example from which the others stem.

The approach to design is started by simply making solid wheels from wood and gradually progressed through iterations involving the removal of material, the development of spokes, the use of multiple materials and the improved techniques as technology itself advanced.

1.2 RACING VEHICLES

It is a characteristic of humans that, in general, we love a good race, and so, it is not too fanciful to imagine that very soon after the wheel was invented, people started racing their 'vehicles'. The first literary evidence of racing wheeled vehicles dates to around 800 BC. The *Iliad*, by Homer, is not only the first literary evidence of racing but also one of the earliest works of Western literature. Given our sporting tendencies and the fact that the Iliad itself came down from a long oral tradition, it is highly likely that people had been racing for dozens of generations before The Iliad was written:

> *The drivers cried out as they set off. They moved very quickly. Each man held on tight, his heart beating with excitement. Everyone cheered their teams. As they raced down the final straight, all the drivers tried to win. One drew ahead, but the next was close behind.*
>
> ***Homer—The Iliad c. 800 BC***

The above quote could apply just as much to a modern F1 race as to a chariot race in ancient Greece even though it was written nearly 3000 years ago. Indeed, the ancient chariot designers faced many of the same design challenges we face today. The quest for reduced weight, more power and more grip were just as relevant then as today. We also know that the sport was plagued with all the difficulties of modern racing.

Ancient chariot teams fed drugs to the horses to boost performance, and some races were fixed. Rival teams were sabotaged, and illegal financial deals were rife. The earliest recorded incident of cheating in wheeled racing was of the Emperor Nero, who bribed judges to declare him the winner of a chariot race. Given that he actually fell out the chariot and failed even to complete the course, foul play must have been fairly obvious to all (Fig. 1.2)!

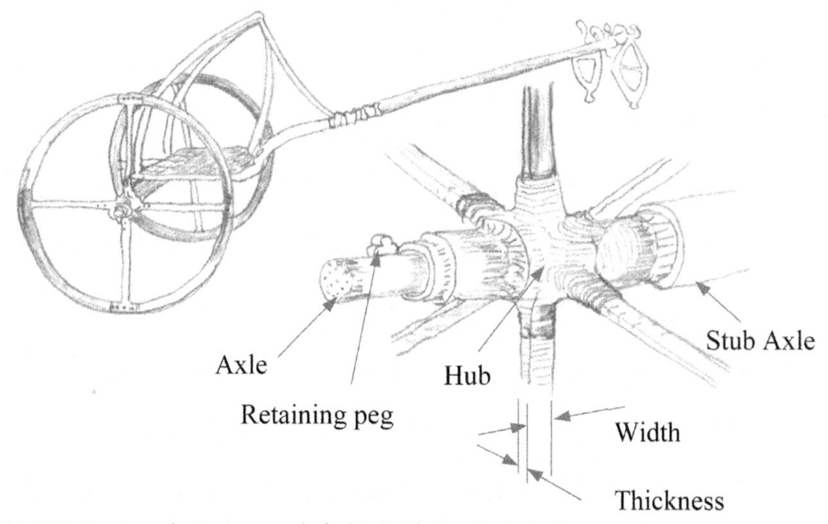

Fig. 1.2 Ancient designers made light rigid structures just as we do now.

The Egyptian chariot above shows many design approaches that are as appropriate today as they were then. The structure is made to be as light as possible yet stiff at the same time. Many different kinds of wood were used, each appropriate for the conditions pertaining to the location of its use. Heavy hardwoods are for the rims, but lighter and less rigid softwoods are for the handrails, for example. The axle detail is similarly skilful. The axle mounts the retaining peg that keeps the hub against the stub axle. The spokes have much greater width than depth, making them much stiffer in this direction and so better able to resist the loads caused by cornering. It is easy to look at an artefact such as this and think it's crude and badly designed but that would be completely wrong. The very best of our current designs will no doubt look rather pale in three thousand years time but that doesn't alter the fact that they are the best we can achieve right now.

In exactly the same way that the invention of the wheel prompted people to go racing with chariots, the invention of the internal combustion engine prompted them to go motor racing. In 1894, an international race was organised in France, and in the following year, Chicago hosted the first American motor race.

Many races were held on the open road and were very dangerous affairs. Indeed, all the accidents led them to be banned. The Milwaukee Mile was a 1 mi horse racing oval and became the first track to feature motor racing in 1903. Wheeled vehicles driven by petrol engines have been raced there ever since. The first purpose-built motor racing circuit was Brooklands in Surrey in Southern England and opened in 1907. The first race was a 24 h event won by Selwyn Edge. Edge was a keen cyclist and had become friendly with David Napier whose company built the car in which he won. Racing performance at the time was actually very good, and it would have been a terrifying experience for most ordinary people to drive the cars of the day.

With today's Formula 1 Grand Prix being considered so much a pinnacle of the racing world, there is not surprisingly some disagreement over which race was actually the first Grand Prix proper. 'Prix' literally means 'price' or 'prize' and, in this context, 'an annual large prize for a race', so confusion over which was the first Grand Prix is inevitable. The French Grand Prix in 1906 is a good candidate for the first ever race and was won by Ferenc Szisz. The race lasted just over 12 h. The cars reached peak speeds of over 90 mph on unmetalled roads featuring pot holes, and the ride must have been unbearable. It was also a very hot day, and the midday temperature reached 49°C. Modern Grand Prix drivers have a much easier time of it! Ferenc's car featured a revolutionary approach to wheel design that allowed a wheel change in only 4 minutes instead of the customary 15 for the rest of the field. Given that the car suffered eight punctures, this must have played a part in the win. Astonishingly, Szisz remains the only Hungarian ever to have won a Grand Prix.

Technical progress was rapid, and within two decades, the cars had power-to-weight ratios of nearly 200 bhp/t. Circuit design was poor, with spectators placed on the outside of bends. Crash structures and barriers were not required, and accidents increased as time went by and speeds increased. Then in 1955, at the LeMans race, motor racing had its worst accident ever when a car crashed through the ineffective fencing on the edge of the track at 120 mph and entered the crowd before finally coming to rest. Nearly a hundred people, including the driver, were killed and around a hundred more injured. The accident prompted new approaches to safety.

Today, of course, things have moved on, and modern Grand Prix cars routinely exert over 4 g and have power-to-weight ratios of around 1500 bhp/t. The cars reliably perform week in week out. Safety is a primary consideration, and the modern record is in a different league when compared with the past. Crashes still happen, but serious injury and death are extremely rare.

People often think that the technical developments made a century ago were much more rapid then than now, but it's not true. It is true to say that one can look at a very well-designed racing car from some decades ago and see what needed to be improved to make it better, but one can't somehow do the same with today's cars. In the last 25 years or so, for example, the time racing drivers spend using the brakes has been halved. Modern developments are incremental, and big advances are made by completing many little improvements. It may be difficult to look at a modern racing car and see where the improvements will come from, but they are all in there somewhere.

I'm often asked whether there is any social benefit from racing. Are road cars better because of racing? Isn't it all a bit of a waste of time and money? There are two main points to bear in mind here. Firstly, it is a sport. Sport is one of those undertakings that separate humans from all the other animals. The ancient Greeks revered sport. The original Olympic Games were a celebration of sport that spanned over a thousand years. It was believed that no one could win without favour from the Gods themselves. Success in sport was the pinnacle of achievement. In many ways that is still true today. Sport commands enormous television audiences, and no one can deny that special status that goes with competing for one's country. Motorsport certainly is an expensive pastime, but the total amount spent is very small in comparison with plenty of other endeavours, for example, football, where just a handful of clubs can have a worth comparable with the entire F1 business.

In answer to the question about whether motorsport produces worthwhile benefits, the answer is a resounding 'yes'. The crash performance of road cars is vastly better now than it was largely due to improvements made mandatory in motor racing. The advanced braking system, ABS, was invented for racing, and there are dozens of other examples. Tyre developments for racing have resulted in a marked decline in the number of blowouts whilst driving, greatly increasing road safety. It is often the case that this happens behind the scenes with neither side seeking publicity, only expertise at the cutting edge in return for material help. Any endeavour that pushes back the limits of understanding will always move knowledge and performance forward generally and so drive-up standards. Of course, motorsport produces technical benefits, lots of it.

1.3 ROAD VEHICLES

If racing is a very high-profile use of wheeled vehicles, transport is a vastly more prevalent one. In the late 19th century, when for the first time the internal combustion engine became the prime mover for wheeled vehicles, we entered a period of exceptionally rapid technological development, just as happened for racing cars. In the 1880s, vehicles, such as there were, offered nearly 10 bhp/t and were capable of over 10 mph. The design of these vehicles drew heavily on the horse-drawn carriages that preceded them, and they were even called *horseless carriages* (Fig. 1.3).

The design approaches taken by early engineers may seem very poor by modern standards but in fact were inspirational and revolutionary at the time. Scientific papers were published on all aspects of car design. For

Fig. 1.3 The horseless carriage really was literally a horseless carriage.

example, Frederick Lanchester [1], the founder of the Lanchester Engine Company that sold cars, published over 50 papers including one that referenced a particular point on the chassis associated with the suspension, something he called the *sideways location*. This was the earliest reference to the roll centre.

A similar process played out for road cars as was experienced in racing, although the increase in performance as the decades went by was not matched by a proportionate increase in safety. Approaches that are unthinkable now were common practice then, for example, no seat belts, no joints in the steering column and no rollover structure in the roof. In 1965, a book was published by Ralph Nader [2] *Unsafe at any speed* that altered the course of road car design. The book detailed the dreadful design practices being used and focused public attention on the fact that cars could very easily be designed to be much safer. Consumer behaviour changed, and safety became a major issue. Seat belts become compulsory. A graph of deaths per year on British roads shows that (excluding the data for WW2) deaths peaked at around 8000/year in the mid-1960s and have fallen to around 1700/year at present, despite the fact that the number of cars on the road has increased by a factor of six over the same period. Overall, this makes cars around 30 times safer now compared with then. The buyer of a new car now has a vast range of very good designs to choose from. Dynamic performance is at an all time high and luxury similarly so.

1.4 TYRES AND RUBBER

The word tyre was originally a shortened form of the word *attire* meaning *the clothing worn* or *to be clothed in*. Its shortened form, *tire*, was used to name the rubber 'dressing' on the metal rim of a car, and for this reason, the American spelling is etymologically more appropriate. In the United Kingdom, the spelling 'tyre' has been normal since the early decades of the 20th century and, like so many other words, has a US and UK version in common use.

Rubber was originally extracted from rubber trees in South America. For thousands of years, the native South Americans extracted the latex and used it to produce a basic watertight shoe. It was even formed into spheres and kicked around as a sport. When the European invaders arrived, the material was taken back across the Atlantic and so began a long story. Initially, the public were fascinated by the new material, but before long, the novelty wore off, and people are tired of the sticky material that lacked any obvious practical application.

However, an entrepreneur called Goodyear became almost obsessed with the material and set about trying to heat-treat and chemically modify the material into something useful. He eventually managed to produce a vulcanised rubber very similar to its modern counterpart, but when he patented it, he found he was beaten to the UK patent by a rival. In the end, neither he nor his family gained any material benefit from the vast Goodyear Company that was named in his honour. In the 1920s, synthetic rubbers were developed (though natural rubber is still widely used in tyre manufacture), and modern tyres now make use of a wide range of polymers and reactants to produce the compounds required. When rubber is collected from a tree, it is a very sticky and adhesive material that requires significant chemical additions and heat treatment to form it into the useable rubber we know in tyre form. An excellent text with much more on this amazing story was written by Paul Haney [3]. People often talk about *compound* when discussing tyres. In essence, the choice of compound is simple. The softer it is, the better the grip will be, but the shorter the life and vice versa. Each manufacturer has their own set of compounds and is able to provide customers with advice about which compound will suit which application.

1.5 VEHICLE DYNAMICS—THE APPROACH TAKEN IN THIS BOOK

And now, we get to the important bit, the engineering. This book is for anyone wanting to understand, in a deep and meaningful way, the dynamics behind cars. It might seem impossible to write down an equation that describes how a car goes around a corner, but it isn't impossible, and in Chapters 3–5, we shall see how to do exactly that. You really can calculate things like how stiff an antiroll bar should be in order to get a car around a corner as quickly as possible. It is the word *calculate* here that is so important. There are plenty of mustard keen car enthusiasts wanting to make great cars but who know nothing of the language of mathematics; they may say things like *if we stiffen the rear we'll get more grip* or *less damping for straights, more for corners*. They may or may not have a point. The issue is that until you put numbers on things you're wasting your time, you're just playing about.

Here is an example I often use with my students. Suppose you are at a race, as a race engineer for a world-class driver. The driver comes into the pits and complains that the car is poor on corner turn in. The first approach you could take is to apply a general, nonmathematical knowledge of cars based on magazines, hearsay and pit-lane small talk. You might, for example, stiffen the dampers, maybe one or two clicks on the high speed setting, or perhaps adjust the antiroll bar, even make changes to the tyre pressures front to rear. Maybe these things will help but maybe they won't. How could you know?

Instead, there is a vastly better second approach you could take. You could use a thorough understanding of cornering dynamics to determine the theoretical best turn-in response that this particular car is capable of achieving. After this, you could make use of telemetry to measure what yaw rate is actually resulting from the steering input. If you find that the driver is failing to produce this best theoretical response, then you need to have a word with him or her because it isn't the car that is the problem, it's the driver. If, however, the driver is on the limit, then nothing can be done with this car, and only redesigning and making a new one will offer improvement. (This assumes that all adjustable settings on the car are at their optimal position, which of course, using dynamics, you will have arranged to be true.)

In the first approach, there is no science, no numbers, its all just guesswork. In the second, we see the rational application of engineering theory and its validation in measured data, which is clearly the proper way to proceed.

One of the first engineers to really bring analysis to bear on vehicle dynamics was Maurice Olley [4]. His analysis and approach was to consider the car as a whole, and he produced equations for the performance of the whole car. Another notable contributor was Leonard Segal [5], whose paper on the response of a car to steering control input brought the mathematical analysis that gave us understeer and oversteer. The Institute of Mechanical Engineering awarded Segal the Crompton-Lanchester Medal for the paper, and Segal became the first person outside the United Kingdom to win it.

The aim of this book is to impart all the understanding and knowledge you need in order to be able to model and accurately predict performance so that you can rationally determine changes that will bring improvement. There's nothing in here about the latest racing technological must-have fashion or the current state of the art for electric vehicle batteries, etc. This text is about the underlying truths that will be used by vehicle designers hundreds of years from now. Vehicles obey Newton's laws of motion [6]. Unless we get cars up to a very significant fraction of the speed of light (not really practical), we simply don't need the refinements brought by relativity, so Newton it is from here on.

We start in earnest on vehicle dynamics with Chapter 2 on tyres. These are such simple devices in principle and yet so complex in practice. They are responsible for generating all the forces that accelerate the car, longitudinally in acceleration or braking, and also in cornering. Once one understands tyres, a knowledge of vehicle dynamics is used to ensure that the chassis provides them with the ideal conditions so that the forces they generate are as large as possible and, for road cars, the ride and tyre life are as good as it can be.

Chapter 3 deals with weight transfer, the name given to the change in vertical loads that tyres experience as a result of acceleration. It's vital to understand it, and it compromises performance (except for one very special situation). Ideally, we would get rid of it, but we can't. So, we shall have to settle for understanding it and doing the best we can.

In Chapter 4 on straight-line acceleration, we begin mathematical modelling in some detail and learn how to produce mathematical models that offer excellent performance prediction.

Chapters 5 and 6 extend this to cornering and braking. Cornering particularly requires complex analysis, and there is plenty of maths here. It's a much more complicated process than most people think.

Chapters 7 and 8 deal with suspensions. This is a large topic, and it is divided up into suspension kinematics in Chapter 7 and dynamics in Chapter 8. The kinematics involves much less maths and is about how suspensions articulate as the vehicle rides over terrain or accelerates and corners. Chapter 8 deals with the optimisation of suspension parameters such as springs and damper rates. This is a complex task, and there is a tremendous amount of computer modelling that comes out of this chapter. Overall, we have two goals, to optimise the dynamic loads on the tyres so that they return the best possible horizontal forces and to reduce the communication of vibrations originating from the road into the vehicle body.

Lastly, in Chapter 9, we see how all this knowledge is brought to bear on computer-based lap-time simulation, vehicle manoeuvre simulation and full-vehicle simulation. This book is by no means a complete treatment of these topics; that would be too much for a text such as this. In many ways, these topics are an end point for vehicle dynamics, something to be undertaken once the subject is understood, and so, it makes sense to finish with them.

CHAPTER 2
Tyres

2.1 INTRODUCTION

In this text, we are concerned with vehicle dynamics that is the study of vehicle motion. Necessarily, the motion of a vehicle results from the forces that act on it. The vast majority of the forces that act on a vehicle originate from the tyres. The few that don't are the gravitational load, aerodynamic downforce and drag. So, it is that most of the forces acting on a vehicle come from the tyres and if we are to get the best out of the vehicle that means getting the most out of the tyres.

A good way to think of a tyre is to consider it to be a *six-degree-of-freedom* element that acts between the road surface and the stub axle to which the wheel is mounted. Since tyres are elastic elements, the forces they develop depend on the deflections they sustain in each of these six degrees of freedom. It is easy to see that a tyre acts like a spring element in the vertical direction, a load on the tyre causes it to deform, and as the deformation proceeds, there develops a force to resist it. The circular nature of the tyre means that as the deformation proceeds the contact patch grows in size and the vertical load increases much more than linearly with deformation. Similar effects apply in plunge when the stub axle is deflected linearly along its axis and also forward and aft under the action of tractive and braking effort. On top of these three linear deflections, there are three rotational ones. Imagine a tyre with a tractive effort being supplied; clearly, the tyre will deflect rotationally under this torque. Further to this, a tyre can be rotated about a vertical axis, as for steering. When this happens, the tyre contact patch will deform and, as we shall see, will produce a torque to oppose this impressed motion. Lastly, in the front view, a tyre may rotate about the contact patch and lean in or outwards. Again, there will be elastic deformations in the contact patch, and these will serve to resist the overturning. Taken together, the tyre therefore offers compliance in all six degrees of freedom.

Performance Vehicle Dynamics
http://dx.doi.org/10.1016/B978-0-12-812693-6.00002-X

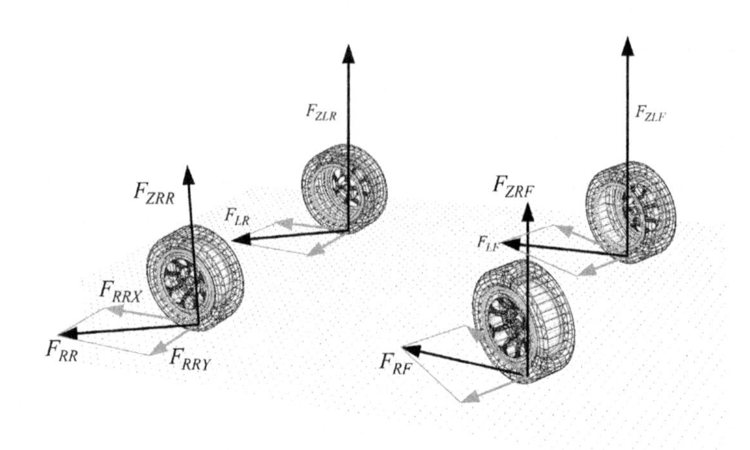

Fig. 2.1 Forces generated by the tyres of a vehicle.

Since the forces that result from displacements in these six degrees of freedom are responsible for accelerating the vehicle, it is this very acceleration that we seek to optimise and control; it is clear that we must understand the origin of these forces and so understand what must be done in order to get the best from the tyre.

We shall start with Fig. 2.1 that shows the forces generated by the tyres of a vehicle. It shows a vehicle that is both braking and on its way into a right-hand corner. The front wheels clearly show the vehicle is turning towards the viewer. The vehicle weight is supported by the four vertical forces. These are all labelled with an F for force and then subscripts that start with Z for the vertical direction followed by an R or L for the *left* or *right* of the vehicle and finally an F or R for *front* or *rear*.

In the figure, we can see that the vertical load is larger on the outside wheels than on the inside as we should expect for a car in corner where the chassis rolls towards the outside under centripetal force. This progresses until compression in the outside suspension springs, and extension in the inner ones brings the chassis into roll equilibrium. There is a larger total load on the front axle than the rear, and for many road cars, with engines in the front, this is normal. However, in this case, with the vehicle braking, we have extra load on the front axle, and the car body will be pitching forwards and so compressing the front springs and extending the rear ones. Again, this will proceed until equilibrium in pitch is restored. In addition to this, we can see horizontal forces generated in the plane of the road. These are labelled F with a subscript starting with R or L for right or left, followed by an R or F

for front or rear. We notice that these four forces, F_{RR}, F_{LR}, F_{RF} and F_{LF}, are neither pointing in the forward-aft or the left-right direction. If the vehicle were only decelerating and not cornering, then we should expect all the forces to point rearwards. If it were cornering in the sense shown and neither accelerating nor braking, then we would expect the forces to point towards the corner centre.

If we think through the situation in Fig. 2.1, it is clear that the forces being generated must be changing rapidly as the cars moves through the corner and conditions change from heavy breaking, through the corner apex, where the forces perpendicular to the cars axis will be greatest, and on to acceleration out of the corner. The horizontal force can be divided up into two components, one longitudinal and one transverse. These are shown for all four wheels and labelled on the rear right. These are F_{RRX}, which acts in the '*X*' direction, and F_{RRY} that acts in the '*Y*' where *X* and *Y* are inline with the tyre and perpendicular to it, respectively. This is a useful step because the mechanisms by which the tyre generates each component are different. It is also helpful because, as we shall see throughout this text, the dynamics of cornering, acceleration and braking are all different. These are all blended into one response for the whole car, but each is separate. Given this view, we can clearly see that if the car is to perform well we want the four forces, F_{RR}, F_{LR}, F_{RF} and F_{LF}, to be just as big as can be arranged.

This viewpoint is also very informative in our consideration of vehicle dynamics. These four contact patches are the only place where the forces that accelerate the car, and so define its performance, are created. In a very real sense, performance vehicle dynamics is all about arranging for the tyres to offer this collection of forces in the best way that they can. In the rest of this chapter, we shall understand how the tyre generates these forces, and in the later chapters, how the forces result in the vehicle motion and therefore how to arrange them for the best performance.

Fig. 2.2 shows the force and moments acting on a tyre. Given that the tyre is composed of inflated rubber and is compliant in all three displacements and rotations with respect to the contact patch, it is clear that it is a *six-degree-of-freedom* component. Movements in these dimensions can be surprisingly large (Fig. 2.3).

In the picture above, the Oxford Brookes University Formula Student car can be seen taking a corner hard with a lateral acceleration of around 1.3 g as measured by the on board telemetry at the time. The tyre on the right of the picture is seen to be very distorted at the contact patch, and the stub axle had moved nearly 30 mm in plunge along the spin axis. We

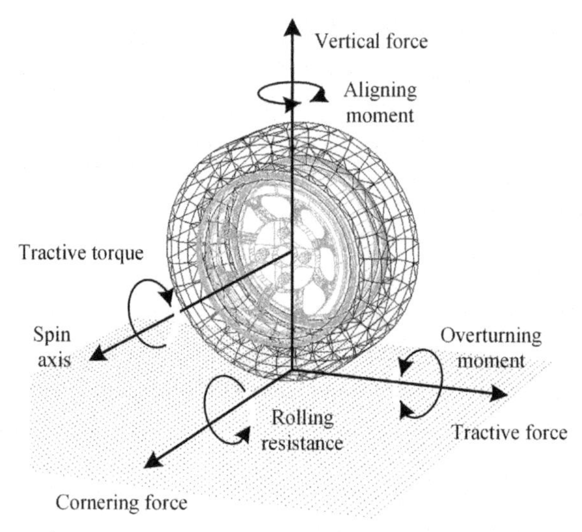

Fig. 2.2 Force and moments acting on a tyre.

Fig. 2.3 The Oxford Brookes University Formula Student car cornering hard. Driver: The late Craig Dawson, a very fine driver indeed.

can also see that this tyre is very well placed with an almost vertical alignment. In fact, the top of the tyre in the picture was leaning inwards towards the car by around one-and-a-half degree despite the cars distinctly rolled position in the opposite direction urging the top of the tyre outwards. As we shall see, this is a very good level of camber control. The tyre on the left of the picture is not nearly so well arranged. The contact patch is visibly smaller, and daylight can be seen under the tyre at the edge on the inside of the corner. In addition, it is clearly leaning with the top much nearer the car body than the bottom making the contact patch much worse. Rather than poor design, this is in fact it's very good design. The tyre on the outside is carrying almost all the vertical load of the front axle at this moment and so is producing almost all the lateral force. The inside tyre has virtually no load; the driver is at the limit, and if the speed had been increased at all, firstly, the inner tyre would have lifted off, and then, a little later, the outer one start to skid instead of holding the road. We can also see extensive wear happening to the tread of all the tyres. This is not in fact quite as good as it might be; the wear is more pronounced on the inner edges of the front tyres than the outer, and if it were more even, it would witness a better loading on the tread. It is however, pretty good, and we were pleased with the situation. One must remember that the designers of the car, very sensibly, used *static camber* to ensure that, in corners, the outer tyre is optimised. *Static camber* means having the front tyre leaning inwards at the top when lateral acceleration is zero. Consequently, under roll, the outer tyre adopts an ideal and nearly vertical position with the tyre top leaning inwards towards the corner centre. An additional consequence is that, when running straight, the inner part of the tread is contacting the ground and this therefore wears more quickly. All this vividly shows how much the tyres deform and move around under the loads they experience. In service, they are nothing like the relaxed toroid that they adopt when stationary.

In the development of our understanding of tyres, we shall therefore have two tasks. The first is to understand how the rubber-road material pair develops the very high coefficient of friction, *Mu*, that it does. In Table 2.1, the *Mu* values for some common material pairs are shown. We see that a racing tyre offers a very high value.

The second task is to consider, in turn, each of the six forces and moments that act between the contact patch and the stub axle and

Table 2.1 *Mu* values for some example material pairs

Material pair	***Mu***
Teflon-teflon	0.04
Graphite-graphite	0.1
Most metal-metals	~0.3
Brake pad-steel	~0.4
Wood-wood	~0.5
Road tyre-asphalt	0.8
Soft compound racing tyre-asphalt	~1.6

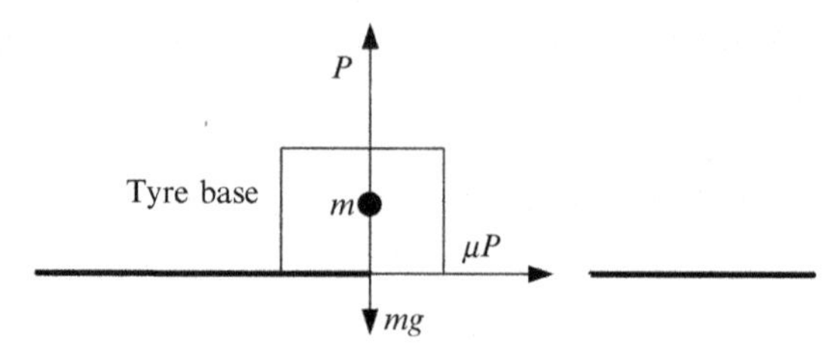

Fig. 2.4 The importance of *Mu.*

understand how the tyre produces these forces and what influences it so that we able be able to make a reasoned case for the maximisation of performance.

Before moving on to look at the mechanisms of grip, it is helpful to consider the importance of the coefficient of friction, *Mu*, by looking at Fig. 2.4. In the figure, a tyre is resting on a flat surface. The weight force, mg, is balanced by a force P caused by the elastic deformation of the ground, and the block is in equilibrium.

The largest horizontal force that can be generated by the tyre is given by μP, and this might be either a cornering, tractive or braking force or indeed a combination of these. The horizontal force will accelerate the car, and if it has a mass, m, the resulting maximum acceleration $\hat{a}$ will be given by

$$m\hat{a} = \mu P$$

and so,

$$m\hat{a} = \mu mg$$
$$\Rightarrow \hat{a} = \mu g$$

Thus, the largest acceleration a car can sustain is determined by μ times g, making the tyres central to the determination of performance.

2.2 MECHANISMS OF GRIP GENERATION IN TYRES

Friction occurs between two surfaces that are in contact and being urged to slide over each other in the plane of contact. Strictly, there are two types, *static* and *dynamic* friction. Static friction describes the force necessary to get the relative motion started, whilst dynamic friction describes the force necessary to keep the relative motion going once started. In general, dynamic friction is less than static friction, and if you observe how someone moves a heavy object, such as a filing cabinet, over a carpet, you can see this in practice. Normally, the person will jerk the cabinet suddenly to get it

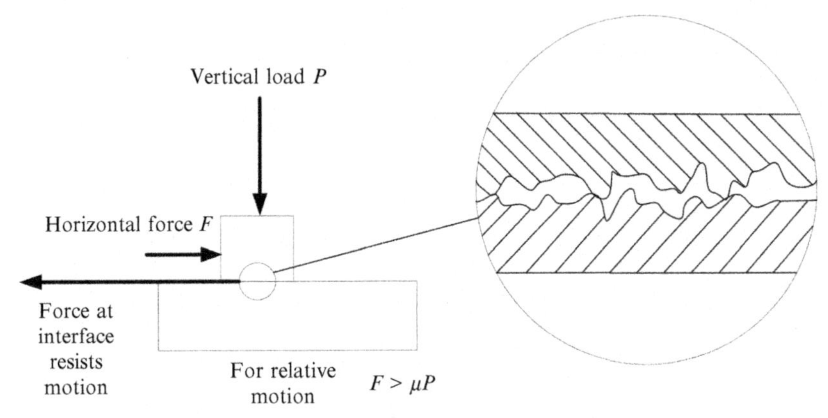

Fig. 2.5 Traditional model of friction.

moving, after which, a smaller constant force is supplied to keep it moving. The person moving the cabinet keeps the motion smooth, being sure not to let the cabinet come to rest even for an instant to avoid needing another strenuous jerk.

Under a microscope, even very smoothest surfaces show irregular shapes in the form of local peaks and troughs. This means that only a very small part of the two objects is actually in contact as is shown in Fig. 2.5.

Several frictional observations can be made from this. The idea of static friction is explained since it will take more force to make the surfaces to move in the first place than it does to keep them moving. This is because the surfaces must be forced apart until the peaks can pass over each other. Once there, keeping the surfaces moving is easier, since when in motion, the upper surface will rumble along and not come to a minimum between each peak since the momentum keeps the bodies in motion. For this reason, dynamic friction is less than static. Surface roughness can be related to friction, and even the rate of particle production (which results from the removal of peaks as one smashes into another) can be modelled.

Unfortunately, this simplistic model of friction does not account for the complex interaction between a tyre and road surface.

Fig. 2.6 shows how a tyre contacts a road surface. The tyre is relatively soft and locally deforms enough to make contact significantly with protruding parts of the hard aggregate. This aggregate is usually made from crushed rock and is constrained by the much softer asphalt. Some of the aggregate is always standing proud of the main surface of the road because, being much harder, it wears more slowly. As it stands proud, the rate of wear increases,

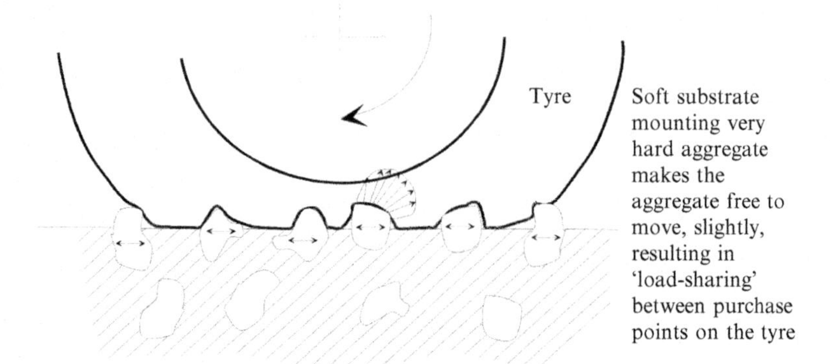

Fig. 2.6 The mechanisms of tyre grip are complex.

and so, a stable situation is reached where there is always roughly the same amount of aggregate protrusion.

- Mechanical purchase: As the aggregate indents the tyre, it engages with it in a way that is similar to a pinion engaging on a geared rack. This allows additional force to be generated at each indentation of the tyre. In the diagram, the forces acting on a single piece of aggregate can be seen, and the net effect is a reaction force pushing the tyre forwards, that is, to the right in this case. Under the tractive effort, any individual piece of aggregate will move a little in the compliant asphalt. This means that each neighbouring piece of aggregate will have load applied to it and so on. Thus, the tractive load is *shared* between all the protrusions of aggregate rather than just one, as would happen if the structures involved were very stiff. In addition to this mechanical grip, there is also the grip that results from the tradition friction model described above applied to a rubber-road material pair. This mechanical purchase provides around a third of the maximum grip and, importantly, is independent of the friction between the rubber and the road and still operates when this is low, for example, in the rain.
- Chemical bonding: As the wheel rolls forwards, new stones press into the tread rubber. In an ideal situation, the stone breaks into fresh rubber that has, so far, never been exposed. Fresh rubber in this condition is chemically reactive and bonds to the stone surface; this adhesive force is very strong, so much so that rubber is left on the track once bonded to it as can be seen on any racing circuit. In turn, this means that a much higher torque can be applied to the tyre before it starts to slide over the stone. Naturally, this also means that the next time that part of the tyre is indented by a stone, the rubber it meets will already have been exposed to the

aggregate and it will not form quite such a strong bond, mainly due to impurities having entered the rubber from the previous deformation. However, as the rubber wears away, each new penetration of aggregate will indeed meet new rubber even if this is just at the very tip. For this reason, we accept the penalty of rapid tyre wear since this progressively exposes new rubber. Thus, tyres for racing are designed to wear down much faster than those for road use. A much higher rate of wear is tolerated in order to get the extra grip. For a road car, this is not desirable since the tyres would wear out in an unacceptably short time. In any case, most road driving is nowhere near the limit, and so, the advantage conferred by a high friction coefficient is only an advantage for a very small fraction of the time. In racing, the car is at the limit all the time, and so, the advantage derived from increased grip is overwhelming.

2.3 FACTORS AFFECTING GRIP GENERATION

2.3.1 Temperature

Considering the two mechanisms of grip generation above, it is not surprising to find that temperature has an effect on grip. An increase in temperature will make the rubber softer and so allow it to form around the aggregate more easily, thus increasing mechanical grip. In addition, a warmer tyre will be more easily be penetrated by the aggregate, and the chemical reactions governing the bonding will happen more quickly making for increased chemical grip. However, increasing the temperature will also reduce material properties, and in particular, the shear modulus and yield stress in shear will drop meaning that the tyre will fail under lower load and so performance is reduced. At modest temperatures, for example, 50 or 60°C, the former effects dominate, and an increase in temperature results in better grip. Much beyond 100–120 degrees, the latter effects dominate, and an increase in temperature is accompanied by a reduction in grip. The typical performance of a tyre is shown in Fig. 2.7.

The effect is pronounced, and in Fig. 2.7, we can see that a change of only around 10 degrees will cause a drop in grip of around 10%. A cold tyre may have as little as half the grip of one at running temperature, and a racing car that can only develop half the grip of the competition will soon be lapped. On very cold days, it can be impossible to get the tyre warm enough, the load on the tyres doesn't result in enough heat generation, by hysteretic deformation of the rubber, to outweigh heat loss, and operating temperatures are too low.

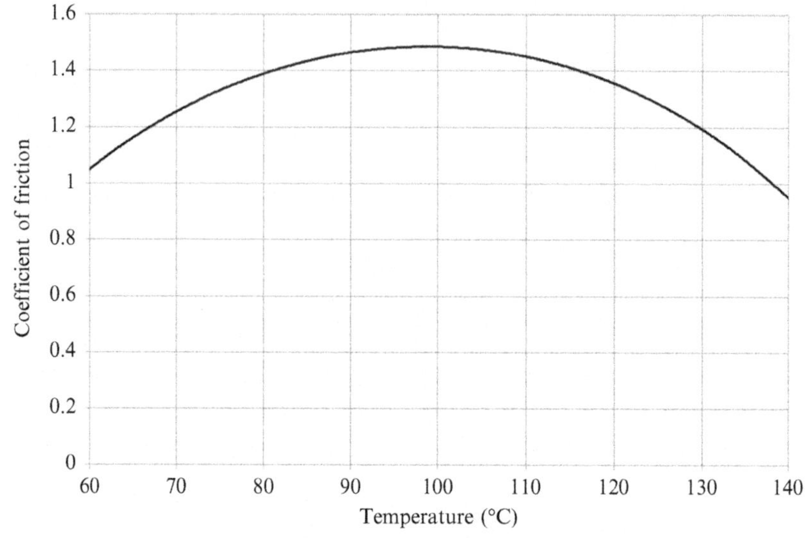

Fig. 2.7 Effect of temperature on grip.

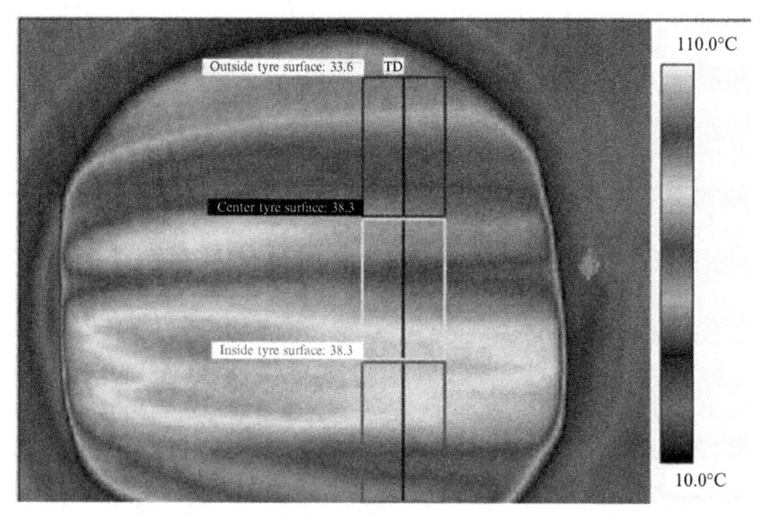

Fig. 2.8 Infrared image of the Oxford Brookes Formula Student car whilst racing.

Temperature is so important that whenever a formula allows it, teams will use tyre warmers consisting of electric blankets formed into tyre-shaped jackets that are applied to the car on the grid and removed as late as possible. Tyre temperatures may be measured using an infrared camera.

Fig. 2.8 shows an image taken of the rear left tyre of the Oxford Brookes Formula Student car whilst racing. The centre can be seen to be hotter than the edges, but given that the temperature gradient is only around 8 degrees across the tyre and it was a very cold day, the situation could have been worse!

One popular race-engineering technique when a car returns to the pits is to use a surface probe to measure the temperature profile across the tyre as quickly as possible. Clearly, this is less sophisticated than in instrumented approach but nevertheless gives a reasonably good indication of how evenly the tyres are being heated and a rough indication of the temperature they are reaching on the track.

Grand Prix drivers have a long tradition of swerving from side to side on a formation lap to try and generate heat in the tyres. The time stationary on the grid is also minimised in order to keep the tyres warm.

2.3.2 Inflation Pressure

An overinflated or underinflated tyre will compromise performance, and in this case, the reason is very simple; unless the tyre is correctly inflated, the carcass will be out of shape, and the contact patch area reduced.

In Fig. 2.9, the normal inflation pressure is shown on the left. Racing tyres are in general rather larger than the road-going equivalents per unit weight, and so, inflation pressures are proportionately lower, often between 8 and 14 psi. Overinflation leads to the position shown in the centre of the figure with the tyre running only on the centre of the contact patch. In response, the central region gets overheated since a level of shear force sufficient for the whole tyre passes though a reduced region. This reduces the

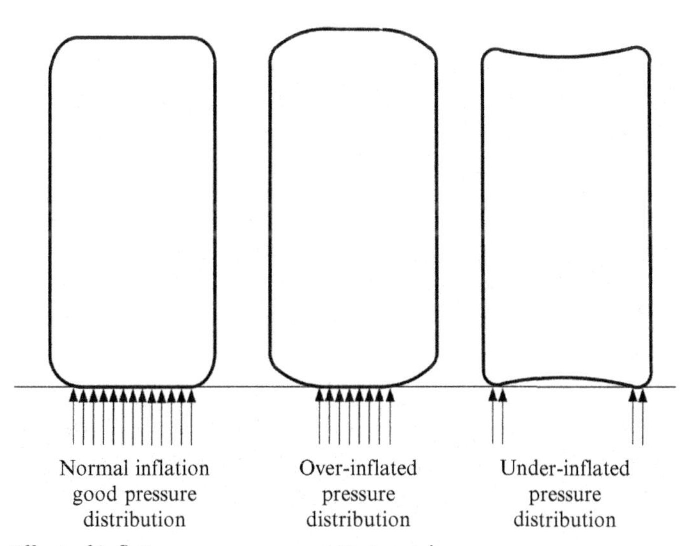

Fig. 2.9 Effect of inflation pressure on contact patch.

level of grip beyond that loss expected from the ratio of areas involved. The situation for underinflated tyres is shown on the right where the majority of the load is carried on the outer regions of the contact patch and similar problems are encountered.

2.3.3 Ambient Conditions

The nature of the surface the car is used on can have a significant effect. If, for example, at the start of a racing session, the track is dusty and devoid of any rubber laid down from previous laps, then the first cars out will be at a considerable disadvantage and will clean the track up for the later cars who will no doubt be much faster. Rain and temperature change are also clearly important, and often everyone is out together the situation is equal for all. However, there are many forms of racing where contestants use the track sequentially, such as drag racing or hill climb and here changes in conditions, and a team's ability to affect when they go out can be significant. In road cars, conditions also have a marked effect; the *Mu* value on a cold road in winter is very different to high summer. Further, the situation of the so-called *split-Mu* can pose problems. This is when one side of the car experiences a significantly different *Mu* value from the other; for example, a car on a country road in winter when the nearside tyre may be in icy water, whilst the offside is on good tarmac.

2.3.4 Tyre Carcass Design

The internal construction of modern tyres is complex and is the result of decades of development and competitive improvements (Fig. 2.10).

Working from the inside out, the tyre starts with an airtight inner liner to contain the air that inflates it when in service. Over this, a rubber layer is placed to distribute the load. The beads at the base of the tyre form a firm anchorage for the whole tyre against the wheel rim. In service, the internal pressure forces them against the wheel rim, and friction transmits the braking or tractive torque from the wheel rim into the tyre itself. The body plys consist of steel wires that curl around the beads for anchorage and run from one side of the tyre to the other. Viewed from the side, these wires run radially, and it is from this part of the construction that the *radial* tyre takes its name. A layer of latex is normally placed over the body plys, and on top of these, the shoulder pads start to build up the familiar shape of the tread region. Belts of steel of Kevlar are placed on top of this layer, and these serve to stabilise the contact patch and reduce the amount of defection it experienced under

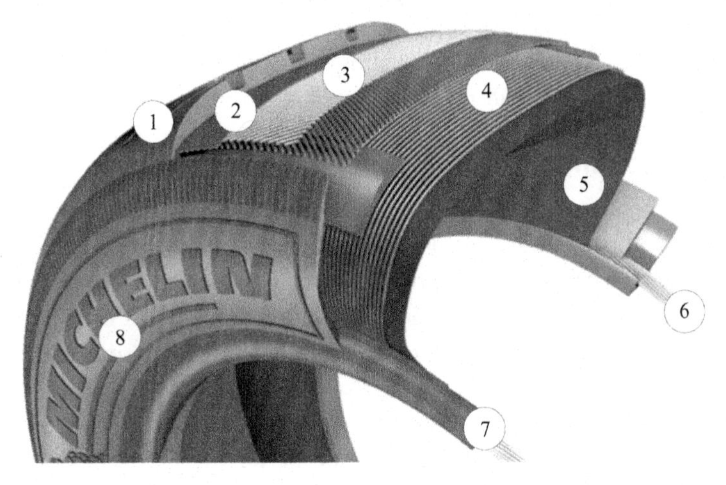

1: Tyre tread
2: Crown belt
3: Crown ply
4: Casing ply
5: Inner liner
6: Bead wire
7: Bead section
8: Sidewall

Fig. 2.10 Internal construction of a radial tyre. *Reproduced with kind permission of the Michelin Tyre Company.*

lateral loads. The tread section is put in position last of all. Once assembled, the whole construction is placed inside a mould bearing the tread pattern and inflated internally, whist the temperature is raised to around 160°C. This process causes the many internal layers to bond together and become one coherent structure. The practice of placing the plys radially was not always followed, and up until the mid-1970s, the body plys were normally inclined at around 45 degrees in the side view, and at least two layers were used crossing over each other in opposite orientations. Thus, the plys were *crossed* giving rise to the name *cross-ply*. Cross-plys offer better grip at very low speeds and a better ride on rough surfaces because the tyre carcass flexes more easily. In all other respects, rolling resistance in particular, the radial construction is considerably superior and now accounts for almost all new tyre construction. Some old road vehicles and historic cars still use cross-plys.

The main design aims for the tyre manufacturer may be summarised as follows:

- Maintaining the static contact patch unchanged under tractive, braking or lateral force conditions
- Controlling damping so that internal friction generates the desired working temperature

- Using the tread pattern to reduce noise generation
- Isolating the rim from vibrational input from, or contact with, the road
- Arranging for the tread to expel water when in rain

2.3.5 Compound

The properties of rubber can be altered and controlled by heat treatment and chemical additions. There are also synthetic rubbers that can be used alongside ones with natural origins. This may either be as a mixture or by having different tyre components from different sources. The most obvious region where the material properties play a major role is the composition of the rubber tread in contact directly with the road. When designers refer to the *compound* of the tyre, this is the bit they mean. In essence, the choice of compound is simple. The softer it is, the better the grip will be, but the shorter the life and vice versa. Each manufacture has their own set of compounds and will provide advice on which compound will suit which application.

2.4 FORCES AND MOMENTS GENERATED BY TYRES

As we have seen, the tyre is a complicated component that can produce forces in all three directions and moments about each of these. The forces generated are important to the vehicle dynamicist because it is these forces and moments that result in the motion of vehicle. We must now turn our attention to how these forces and moments are produced so that we can design a vehicle to provide the tyre with the ideal conditions for the maximisation of their useful components.

2.4.1 Cornering Force

A tyre can produce a force that is perpendicular to the direction in which it points. This is the force that allows a car to corner. There are two important methods by which a force in this direction can be generated. These are through the development of a *slip angle* and through *camber thrust;* both are discussed in turn below.

2.4.1.1 Slip Angle

The first mechanism of lateral force generation is that of slip angle. Most people think that when vehicles go around a corner, the tyres point at a tangent to the corner and the process is much like that of a train on tracks. In fact, this is not the case at all. To understand the mechanism at work, we start

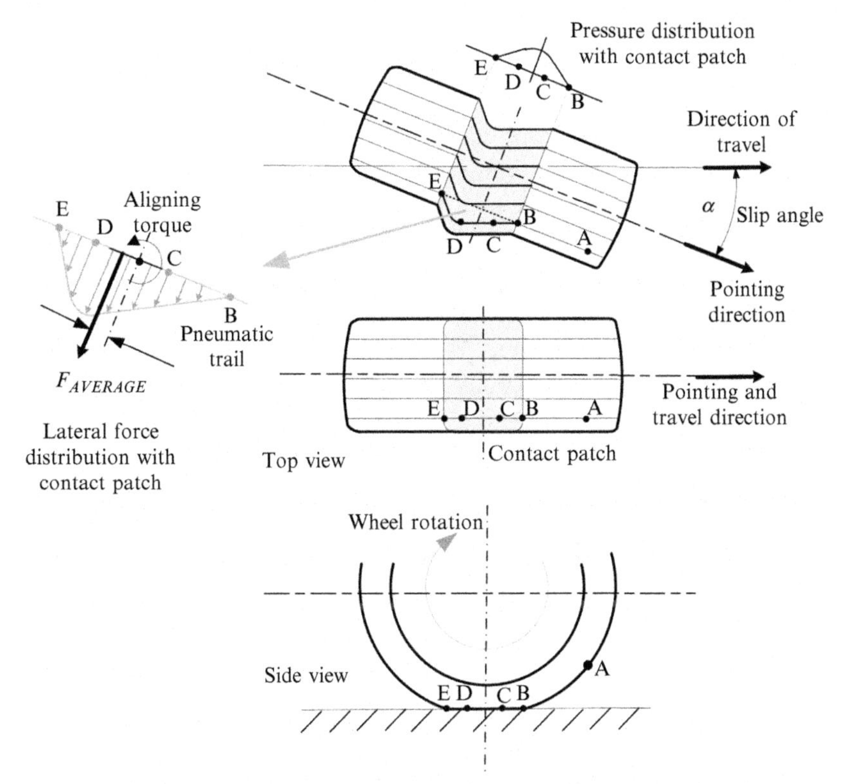

Fig. 2.11 Lateral force generation with slip angle.

by looking at Fig. 2.11. At the bottom, we see a tyre viewed from the side and rotating clockwise and so moving to the right of the figure. We shall follow an element of rubber on the surface starting at position at 'A'. In this position, the element is in the air stream and experiences no forces to displace it either left or right (that is in or out of the plane of the picture) from the road. When the tyre has advanced, the element will reach the point 'B', and this marks its entrance into the contact patch.

Under the vertical load of the vehicle weight, the contact patch is a flat region of sufficient area that the elastic deformation, together with the much larger inflation pressure forces, is enough to balance the weight force.

Rubber within the contact patch is in direct contact with the road surface, and so, the mechanisms of grip described above apply. It is clear that on the passage from 'B' to 'E' there has been considerable elastic deformation and recovery. As the element moves to position 'C', the vertical load on it increases. This will continue until around the midpoint after which the vertical load will decrease until it reaches zero at point 'E' where the element leaves the contact patch. Because rubber is an elastomer and has an hysteretic force displacement curve, there will be less energy recovered in the unloading than expended in the loading, and heat will therefore be generated and energy lost leading to rolling resistance.

Moving now to the middle part of the figure, we can see the tyre viewed from above. The tyre is still rolling in the direction that it points in. The points of interest, 'A' to 'E', are marked. The important part of the lateral force generation is shown as we move to the top of the diagram. In this situation, the tyre has been steered. It is essential to realise that it is still travelling to the right in the diagram even though it is not pointing in this direction. The pointing direction and direction of travel are clearly marked. If we now consider the passage of the element, as before, but in this now rotated orientation with respect to the road, we can see how lateral force is generated. Starting at point 'A', the element is still in the air stream and not experiencing forces from the road. As it reaches point 'B', it makes contact with the road, bonds to it and is therefore constrained to stay at rest with respect to that particular point on the road. As the tyre moves to the right and the element starts on its journey from 'A' to 'E', we can see that in the absence of the road, it would take the dotted path as before; however, because the element is bonded to the road, it must move to position 'C'. Clearly, the element has been displaced laterally, in a direction at right angles to the pointing direction of the tyre, and this therefore creates an elastic force urging the tyre in this direction. As the element moves rearwards, two effects are at work. The first is that the displacement increases and so therefore so does the lateral force. In addition, the vertical load on the element increases as it is squashed harder and harder into the road surface. The pressure distribution within the contact patch is shown above the tyre.

The increase in vertical load is enough to outweigh the tendency of the element to return to its undisplaced position, and it remains firmly adhered to the road surface. However, the situation starts to change when the element passes the midway point around position 'D', and the vertical load on the element begins to decrease as it gets closer to the exit of the contact patch and moves to its radially undisplaced position. However, the lateral displacement continues to increase. Thus, the force urging the element back into position is increasing, and the force between the element and the road resisting this tendency is reducing. It soon reaches a point, very shortly after 'D', when it changes direction and starts to move back into the rest position. Given that the element is still in the contact patch at this point, it is clear that there must be some relative movement of the rubber laterally with respect to the road, and it slides back adopting a position during the process where the restoring force is met by the frictional one. By the time the element reaches position 'E', it exits the contact patch, and the process is over.

The insert on the left shows in more detail the situation that pertains as the element moves rearwards through the contact patch. In particular, we can see the lateral forces that initially increase, reach a peak and then decrease. Scrutiny and consideration of the figure will show that the force has not reached a peak at the contact patch centre; this happens afterwards. Thus, the average of all the forces from all the elements within the contact patch will lie after the centre of the contact patch. In the insert, this is labelled $F_{AVERAGE}$. Since this force lies behind the centre line of the contact patch, it generates a torque about it called the *aligning torque*. This torque serves to pull the steered wheel back to the straight ahead position, realigning the wheel heading, hence its name. This provides a form of feedback to the driver on how much steering is applied to the tyre. The harder a tyre is made to corner, the more it resists the impressed steering, and even in the absence of visual input for the steering wheel position, drivers generally know the current heading and don't need to look at the steering wheel to know the tyre heading; they can feel it. Most people are familiar with the tendency of steering wheel of a car to return to the straight ahead position; indeed, if released, this is exactly what it does, and the aligning torque is a major source of this effect. The distance between the contact patch centre and the average lateral force is called the *pneumatic trail*.

From above, it is clear that no lateral force can be generated when the slip angle is zero. Necessarily, if a tyre is generating a lateral force, there must be a slip angle present. We can also visualise some other effects that are at work. For example, as the slip angle increased from an initially very small value, we should expect the process to work just the same but simply with larger displacements, larger forces and, with them, a larger total lateral force, which is indeed what happens. However, as the slip angle increases, the displaced element will reach its maximum lateral displacement earlier and earlier. This in turn will mean that the point at which it starts to return to the undisplaced position will become earlier too. This will serve to make the region between 'D' and 'E' larger. This is an unproductive region for lateral force, and so, the lateral force will decrease, and indeed, this is observed in practice. Further to this, the average force, $F_{AVERAGE}$, will move forward and eventually be ahead of the midline of the contact patch. This means that the aligning torque actually forces the steering input to increase. This is very undesirable and makes a car very difficult to drive since, when a steering input is supplied, the steering system urges it to get bigger making steering unstable.

2.4.1.2 Lateral Force Characteristic

With the understanding developed above, we can now make an informed consideration of the expected characteristic for the lateral force dependency on slip angle. Clearly, at zero slip, there will be zero lateral force. As the slip angle increases, we should expect that, to begin with, the lateral force would also linearly increase. The gradient would depend on the actual tyre parameters, a wider or larger tyre giving a greater gradient, as would a compound with a higher coefficient of friction. After some time, however, this linear increase will start to tail off. As the point of maximum lateral deflection moves forward in the contact patch, so the maximum lateral force decreases, and a point is reached where the increase in lateral force due to the increasing slip angle is more than offset by the advancing point of maximum displacement causing a reduction in total lateral force. After this point, the lateral force will decrease until the slip angle becomes so large that the entire contact patch is sliding laterally. This characteristic is shown in Fig. 2.12.

The characteristic shown in Fig. 2.12 is typical for a tyre of any kind. The actual values will naturally vary, and the point of maximum lateral force moves with tyre design and vertical load, but the general shape holds.

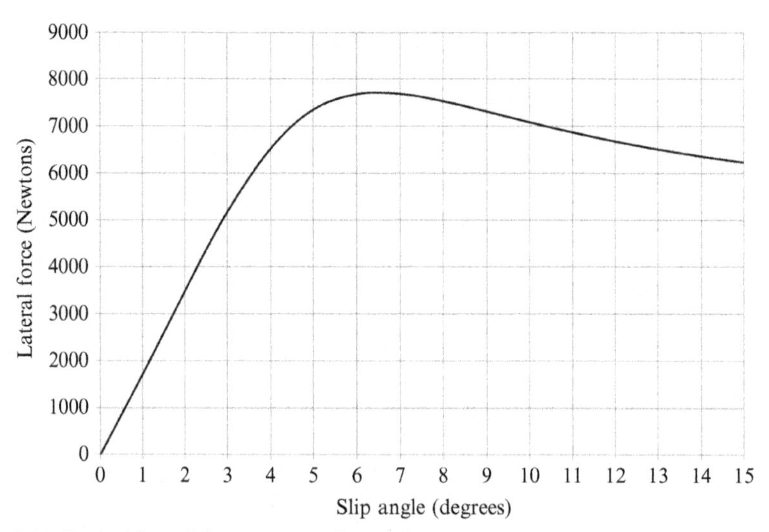

Fig. 2.12 Typical lateral force versus slip angle.

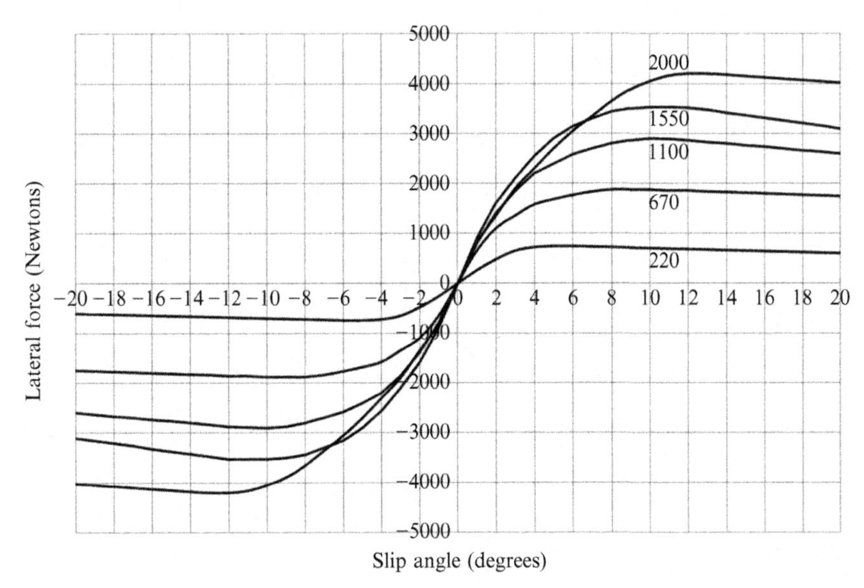

Fig. 2.13 Lateral force versus slip angle for various vertical loads.

The effect of vertical load can be seen in Fig. 2.13. In this figure, data are presented from measurements taken from the formula student car at Oxford Brookes University and relate to a 13 in rim tyre.

In the figure, we can see the tyre exposed to a very wide range of loads. The first line relates to a vertical load of just 220 N. For a tyre suitable for a 180 kg car, this is clearly a very underloaded condition. Nevertheless, the characteristic expected above is present. The next line shows a vertical load of 670 N, and in this case, we see that the line has moved upwards with a larger peak lateral force and that the slip angle at which the lateral force peak occurs has moved to the right. The next two lines corresponding to vertical loads of 1100 and 1550 N show similar movement. The last line corresponds to a vertical load of 2000 N, and this is a heavily loaded tyre. It is the sort of response the tyre would give if the car were braking and turning so hard that almost all the vertical load was carried by just the one tyre. We can see that the peak force has again moved to the right and increased. In the end, the increase will stop as the carcass deformations become so large that the area of contact is significantly reduced. However, even at this value of loading, the tyre will rapidly overheat and become damaged.

We saw above that the coefficient of friction is the parameter that dictates the maximum acceleration a vehicle can sustain, and it is therefore helpful to represent the data in Fig. 2.13 in terms of *Mu*. This is presented in Fig. 2.14.

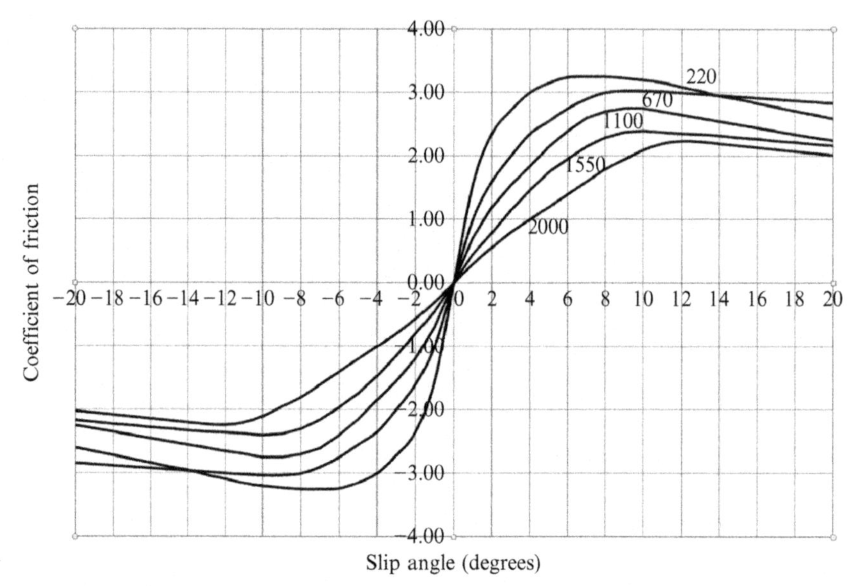

Fig. 2.14 Coefficient of friction versus slip angle.

In Fig. 2.14, we see that it is the lowest vertical load that offers the highest coefficient of friction. It is obvious that as the vertical load on a tyre is increased the lateral force that it can generate will also increase; this is normal frictional behaviour. However, what is not so obvious and is made clear in Fig. 2.14 is that the coefficient of friction gets worse as the load increases. It is true that the lateral force increases with increased vertical load, but it does not keep pace. A doubling of the vertical load does not result in a doubling of the horizontal load. Thus, a well-designed performance vehicle will have the largest tyres that are practical. For example, a road-going hatchback has tyres that are around 40% of the vehicle height; for an F1 car, they are over 60%. In the end, increasing the tyre size brings disadvantage, and there is a limit. Issues such as rolling resistance, aerodynamic drag and rotating weight mean that, after a while, the disadvantages of increasing the tyre size start to outweigh the advantages and a limit is reached.

The vertical load acting on a tyre therefore affects the maximum lateral force it can generate. The graph below shows maximum lateral force against vertical load, and we can see that it is not a straight line. Close but definitely curved (Fig. 2.15).

The effect of this is particularly important when we consider weight transfer in Chapter 3. In the meantime, it can be made clearer if we consider the graph of coefficient of friction against vertical load as shown below (Fig. 2.16).

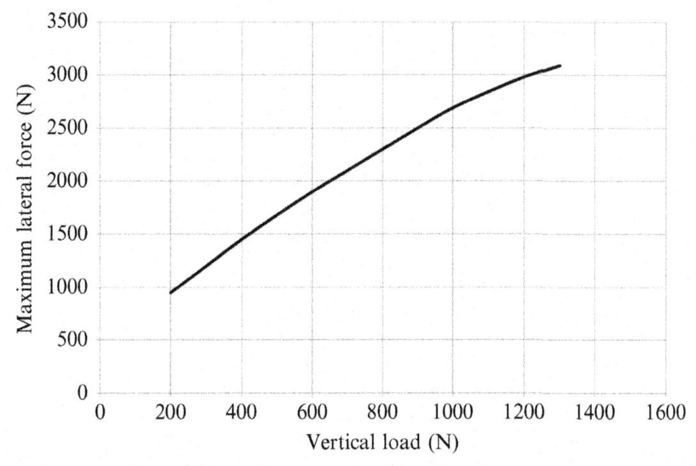

Fig. 2.15 Maximum lateral force versus vertical load.

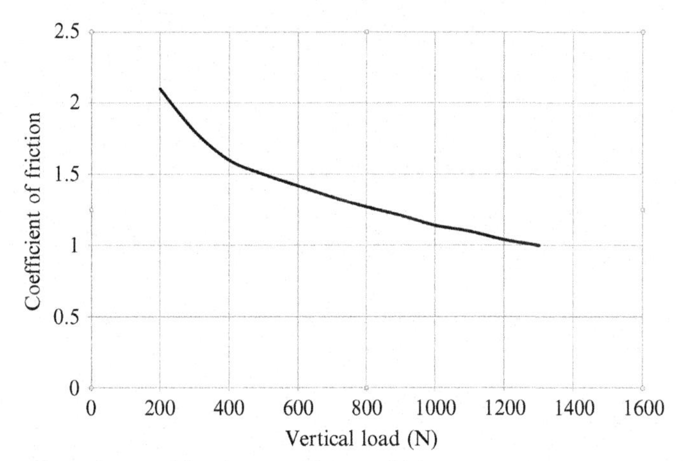

Fig. 2.16 Effect of vertical load on coefficient of friction.

In this graph, we see that the lower the vertical load, the higher the coefficient of friction and so the faster the car will be. For this reason, the first question a good vehicle dynamicist should ask when invited to design a new racing car is *how big are the tyres allowed to be?*

2.4.1.3 Camber Thrust

The second source of lateral force generated by a tyre is *camber thrust.* The term *camber* has three meanings in automotive engineering. It refers to the asymmetry of an aerofoil and is also the name given to the convex curve built

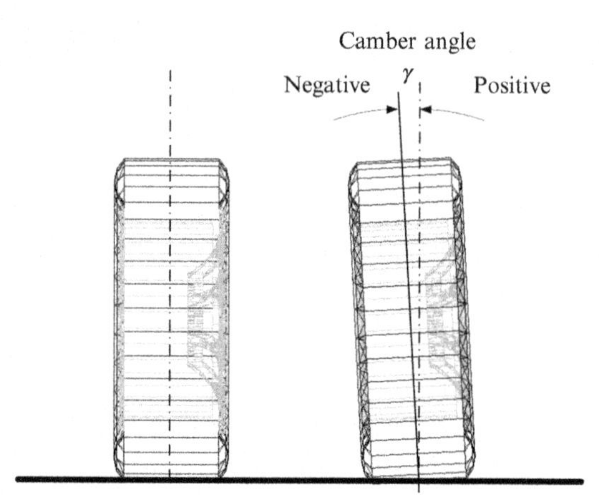

Fig. 2.17 Camber angle.

into road surfaces so that water drains away quickly. A cambered road surface is also provided for safety since if a driver falls asleep at the wheel, the car will tend to crash into the nearside road edge rather than into oncoming traffic. The effect that we are interested in here is the inclination of a wheel when viewed from the front.

Fig. 2.17 shows the camber angle and the sense in which it applies (the corner centre is to the right in the figure). If the top of the wheel leans inwards towards the vehicle body, then the wheel is said to be under *negative* camber. The symbol gamma, γ, is usually used to denote camber. To understand how the inclination of the tyre results in the generation of a lateral force, we shall follow a similar approach to that taken above. Fig. 2.18, in part one of the diagram, shows a cambered wheel from above. Just as when we considered the effect of slip angle, we shall follow the passage of an element on the surface of the tyre. Starting at 'A' where the element is in the air stream and experiences no force from the road and so is at elastic rest with respect to the tyre carcass.

As the wheel rotates, the element reaches point 'B' where it enters the contact patch and so adheres firmly to the road surface and must from now on remain at rest with respect to the road surface. If the tyre were in free space and the road not present, then the passage of the element would take as the dotted path from 'B' to 'E'; instead, because the tyre is indeed in contact with the road, the element must take the straight path shown with the solid line since it is this path that allows it to remain at rest on the road surface as the wheel rolls on. In the diagram, these two paths are very close together and hard to discern. Part two of the diagram shows the path enlarged for

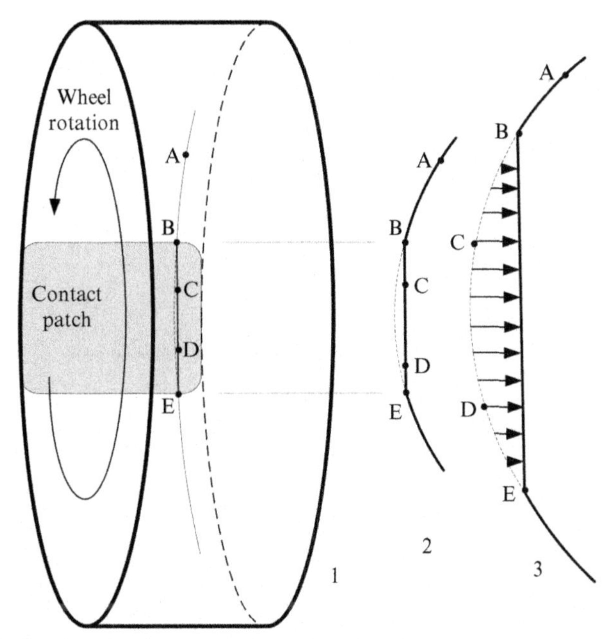

Fig. 2.18 Mechanism for the generation of camber thrust.

clarity. Consideration of the paths shows that at, for example, point 'C' the element will have been displaced to the right in the diagram and will therefore be exerting an elastic force on the tyre in the direction shown in part three of the diagram. Indeed, part three shows that the elastic forces will reach their maximum at the contact patch centre. The pressure distribution similarly reaches a maximum at this point, which helpfully ensures that the elements experiencing the largest lateral force are also experiencing the largest vertical load pressing them into the road surface and ensuring that the elastic forces within the carcass do not overcome the frictional force between the rubber and the road surface. Fig. 2.19 further demonstrates the generation of camber thrust.

In the left hand part of the figure, the deformed and undeformed shapes of the tyre under camber are shown. An element that is forced to be at position 'B' in its elastic equilibrium position would be at position 'A'. In the right hand part of the diagram, showing the situation enlarged, it is clear that this displacement will give rise to a vertical force that supports the vehicle load and a horizontal force in the same direction in which the tyre leans, and this is the camber thrust. Camber thrust will clearly vary with camber angle, and the dependence is shown in Fig. 2.20.

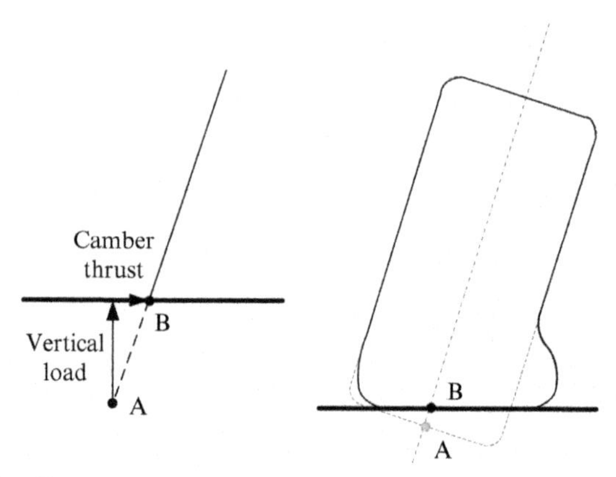

Fig. 2.19 Camber thrust.

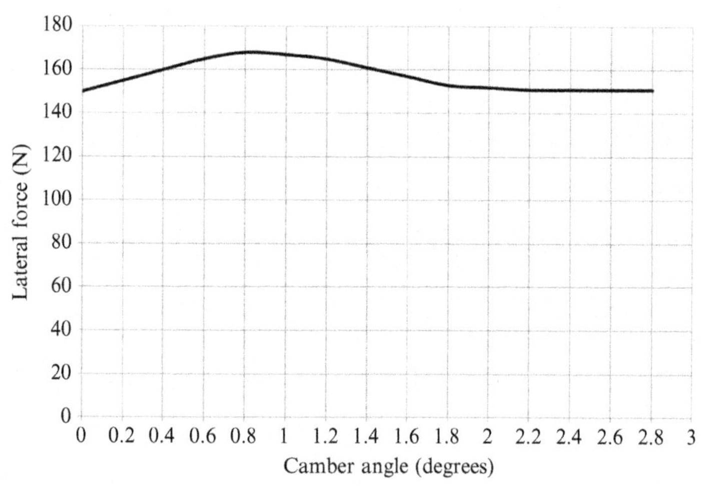

Fig. 2.20 Camber thrust vs. camber angle for a given slip angle.

2.4.2 Longitudinal Force

To understand how a tyre may generate tractive or braking effort, we shall make use of the brush model. The brush model has been in general usage for some time, and many references can be found to it in the guided reading. Pacejka [7] is a good place to start. In this representation of the tyre, elastic elements are imagined to connect the tyre to the road surface; this is shown in Fig. 2.21. The tyre is travelling to the right with velocity V_X and is rotating with angular velocity Ω. As before, we shall follow an element on the surface of the tyre from a point before it enters the contact patch. This is

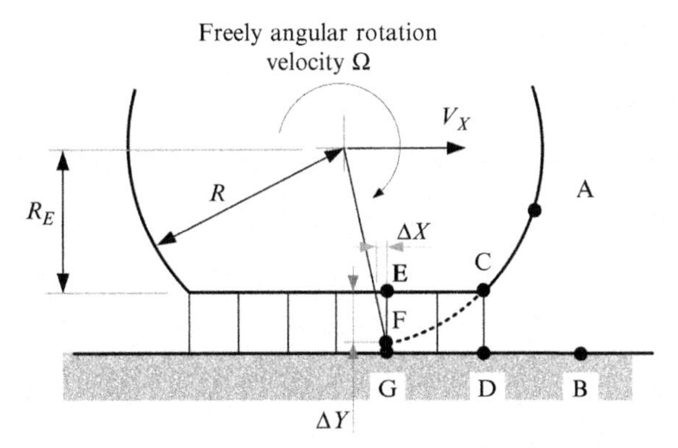

Fig. 2.21 Brush model for the free rolling tyre.

the situation at point 'A'; point 'B' is the point on the road surface to which point 'A' will make contact when time comes. Positions 'C' and 'D' show this. In the bush model, a line element is used to joint from 'C' to 'D', this being an elastic connection that permits force to be applied between the two points it connects in response to the deformation it experiences.

In the case of the free rolling tyre, the horizontal displacement of 'C' with respect to 'D' does not change as the tyre rotation advances. This is because an element in the contact patch has two components of velocity that cancel each other out. The first is a forward velocity given by the vehicle speed V_X. The second is a rearward velocity given by the angular velocity, Ω, and multiplied by the effective rolling radius, R_E. For the free rolling tyre, these two velocities cancel each other, and there is no tractive or braking torque developed. As the tyre rotation progresses, an element on the tyre surface is forced to follow the flat path through the contact patch rather than the dotted path corresponding to the undeformed tyre surface. Consideration of the line from 'E' to 'G' shows that this will result in the deformations ΔY and ΔX. The sum of the vertical forces caused by the vertical deformations ΔY gives rise to the vertical force that supports the corner load of the vehicle on the tyre. For the first half of passage through the contact patch, the displacement ΔX will be in the direction of travel, whilst in the second half, it will be symmetrically in the opposite direction. Thus, the sum of the forces caused by the displacement ΔX will add up to zero. However, as a result of these forward and aft displacements, there will be work done within the tyre carcass, and this is one source of rolling resistance. The brush model can equally be applied to the generation of lateral and longitudinal forces.

2.4.2.1 Braking Condition

Having established the brush model, we can now apply it to the braking case, and this is shown in Fig. 2.22. The figure works in the same way as before with one important difference. Under braking, an element on the surface of the tyre entering the contact patch has some forward velocity with respect to the road surface. In the figure, we can see that as the element moves rearwards from position 'C' to 'E', the point moves forwards and a displacement in the X direction develops. Tension in the brush element therefore develops as the element moves rearwards through the contact patch, and component of this tension in the X direction exerts a force in a rearward direction serving to decelerate the vehicle and hence *brake* its motion. The difference in velocity of point 'C' with respect to 'D' is given by $V_X - \Omega R_E$, and for braking, this is positive quantity.

If the difference in speed is small, then the brush element will be able to extend throughout the passage through the contact patch. If this were the case, then the brush element lines in the figure would become progressively more steeply inclined as they move rearwards as shown between positions C–D and E–F. However, if the difference in speed is sufficient, the situation becomes more yet complex, and this is shown in the figure. If the elastic force within a brush element urging the element in contact with the ground forwards exceeds the frictional force it can sustain at the contact patch, then it will slide forwards, reducing the tension in the brush element until the frictional force can resist the tension force. This process is seen to be

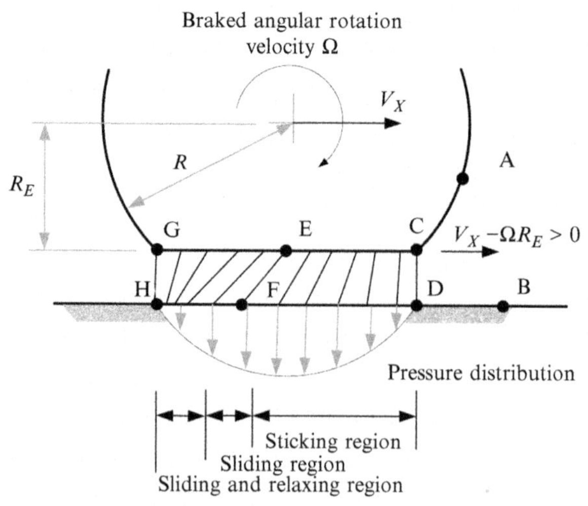

Fig. 2.22 Brush model for braking.

occurring as we move rearwards from element 'E–F'. By the time the element reaches the rear of the contact patch where the vertical load is zero, the brush element is once again vertical, as it was upon entry to the contact patch. There are therefore three regions within the contact patch of a braking tyre. Firstly, a *sticking* region in which the tyre elements remain in good contact with the road and the brush elements extend. Secondly, a *sliding* region where the reduction in vertical load allows the tension in the brush elements to reduce. Finally, there is a *sliding and relaxing* region in which the brush elements return to a vertical orientation in readiness for exiting the contact patch when they must be fully relaxed.

2.4.2.2 Tractive Condition

For the tractive condition, the situation is reversed, and $V_X - \Omega R_E$ is a negative quantity. Point 'C' would move left in the diagram with respect to 'D', and the brush elements would slope the other way. There is still a *sticking* region and following this a *sliding* and then a *sliding and relaxing* region. Naturally, the forces in the brush elements are urging the tyre forwards, and a tractive force is developed, which is the sum of all the forces in the individual brush elements.

2.4.2.3 Longitudinal Slip Ratio

Just as we saw, there is a lateral slip angle, so there is a longitudinal equivalent, the *slip ratio*. In the longitudinal case, the slip ratio involves the ratio of forward velocity to the velocity of a point on the surface of the tyre at the effective rolling radius. To produce a number that varies between zero and one, the usual form is given by

$$SR = \frac{\Omega R_E}{V_X} - 1$$

The typical characteristic for longitudinal force versus longitudinal slip is shown in Fig. 2.23. The graph shows the normalised longitudinal force that simply means the value is divided by its maximum value to obtain a generic graph that applies to all tyres. The longitudinal slip ratio is as defined above.

As expected from the brush model, the graph shows that at no slip there is no force. After this, the longitudinal force increases linearly until after sometime, the slipping and sliding region increases significantly and linearity is lost. A peak is reached, and from there, the longitudinal force decreases until, at a longitudinal slip ratio of 1, corresponding to sliding throughout the contact patch, a reduced value of around three quarters of the maximum is

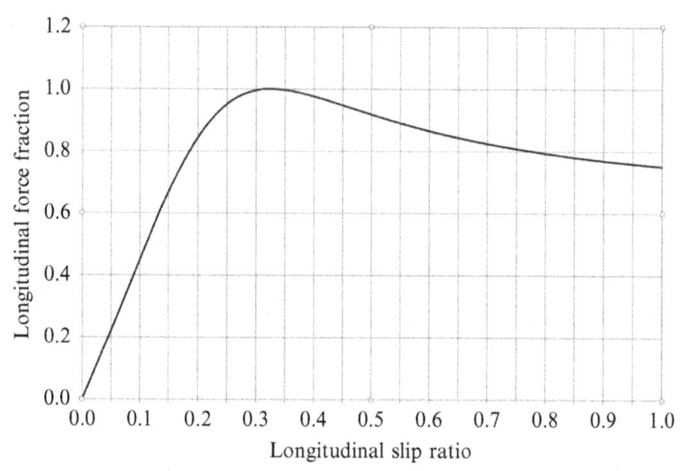

Fig. 2.23 Longitudinal force versus longitudinal slip ratio.

reached. The details of the characteristic vary between traction and braking, but they are both broad as shown.

2.4.2.4 Further Developments of the Brush Model

Having seen the brush model applied above to the tractive and braking case, it is straightforward enough to see that models for the total tractive or braking effort could be developed by forming expressions for the deformations in the brush elements, integrating over the duration of contact and then multiplying to account for the width of the tyre. Such a process would result in expressions for the tractive and braking effort that can be generated in terms of tyre parameters such as coefficient of friction, width, radius and pressure. In this text, we shall stop at this point since this is a text on vehicle dynamics and not tyre modelling and it is sufficient to make use of published curves that provide the characteristics we require without going further into the origin of their shape. The bibliography offers sources for further enquiry.

2.4.3 Vertical Force

Having given a detailed consideration of the lateral and longitudinal forces that a tyre can generate, we now turn our attention to the remaining four forces and torques. In the case of the vertical force that supports the load on each of the corner of the car, the situation is reasonably straightforward. Movement of the rim vertically downwards deforms the tyre carcass compressing the bottom and extending the top sections. These deformations give rise to forces that oppose them, and a vertical force is generated. The larger the displacement, the larger the force. Typically, tyre vertical stiffnesses are significantly greater than the suspension stiffness varying from a factor of around 2 up to 5 or 8. In racing, the lower end is used, and we can see immediately that there will be a roughly equal amount of movement in the tyre as there is in the suspension, making the tyre an integral part of the suspension. As the vertical load is increased, the displacement increases, resulting in the tyre bulging more at the contact patch. This in turn means that more heat is generated, since on every passage through the contact patch, the rubber is flexed back and forth more and more. Eventually, the deformations become so large that the contact patch is compromised, losing its desired shape and pressure distribution and so impairing performance. In road cars, the tyre stiffness is generally at the larger end of the range above, and these issues are consequently less significant.

2.4.4 Aligning Moment

The aligning moment can be seen in Fig. 2.2 to be the torque that urges the tyre to steer. The torque that causes this was described in above when considering the slip angle and lateral force generation and is shown in Fig. 2.11; there are no new sources that we need to cover. However, we have assumed that the tyre was rotating about a vertical axis passing through its centre. In fact, the axis about which the tyre rotates in the top view depends on the suspension geometry, and it is common for this to be several degrees away from vertical, influencing the total aligning torque. To distinguish the torque that comes from these two different sources, the *pneumatic trail* refers only to the distance shown in the diagram and not from suspension geometry that we shall consider later in Chapter 7—Suspension Kinematics.

2.4.5 Overturning Moment

There are two main sources of the overturning moment. The first is deformations in the contact patch, the result from camber. Fig. 2.19 shows a heavily cambered wheel. Clearly, the compressed rubber on the inside will serve to urge the tyre back to the vertical position resulting in aligning torque. The second mechanism is shown in Fig. 2.23. In this situation, the tyre is cornering heavily, and the contact patch is shown in its deformed shape. As a result of the deformation, the vertical load that was 'P' has moved to position 'R' and is offset from the midline of the tyre. As a result an overturning moment is generated about the tyre centre. The overturning moment is resisted by the suspension members. The value of the loads can be estimated

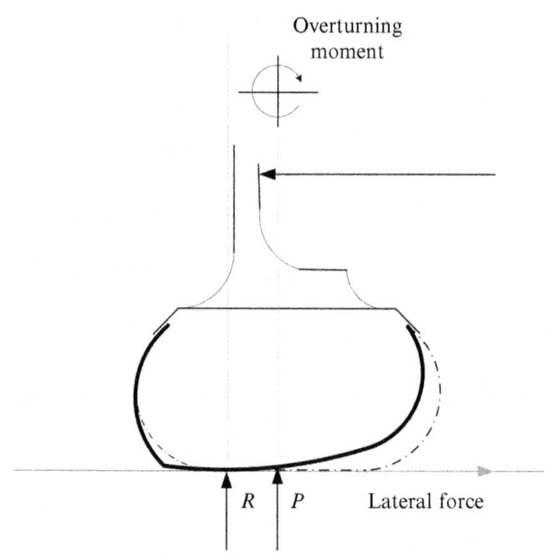

Fig. 2.24 Overturning moment under lateral force.

for the steady-state case by using the knowledge of the forces and moments that apply from equilibrium, and these make a useful first estimate of loads. However, estimates prepared that way will inevitably be a distinct underestimate of the real loads, and it should be remembered that the loads caused by road bumps and excursions such a kerb strikes will be much larger. This is treated in Chapter 8—Suspension Dynamics (Fig. 2.24).

2.4.6 Rolling Resistance

A tyre in motion offers resistance to the imposed motion in a number of ways. The first results from the movement of the vertical reaction to the load on the tyre.

In Fig. 2.25, we see a tyre under tractive effort. A torque applied at the axle serves to rotate the wheel clockwise in the figure. Bearing in mind the brush model developed above, it is clear that under the action of the torque, the centre line of the axle will move rearwards with respect to the contact patch, therefore distorting the carcass into the shape shown with the dotted line. As a result, the vertical reaction force R will be in front of the wheel axel by a distance k. The product of Rk gives a moment resisting the tractive torque, and this is one source of rolling resistance. Under braking, the situation is reversed, and the torque applied at the brake wheel is opposed. For a 13 in tyre with a vertical load of 80 kg offset by 5 mm, this gives a torque of

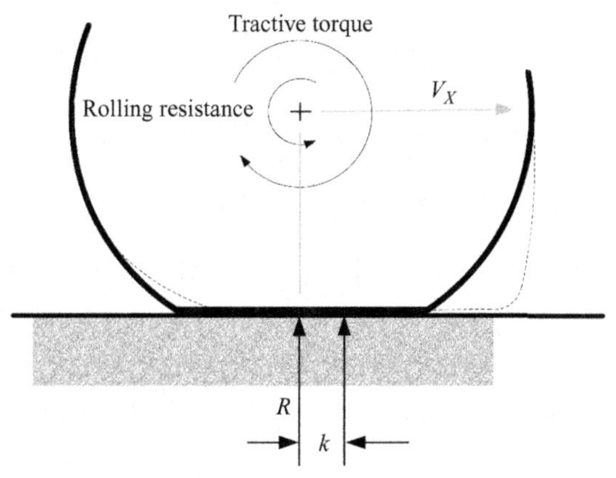

Fig. 2.25 Rolling resistance from vertical load migration.

3.8 Nm with an angular velocity of 170 rads/s. This corresponds to around 650 W or nearly a horsepower all of which is dissipated within the tyre causing its temperature to rise.

In addition to this, it must be remembered that rubber exhibits considerable hysteresis on bending and the force applied to cause a distortion is not recovered when the force is relaxed. As a result, every instance of bending results in energy lost and so causes rolling resistance. For example, an element on the tyre surface will be bent on entry to the contact patch, compressed and relaxed on its passage through it, and then bent again on exit. Indeed, elements throughout the carcass and even in the tyre wall will experience a similar process as they pass through the stationary bulge that results from the vertical load.

Together, these sources of internal heating serve to increase the operating temperature of the tyre. This internal heating brings the advantage that the coefficient of friction is increased and with it maximum acceleration. However, the benefit comes unevenly; ideally, the tyres would have increased grip at all speeds and times, but in fact, it depends on speed in a very nonlinear way. Rolling resistance and with its internal heating can be approximated to the cube of forward speed. Further, increased grip brings reduced life, and tyre design, like most design, is a compromise.

2.5 THE FRICTION CIRCLE

In the above treatment of the force generated by a tyre, we have seen the longitudinal force that results from the longitudinal slip and the lateral force

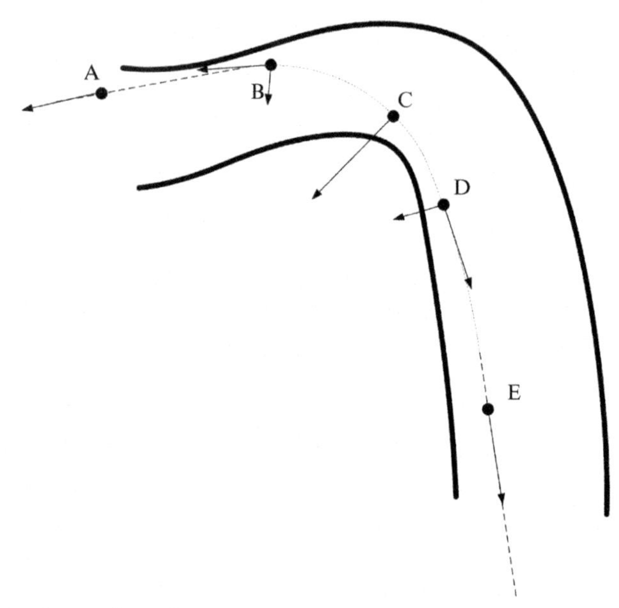

Fig. 2.26 Tyre forces through a corner.

that results from the slip angle. In general, a tyre will be experiencing both of these conditions together. Fig. 2.26 shows this. As a car approaches a corner, at position 'A', the brakes are applied, but no lateral force is developed, and the steering has not yet been applied. The driver passes the 'turn-in' point when steering is first applied and the situation shortly after is shown in position 'B' where both a modest lateral force and significant braking are simultaneously applied. At the apex of the corner, position 'C', the driver is not applying longitudinal force, neither the brakes nor the accelerator is applied, and instead, the entire capacity of the tyre to generate force is expended in the generation of lateral force. At position 'D', the corner exit is well under way, the lateral force is reduced, and the car is travelling in a much larger radius. At the same time, tractive effort is supplied, and the car is accelerating. By position 'E', the corner is over, the car is accelerating only in a longitudinal direction, and the driver will be giving thought to the next corner where the whole process must be repeated.

A good driver will smoothly blend the lateral and longitudinal accelerations and will always be generating the largest values possible. Clearly, the two components can be added, and this is the idea behind the friction circle shown in Fig. 2.27. The plot shows longitudinal and lateral acceleration on the two axes. On this, a circle representing the largest value of acceleration,

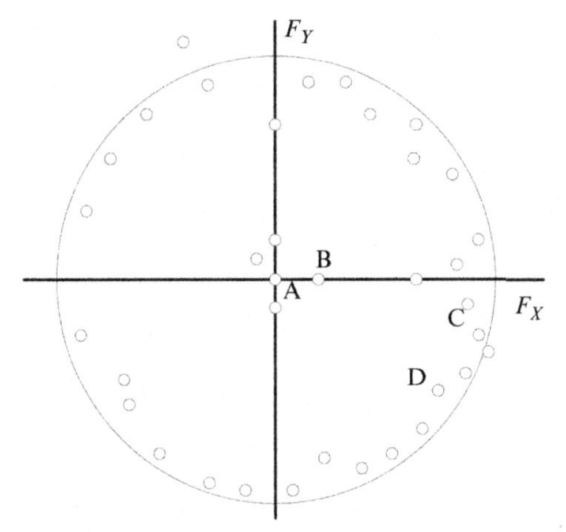

Fig. 2.27 The friction circle.

either longitudinal or lateral or any combination is shown. The concept is very useful when performing lap-time simulation because it allows the determination of the maximum longitudinal and lateral force that can be applied at any time under combined conditions. The diagram is also useful for drivers, and this is the explanation for the dots on the diagram. Each one shows the longitudinal and lateral acceleration recorded by the cars telemetry at a given time. The point at the origin, 'A', corresponds to the start, and the car pulls away to position 'B' some moments later. By point 'C', the car is under combined lateral and longitudinal acceleration. A driver may be judged on how well he or she keeps the car at the limit, meaning that most points lie on the circle or at least close to it. This will not be possible all the time; for example, on a long straight where the car nears terminal velocity, the acceleration will be much less than the friction limit but is still the best that can be done. In racing, such times are not the norm, and keeping the car at the circle is indeed the goal.

2.6 COORDINATE SYSTEM

There are a number of coordinate systems in use, the ISO system and the American SAE system being two prominent examples. In this text, we shall adopt the system below which is drawn heavily from the SAE system.

The tyre coordinate system shown in Fig. 2.28 defines the quantities and directions we shall use when referring to tyres. In the top view, the positive slip angle is shown, and this results in a lateral force that acts in the direction

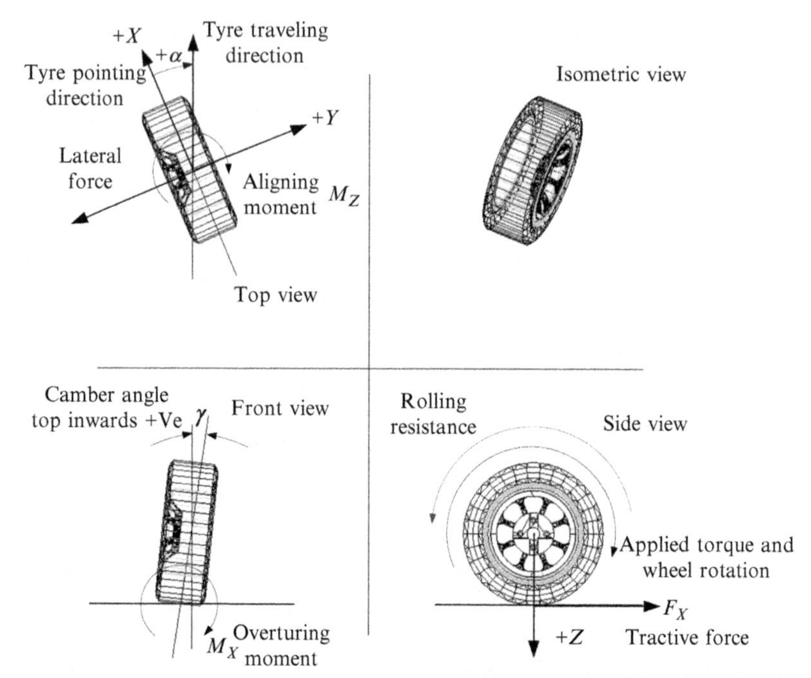

Fig. 2.28 Tyre coordinate system.

shown. In our coordinate system, the positive Y-axis is in the opposite direction, and as a consequence, the cornering stiffness coefficient of a tyre will always be negative. Often, the cornering stiffness is approximated to a constant coefficient, and this will be negative. When, more accurately, it is modelled as a function, it again must be negative. When dealing simply with cornering stiffness on its own, most people ignore the negative sign, and that doesn't matter, but when we apply analysis, for example, in the derivatives analysis in Chapter 5, it is important. Camber angle is defined with positive with the top of the tyre being inboard of the bottom.

2.7 TYRE MODELLING

The lateral and longitudinal forces developed by a tyre result from the complex processes we have seen above. In the practice of vehicle dynamics, we shall clearly need to make use of the knowledge of the values of forces that the tyres in question can develop. For example, if we want to select spring values that result in a given amount of roll, we shall have to determine the moment generated by the suspension and balance this against that produced by the tyres. Alternatively, when doing lap-time simulation, we shall constantly need to know what longitudinal and lateral force the tyre is able to generate and then apply these to the car to determine its local acceleration. From this, the position can be indexed to the next point in the simulation and so on. One way of doing this is to simply refer to tables prepared from experimental data. This is simple and easy and has the benefit of being based on actual data that will give good results. However, experimental data are not always available, and tyre testing is very expensive. For these reasons, one often ends up approximating the tyre curves.

A very practical and reasonably simple formula for this was developed by Hans Pacejka [7] and is shown below:

$$y(x) = D\sin\left[C\ \arctan\left\{Bx - E\left(Bx - \ \arctan\ Bx\right)\right\}\right]$$

In the formula, $y(x)$ is the output variable. This might be the lateral force, F_X, or the aligning moment. All tyre characteristics can be modelled using the formula. X is the input variable, slip angle α or longitudinal slip ratio, for example. The remaining terms are the following:

B = Stiffness factor
C = Shape factor
D = Peak value
E = Curvature factor

where 'D' is the peak value of the curve being modelled and 'B' is the cornering stiffness value that is the gradient of the line in hand at the origin. 'C' is determined from the following:

$$C = 1 \pm \left(1 - \frac{2}{\Pi}\arcsin\frac{y_a}{D}\right)$$

The formula for 'C' will produce two values for 'C', but only the value of 'C' greater than or equal to 1 is used. E is determined from

$$E = \frac{Bx_m - \tan\left(\Pi/2C\right)}{Bx_m - \arctan\left(Bx_m\right)}$$

The formula is applied to simulate a real tyre in Fig. 2.29.

In the figure, experimental data can be seen with the solid line, and the 'magic formula' approximation is shown with the dotted line.

To extend the sophistication and accuracy of tyre modelling, conceptual models such as the brush model above can be used. Once a method of estimating elastic displacement in the contact patch is prepared, integration can be applied to estimate the total value of the parameter in hand. Many such

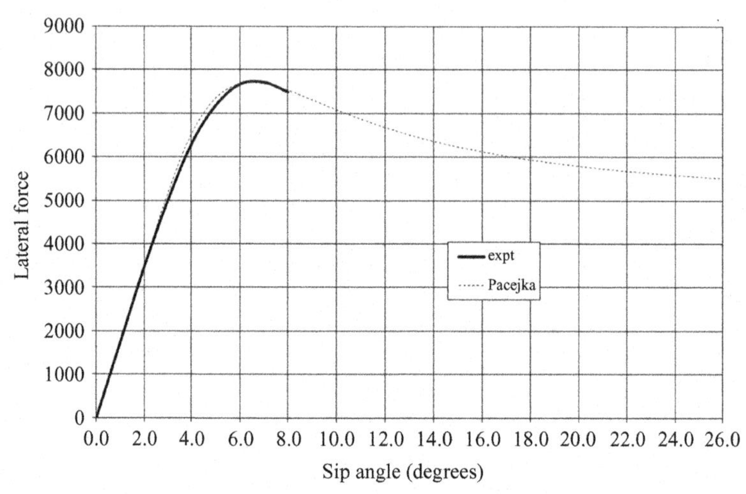

Fig. 2.29 Magic formula approximation to real tyre data.

conceptual frameworks have been proposed over the years, and a literature review starting with the texts listed at the end of this chapter will make a good start.

In addition to analytical models, computer simulations can also be employed. These are sometimes based on a finite element approach in which the tyre is divided into a large number of elements, each small enough that parameters change only by small amounts as one moves from one to the next. Matrix-based computational methods are then used to solve simultaneous equations describing the whole tyre, and results obtained. This approach has the advantage that effects far too complex for any analytical approach can be simulated. For example, the best current tyre simulations include modelling for effects such as the temperature-dependant chemistry of bonding to the road, all sources of hysteretic heating and interaction with water. Other approaches can be used, in the multibody code Adams; for example, the tyre is treated as a six-degree-of-freedom element linking the road to the stub axle, and a separate file is used to define the functions of the force in each dimension.

2.8 TYRE TESTING

Because simulation can only ever be an approximation and because many tyre models require the presence of empirically determined parameters, tyre testing is required. This will clearly involve a significant piece of apparatus.

If one considers an F1 tyre forces that are large, around 5 kN of lateral force, a positional accuracy of fractions of a degree required, all happening under tractive power of around 800 hp, then, it is clear that a substantial apparatus will be needed.

2.8.1 Tyre Testing Machine With a 'Belt'

One generic kind of tyre testing machine is those that involve a belt that is used as the 'road'. The tyre is made to roll along a belt being held in position with an adjustable and very rigid support structure. The camber, slip angles, etc. can all be varied under control, and a complete map of the tyres performance can be acquired. Fig. 2.30 shows the general approach to such machines.

The advantages offered by this method include the fact that the test is conducted in laboratory conditions where the environment is controlled. Temperature, humidity, etc. can all be monitored and, if desired, controlled. A criticism of this method is that the 'road' surface is not actually road. The pragmatic requirements of having a belt that can pass around the drums and simulate the road mean that the coefficient of friction can never be as it really is for the tyre-road interface. On the other hand, tests made between tyres are clearly comparative, and one only needs to establish a 'correction factor' that should be applied to convert measurements of, for example, cornering stiffness, as measured to that that pertains in real life, and this criticism is largely dispelled.

Fig. 2.30 The approach taken for 'belt' tyre testing machines.

There are indeed many testing facilities around the world based on this approach, and the web-based research suggested at the end of the chapter will lead to a selection.

2.8.2 Tyre Testing With a 'Trailer'

Another approach to tyre testing is to mount the tyre under test to a trailer, which itself has tyres vastly larger than the tyre under test. The test tyre, which is mounted underneath the trailer, makes contact with the road surface, and as before, parameters such as slip angle and camber angle can be varied. Again, it is necessary to have some very stiff and substantial structures on which to mount the tyre, and the trailer itself must be sufficiently large and heavy that the lateral and longitudinal forces developed by the tyre under test are negligible compared with those generated by the tyres of the trailer itself. It's important that the tyre under test doesn't steer the trailer (Fig. 2.31)!

2.8.3 The Approach Taken With Direct Wheel Sensors

An alternative approach is to mount a wheel sensor directly between the stub axle and the wheel rim itself. The approach is shown in Fig. 2.32. The stub axle is no longer used to mount the wheel rim; instead, it mounts the sensor to which the wheel is mounted. A rotating part of the sensor then supports the wheel rim. Through this, all the loads between the road and the stub axle are communicated and can be measured. It is made complex by the fact that the wheel must rotate and slip rings are used to make connections between the rotating part of the sensor and the cable outlets.

Fig. 2.31 Approach taken for tyre testing trailers.

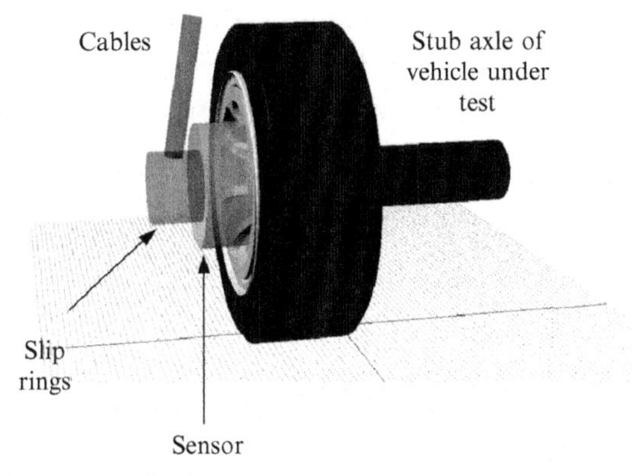

Fig. 2.32 Wheel sensor method.

This method offers the distinct advantage that it measures directly the loads in all six degrees of freedom on a tyre travelling over the actual road surface. It is therefore difficult to improve on measurements made in this way.

2.9 QUESTIONS

1. What is meant by the term *static camber*? Why and when is it a desirable feature of tyre usage?
2. Draw a graph of lateral force against camber angle for a typical tyre.
3. Draw an annotated diagram to explain how camber thrust is generated within a rotating tyre.
4. Use a diagram to explain the two primary mechanisms by which the tyre-road material pair produces the very high value of friction normally found.
5. Sketch a graph with several curves to show how lateral force against slip angle varies with vertical load.
6. What is 'aligning torque'? What causes it and how does the value taken vary as the operating conditions of a tyre change?
7. If the vertical load on a tyre is doubled, by how much will the maximum horizontal load it can generate increase.
8. Draw a diagram of a tyre in contact with a road surface showing the passage that an element of rubber takes as it passes through the contact patch. Use the diagram to explain how tyrc gcnerates lateral forces and why this initially rises with slip angle and then falls.

9. Explain the role that the *aggregate* in a road surface plays in ensuring a good level of grip is produced by the tyre that uses it.
10. Estimate how much cornering speed is lost by a car when its tyres are 10% below the optimal temperature.
11. Explain how the *brush model* explains tyre operation.
12. Discuss the advantages and disadvantages of tyre testing by the *belt* method, the *trailer* method and the *direct wheel sensors.*
13. What are the constructional and performance differences between the *cross-ply* tyre and the *radial* types of tyre construction?
14. What factors combine together and mean that the vertical stiffness of a tyre is not a linear function?
15. Distinguish between *cornering force, cornering stiffness, horizontal force slip angle* and *longitudinal slip.*
16. Why is there a maximum lateral force that a tyre can generate and what conditions must prevail for it to do so?

2.10 DIRECTED READING

Many of the explanations of tyre operation given above are in general usage and form part of the cannon of knowledge vehicle dynamicists make use of. I have presented my own versions of these principles, but it is always helpful to widen one's understanding by reading the work of others. Below are some suggestions:

Possibly, the most prolific author on tyre modelling is Hans Pacejka [7]. The text referenced is a very comprehensive book on tyres within which many further references may be found and followed.

Paul Haney [3] offers another excellent text with much a much less analytical approach and an excellent section on rubber.

John Dixon [8] is a more general text with a good first chapter on tyres.

Milliken and Milliken [9] also have a very good chapter on tyres. The text is excellent for a number of other topics and is referenced elsewhere.

2.11 LEARNING PROJECTS

- Perform a literature review using a university standard library and internet resources such as Google Scholar and research five distinct and separate methods of modelling tyres. For each one, explain what is unique to the model and under what circumstances might an ability to model a tyre in this way confer an advantage in either road or racing car design.

- Perform a literature review to determine the current state-of-the art modelling for tyres that includes temperature and chemical effects.
- Prepare a design for a simple tyre test rig. The rig is to be a piece of laboratory equipment and suitable for a university to build. Start by deciding on the parameters to be measured and the accuracy required and prepare a design specification for the rig. Then, proceed to produce an analysis in a dynamic modelling package such as Adams before proceeding to produce detailed drawings to permit manufacture.
- For a collection of different vehicle cornering conditions taken from racing and road car observations, make an estimate of the coefficient of friction the tyres must have been producing in each situation. For example, using televised data from an F1 race and the knowledge of the car's speed at a particular corner together with the knowledge of the radius of that corner, estimate the *Mu* value for the tyres.
- For any vehicles you may have access to, determine the vertical stiffness of the tyres by firstly measuring the static displacement between the rim and the ground. Then, load the tyre by sitting on the corner of the vehicle and remeasuring the distance. This will allow the determination of a spring rate from a single pair of measurements. If possible, use a range of weights and obtain a curve for stiffness against displacement. Repeat the curves for a range of inflation pressures.
- Undertake an examination of the Pacejka 'magic formal' in a spreadsheet and use it to approximate a real tyre lateral force curve obtained from research.

2.12 INTERNET-BASED RESEARCH AND SEARCH SUGGESTIONS

Using the internet, perform a series of searches collecting useful URL's into a dedicated folder on your PC for each of the following search terms. Make use of searching Google by 'images' and on 'Youtube' as well.

- Calspan tyre testing
- Belt tyre testing
- MTS tyre testing
- Delft tyre test trailer
- On-vehicle tyre testing transducer
- On-track tyre testing transducer
- Tyre testing machine
- Tyre force test

- Tyre testing Kistler
- Tyre manufacture
- Hans Pacejka
- Cornering stiffness
- Camber thrust
- SAE vehicle coordinate system
- ISO vehicle coordinate system

CHAPTER 3

Weight Transfer and Wheel Loads

3.1 INTRODUCTION

In Chapter 2, we saw how a tyre produces a horizontal force at the contact patch. This might be forwards or backwards, for accelerating and braking the vehicle, or laterally for cornering, or indeed any combination in between. Clearly, if we want a car to perform well, we want the largest possible horizontal forces to be produced, and since the amount of horizontal force produced is dependent on the vertical load, we shall start our consideration of vehicle dynamics by determining the vertical loads on a car wheels.

In this section, we shall consider the steady-state loads that act. This is a good place to start since we must understand the steady-state case before considering anything more complicated. The knowledge gained in this chapter and in the previous one on tyres will enable us to proceed in the ensuing chapters to consider the four main vehicle dynamics topics of straight-line acceleration, cornering, braking and suspension dynamics.

3.1.1 Wheel Loads Under Constant Acceleration

We start by considering a car at rest or constant velocity in the side view as shown in Fig. 3.1 The car has a wheelbase, l; the centre of gravity, (CoG), is a distance 'a' behind the front axle and a distance 'b' in front of the rear axle.

By balancing moments, it is a simple matter to determine the static axle loads. These are the static load in the Z direction at the front F_{SZF} and similarly F_{SZR} at the rear. The subscripts stand for *static*, 'Z direction' and F or R for front or rear, and most organisations adopt this or a similar nomenclature. Starting by taking moments about the front axle,

$$
\begin{aligned}
mg \times a &= F_{SZR} l \\
\Rightarrow F_{SZR} &= mg\frac{a}{l}
\end{aligned}
\tag{3.1}
$$

Thus, the static load on the real axel is a fraction of the total weigh in proportion to the dimensions of the centre of gravity and wheelbase only. By a similar process, F_{SZF} may be determined:

Performance Vehicle Dynamics
http://dx.doi.org/10.1016/B978-0-12-812693-6.00003-1

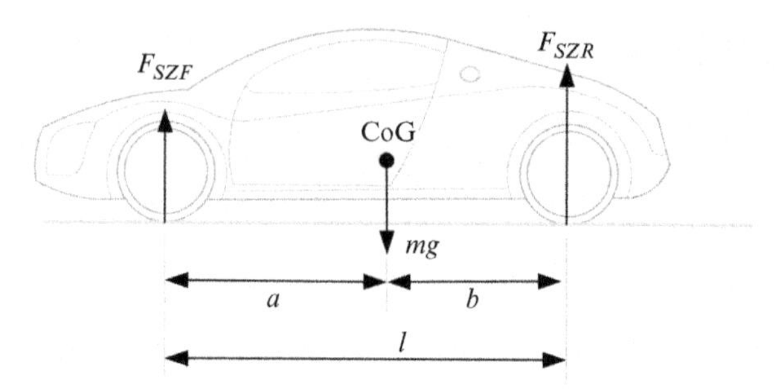

Fig. 3.1 Static wheel loads in the side view.

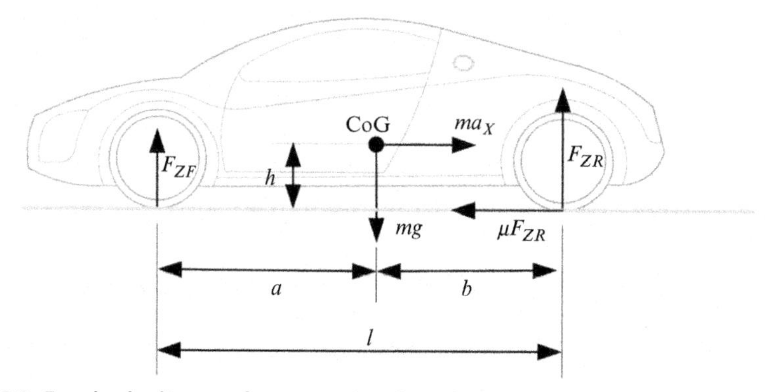

Fig. 3.2 Free-body diagram for an accelerating racing car.

$$F_{SZF} = mg\frac{b}{l} \tag{3.2}$$

When the car is accelerating, a longitudinal force at the driven wheels pushes the car forward. We may consider this force to be reacted at the centre of gravity by the inertia force, ma_X, to produce equilibrium in the side view. The free-body diagram for this situation is shown in Fig. 3.2.

Again, taking moments about the rear contact patch,

$$F_{ZF}l + ma_Xh = mg \times b$$

$$\Rightarrow F_{ZF} = \frac{1}{l}(mgb - ma_Xh)$$

$$\Rightarrow F_{ZF} = mg\frac{b}{l} - \frac{h}{l}ma_X$$

Noticing that the first term on the right is the static force on the front axle, from Eq. (3.2), we have

$$\Rightarrow F_{ZF} = F_{SZF} - \frac{h}{l}ma_X$$

Similarly, taking moments about the front contact patch,

$$F_{ZR} = F_{SZR} + \frac{h}{l} m a_X$$

Thus, when a car is accelerating, the load on the rear wheels increases, and the load on the front wheels decreases by the same amount, given by $h\,m\,a_X/l$. This quantity is often called the *longitudinal weight transfer* since it is a weight force that is *transferred*, adding one end and subtracting from the other during acceleration. Clearly as much must be gained by one end as it lost by the other since the total load must remain unchanged.

Two important points need making here. Firstly, this is an absolutely fundamental effect. It makes no difference, for example, what kind of suspension the vehicle has or what dampers are fitted. There will be weight transfer from the front to the rear under acceleration, and nothing can be done about it. A common misconception is that putting stiffer springs on the rear will change the weight transfer. Stiffer springs will simply result in a smaller deflection being needed to generate the same equilibrium force. The weight transfer comes first; then, the chassis adopts the resulting equilibrium position. The second point is that the rear axle has more load on it, and since the maximum horizontal force it can develop depends on the vertical load, this too will be increased. If the rear wheels are the driven ones, this will increase the acceleration performance in the region where there is sufficient engine torque to spin the wheels, after which the benefit will not apply. Thus, the situation is complex, and to evaluate it properly, we shall have to apply a lot more analysis. This will be the subject of Chapter 4—Straight-line acceleration.

3.1.2 Wheel Loads Under Constant Braking

The situation under braking is very similar and is shown in Fig. 3.3. In this case, however, braking effort is developed at both axles, and with the car decelerating, the load is larger on the front axle, as shown by the larger arrow.

The same analysis above applies, but this time, the front axle gains vertical load, and the rear loses it.

$$F_{ZF} = F_{SYF} + \frac{h}{l} m a_X$$

and

$$F_{ZR} = F_{SYR} - \frac{h}{l} m a_X$$

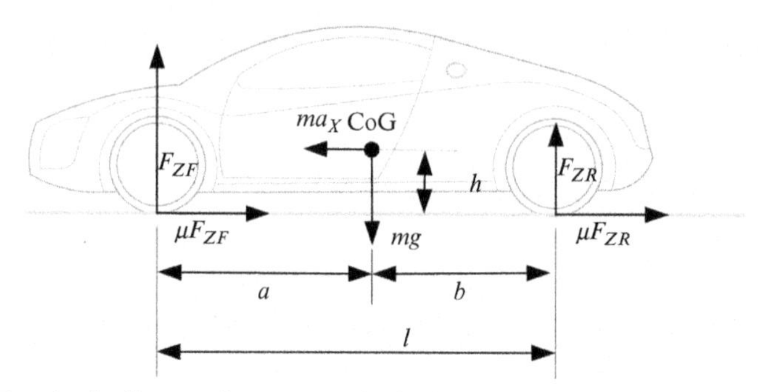

Fig. 3.3 Free-body diagram for a car under braking.

This means that under heavy braking, there is much more horizontal force available from the front wheels than there is at the rear. For this reason, racing cars and many road cars have a brake biasing system that makes the front callipers grip the front discs more than the rears do. In an ideal braking situation, all tyres would be operating at a longitudinal slip ratio to give the largest longitudinal force. Ideas such as this, aimed at optimising braking performance, will be dealt with further in Chapter 6—Braking.

3.1.3 Wheel Loads Under Cornering

3.1.3.1 Rigid Chassis Model

In a very similar way as we saw above for the longitudinally accelerating car, an equilibrium diagram can be prepared for the laterally accelerating car, and this is shown in Fig. 3.4. The centripetal force, ma_Y, is balanced by the lateral forces μF_{ZO} and μF_{ZI}.

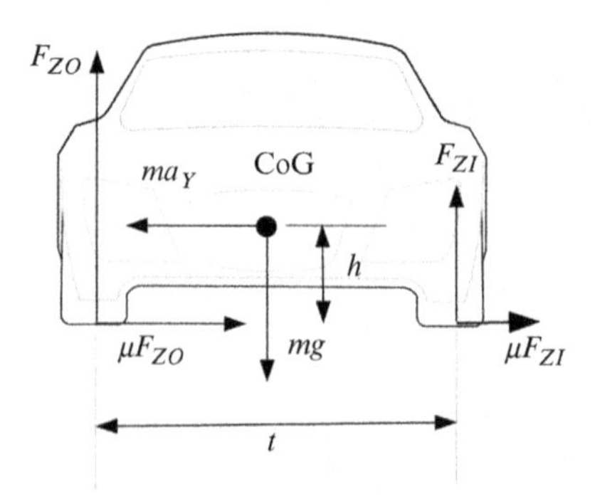

Fig. 3.4 Free-body diagram for a racing car in a cornering manoeuvre.

Taking moments about the inside wheel,

$$F_{ZO}t = mg\frac{t}{2} + ma_Y h$$

$$\Rightarrow F_{ZO} = \frac{mg}{2} + ma_Y\frac{h}{t}$$

Assuming that the centre of gravity is on the midline of the car, we can see that the static wheel load, meaning the load in the absence of any acceleration, on each side of the car, will be given by $mg/2$, and we can write the following:

$$F_{ZO} = F_{SZO} + ma_Y\frac{h}{t}$$

Similarly,

$$F_{ZI} = F_{SZI} - ma_Y\frac{h}{t}$$

Thus, as a car goes round a corner, the outside wheels have more load than the inside ones. Just as with the longitudinal case, the amount depends only on the centre of gravity height-to-track (rather than wheelbase) ratio; the mass and lateral acceleration, suspension characteristics, etc. play no role in the determination of this, steady-state value. The amount of weight transfer is given by $ma_Y h/t$.

3.1.3.2 Torsionally Compliant Chassis Weight Transfer

We have given no consideration yet to the fact that this total weight transfer must, in reality, be divided up between weight that is transferred at the front and weight that is transferred at the rear. If the chassis is very rigid, the wheel rates are the same front and rear, and the centre of gravity lies midway between the axles; then, indeed, the same amount will be transferred front and rear. However, this is rarely the case. Indeed, the total weight transfer doesn't depend on these quantities, and so, whilst the amount transferred at one end may differ from the other, the total must be independent of them, as given above.

Given that we have started to consider this difference front to rear, it may seem odd not to now give thought to it left to right for the cornering car. However, almost every vehicle is symmetric left to right, whilst very few are symmetric front to back. Thus, whilst amount of weight transferred between the front and rear axle under cornering is important, the amount transferred between the left and right under straight-line acceleration is not of interest, being generally equal. In addition, the width is much less than the length and the cross-section much less in the front view, making torsion larger than the side, and so, whilst longitudinal torsional rigidity is significant, lateral torsional rigidity is not.

One simple treatment for torsional rigidity is to introduce the parameter, ρ, which is defined as that proportion of the total weight transfer that is reacted by the front suspension. By definition, the amount reacted by the rear is all the rest. Using this, we may extend the equations above to provide four equations that give the loads on each wheel resulting from cornering.

$$F_{YOF} = \frac{F_{YOS}}{2} + \rho m a_X \frac{h}{t}$$

$$F_{YIF} = \frac{F_{YIS}}{2} - \rho m a_X \frac{h}{t}$$

$$F_{YOR} = \frac{F_{YIS}}{2} + (1-\rho) m a_X \frac{h}{t}$$

$$F_{YOR} = \frac{F_{YOS}}{2} - (1-\rho) m a_X \frac{h}{t}$$

To improve the understanding of the effect of torsional rigidity on weight transfer, we can consider the chassis to be made up of individual segments, each connected to the next with a torsional compliance. This is shown in Fig. 3.5.

In the figure, the chassis is divided up into nine segments. The horizontal cornering force is shown acting at the chassis centre of gravity in segment six.

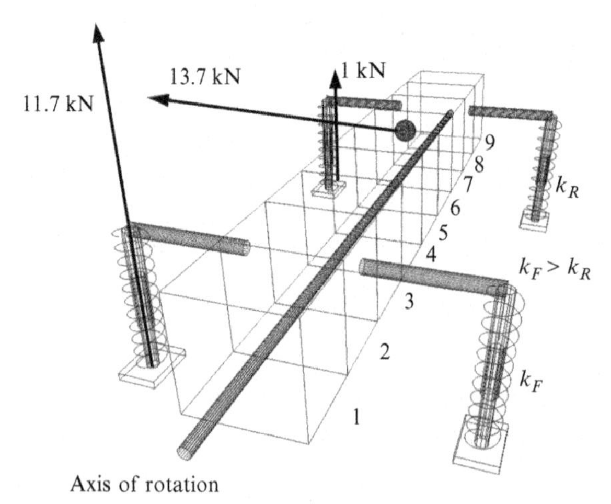

Fig. 3.5 Torsionally stiff chassis under cornering loads.

The front suspension springs are stiffer than the rear ones. The important point is that the torsional rigidity of the chassis is high here and sufficiently high that torsional deflections along the length of the chassis are negligible. Because of this, the whole chassis has to rotate as one, and it rotates about the axis of rotation shown. Because the angular rotation of the chassis along its length is the same, the linear compression of the front springs is the same as that of the rears. However, since the spring stiffnesses are higher at the front, the vertical load on the front wheels is much larger, and the length of the arrow denoting this load is much larger than that at the rear.

We may now compare this to the situation that applies when the chassis is torsionally compliant as is shown in Fig. 3.6. In this case, the situation is the same as above except that the chassis has significant torsional rigidity, meaning that the angular deflections that develop are significant fractions of the dimensions of the structure. The spring stiffnesses at the front are still much higher than those at the rear. In this case, we no longer have the requirement that the angular displacement along the length of the chassis is the same, and so, the deflection at the front springs can be different from the deflection at the rear.

In the case shown, the centre of gravity is near the rear of the car, and there are fewer segments (less compliance) between the centre of gravity and the rear than there are between it and the front (more compliance). Thus, under the action of the cornering force, the front suspension segment rotates until the resulting suspension force produces torque equilibrium, and this involves a different rotation from that at the rear. This results in a larger

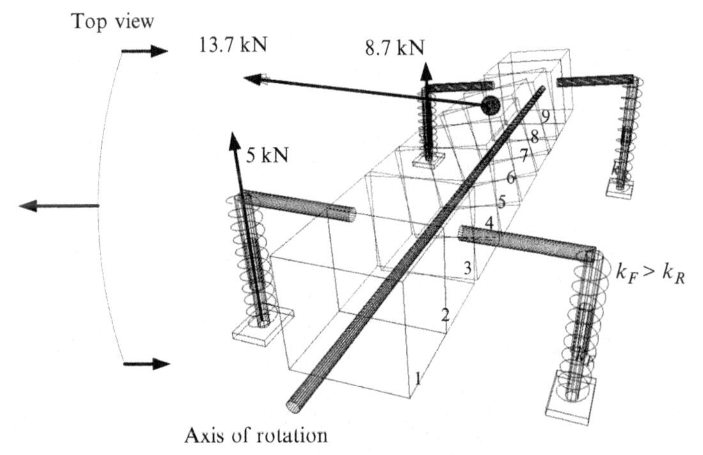

Fig. 3.6 Torsionally compliant chassis under cornering load.

force being developed in the rear suspension than the front, the opposite way around from the case above with a rigid chassis.

We shall see later that this is a very important consideration when dealing with the design of antiroll bar. These devices control the balance of a car and work by altering the roll stiffness of a given axle without altering its vertical stiffness, and so, in turn, controlling the amount of weight transfer at each end of the car. If the chassis is stiff, then this is indeed possible. However, if the chassis is torsionally compliant, then, as we have just seen, the weight distribution front to rear will depend on the chassis compliance and not the antiroll bar setting.

In fact, vehicle dynamicists are concerned with the chassis torsional rigidity for other reasons. For example, consider the chassis in Fig. 3.6 but viewed from above as shown on the left. The centripetal load would serve to bend the chassis into a banana shape as shown. This is an oversteering effect; since under its action, the car will adopt a smaller radius corner than demanded by the steering input. An additional requirement for torsional rigidity comes from the suspension kinematics. As we have seen, tyres are sensitive to very small amounts of camber change; for a wide tyre, as little as a twentieth of a degree is significant. Expressing the need for torsional rigidity in these terms, we would say that the camber change at the tyre as a result of cornering loads displacing the suspension in torsion should be negligibly small compared with the level of camber control designed into the suspension. In addition to all this, there are the load inputs from the wheels as they ride the undulating terrain. These also serve to deform the chassis torsionally, and again, we require that the deflections resulting from these loads are sufficiently small that they don't compromise the suspension geometry.

In order to model these effects, one could produce a full-car simulation, as discussed in Chapter 9, and study the effect of chassis compliance over a range of manoeuvres, characterising the loss of performance. However, the much simpler model shown here not only produces understanding but also can be used simple estimates of the quantities involved in a fraction of the time.

An Adams model based on the idea above is shown in Fig. 3.7. In this model, the chassis torsional rigidity can be controlled, the lateral force applied, and antiroll bars could easily be fitted. Measurements of deflections at the segments mounting the suspensions can easily and simply be made, and these converted into camber change values using CAD.

The model features hinge joints between each segment and rotation axis, which is fixed to the ground. There is a torsional spring between each

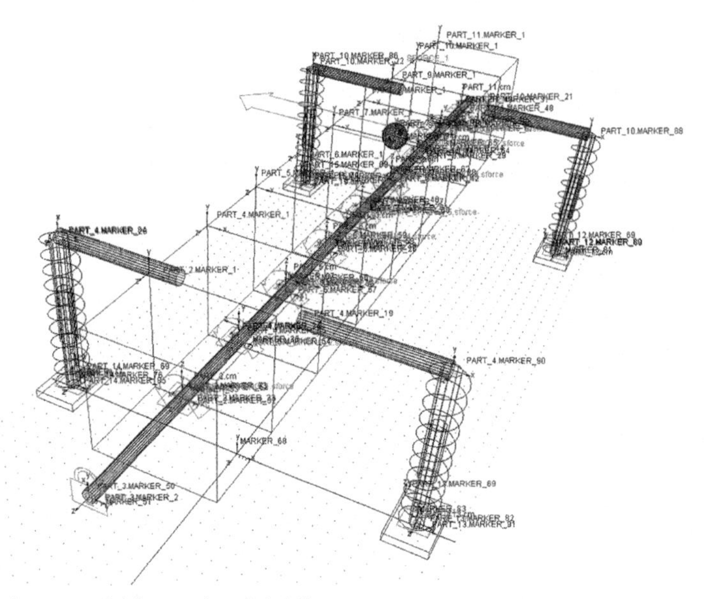

Fig. 3.7 Adams model for torsional rigidity.

segment and the next. The horizontal force is applied to segment six. The suspension springs are mounted between small square plates on the ground and mounting extensions to segments two and nine. The Adams model shown in Fig. 3.7 can be downloaded from the website for the book and viewed in motion.

3.2 TOTAL WHEEL LOADS UNDER COMBINED ACCELERATION

Using the material developed above and the concept of the forward/aft weight transfer coefficient ρ, one can easily produce a spreadsheet to determine weight transfer under lateral and longitudinal acceleration. Such a mode is shown in Fig. 3.8.

In the spreadsheet, cells B4/D16 contain the input parameters including the corner radius, current speed and weight transfer coefficient ρ. In cells B18/D25, intermediate quantities needed for the wheel load calculation are evaluated for the input parameters. In G6/I17, the individual wheel loads are determined from the equations above, and in this case, the aerodynamic downforce is added simply as the total downforce distributed evenly among the four wheels. Care must be taken to ensure that quantities have the correct sign values. Finally, as a check, the total wheel loads are determined by

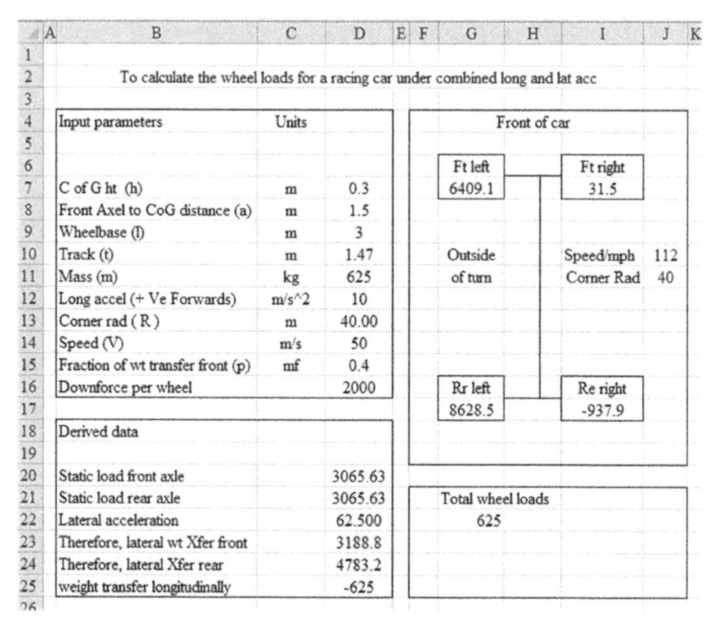

Fig. 3.8 Spreadsheet determination of wheel loads.

adding G7, I7, G17 and I17 that should equal the total wheel load originally input in D11. A file for the above model is available from the website for the book. As an alternative modelling approach, the Adams model in Fig. 3.7 could be adapted to include a longitudinal force and modified to demonstrate weight transfer. This could then be validated against the values determined by the spreadsheet to bring confidence. This is left as an exercise for the reader.

3.2.1 Effect of Aerodynamics

The effect of aerodynamics was considered in the spreadsheet in Fig. 3.8 but was treated very simply. A value was entered for the downforce on each wheel, and this is added to the sum of the vertical load from the other sources. In reality, aerodynamic downforce is more complex. To begin with, it increases with the square of the speed. In addition, the centre of pressure, meaning the location of the resultant of all the downforce components generally, moves forward with speed. This is because the front wing, being exposed to undisturbed air, works more effectively than the rest of the aerodynamic package at high speed. For this reason, the centre of pressure moves forwards with increasing speed for a racing car, and the downforce on the front wheels increases more with speed than it does at the rear. By contrast, the elements towards the rear, such as the rear wing, operate in more turbulent air as the speed increases.

In a road car, designers may choose to design in 'upforce', where the result of aerodynamic forces acts to lift the car with speed. This reduces tyre wear slightly and also brings a very progressive and slow deterioration in performance that has the effect of discouraging excessive speed in a safe way. In a racing car, the increasing downforce can make the car progressively better and better until it very suddenly loses controlled response. Whilst a very skilled racing driver may be able to deal with this sudden onset of poor control, it certainly would not be desirable on the open road.

With this understanding, it would be simple enough to include these effects in the spreadsheet in Fig. 3.8, but clearly, one would need to know the summary of the aerodynamic package of the particular car in order to do so.

3.2.2 Roll Over Limit, Skidding Limit

There are only two ways in which a vehicle can reach the limit of performance as the speed through a corner is progressively increased. The first is for it to roll over; the second is for it to skid sideways. We can determine limits for rollover and skidding as follows:

In Fig. 3.9, we can see the front view of a car going around a corner in equilibrium. The number N is the downforce multiple of the car weight. When N is zero, there is no downforce, and the force acting is simply the weight force of the car. The peak acceleration that the car can sustain in roll is denoted as $\hat{a}_{YROLL}$. Taking moments about the outside wheel,

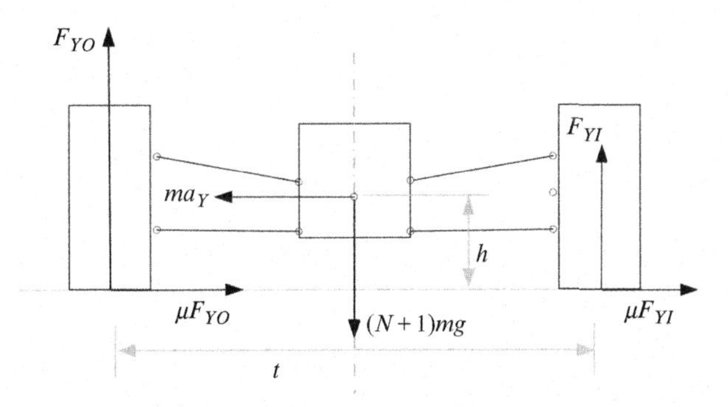

Fig. 3.9 Equilibrium in the front view.

$$F_{YI}t + m\hat{a}_{YROLL}h = (N+1)mg\frac{t}{2}$$

The onset of roll will occur approximately when the inside wheel vertical load decreases zero. Thus,

$$0 + m\hat{a}_{YROLL}h = (N+1)mg\frac{t}{2}$$

$$\Rightarrow \hat{a}_{YROLL} = (N+1)\frac{t}{2h}g$$

This very simple equation shows that, as the aerodynamic downforce is increased, the rollover threshold is too.

We might expect some vehicles to roll over at the limit, a double-decker bus maybe, but others would not, a racing car, for example. We can see why this is by looking at the maximum acceleration that a car can sustain frictionally. Working again from the figure,

$$m\hat{a}_{YSKID} = \mu F_{YO} + \mu F_{YI}$$

$$\Rightarrow m\hat{a}_{YSKID} = \mu(F_{YO} + F_{YI}) = \mu(N+1)mg$$

$$\hat{a}_{YSKID} = \mu(N+1)g$$

Thus, again, we can see that the peak lateral acceleration a racing car can sustain goes linearly with downforce. A car with $N=4$ has four times the vertical load pushing down on the tyres and so can develop four times the lateral force. Since the car mass has not been increased, this results in four times the lateral acceleration. For this reason, some people describe aerodynamics as adding weight force without adding actual weight.

A vehicle will fail at which ever limit comes first as the speed increases. Thus, a vehicle will roll if $\circ a_{YROLL} < \circ a_{YSKID}$. Comparison of the two equations shows that rollover,

$$(N+1)\frac{t}{2h}g > \mu(N+1)g$$

$$\Rightarrow \frac{t}{2h} > \mu$$

If we approximate μ to unity, then to avoid rolling over the track must be more than twice the centre of gravity height. The larger the μ is, the larger this multiple is.

3.3 TRANSIENT WEIGHT TRANSFER

In the above analysis, we have treated the vehicle as if it were a solid object. The diagrams showed the suspension links and the tyres, for example, yet no account was taken of the fact that the suspension will move under the loads applied. For example, as a car rolls in response to corner entry, the body will rotate in roll, the inside tyre will extend in response to the decreased vertical load, and the outside one will compress. Processes such as these can take time to evolve, and the vertical loads on the tyre will not suddenly adopt the equilibrium value obtained above but instead go through a *transient* where the load is changing with time. Indeed, it isn't only the vehicles response to the applied acceleration that goes through a transient rather than simply adopting the final position, the source of the transient themselves takes time to evolve. A steering input is never instantaneously applied, nor is a braking one, nor is the torque response of the engine to a throttle position change. In this section, we shall briefly consider a number of effects that make for a transient weight transfer before the steady-state values are reached; these will be revisited later.

3.3.1 Roll Transient

Looking at Fig. 3.10, we can see that the lateral force at the contact patch, F_1, is reacted by a horizontal force, mV^2/R, at the centre of gravity to produce an equilibrium diagram in the front view. Clearly, the action of F_1 is to produce a torque that acts about the centre of gravity, and this is shown by the torque, T. The exact way that this happens is not at all straightforward; for example, the force F_1 must act on the body at some point by being communicated through the suspension links, but where would this be? Its position is important because it affects the magnitude of the torque T that in turn affects how much the body rolls, which, in its turn, affects the force in the pushrod

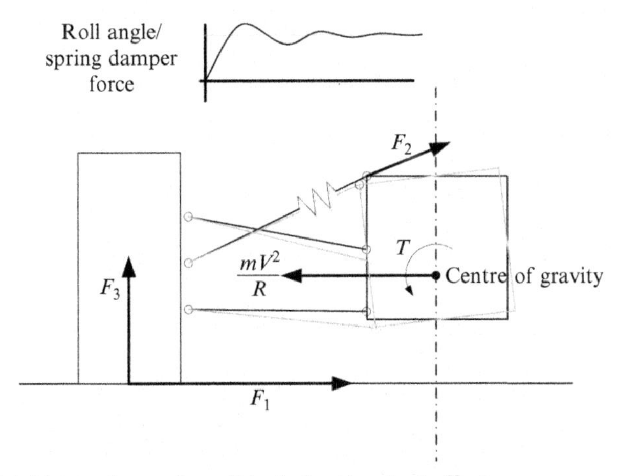

Fig. 3.10 Weight transfer under roll includes transient effects.

and so affects the vertical load on the contact patch. The situation is even more complex since the suspension links are not normally horizontal and given that they can only support an axial force since they have ball joints at either end, there must be a vertical component of this acting on the body. Indeed, there is and this is the force that gives rise to *jacking* where a car body rises or falls on corner entry. A full consideration of this will have to wait until Chapter 7 on suspension kinematics. For the time being, where we are concerned with weight transfer, we shall be content simply to see that the torque T must exist and that, as a result of it, the body will roll.

Under the action of this torque, the body will roll until equilibrium in roll is established. The body clearly has a moment of inertia in the front view. The figure shows a graph of the roll angle against time. The graph evolves in the following way. Firstly, the torque acting on the body starts it rotating; then as its angular velocity increases, it gains momentum. On top of this, the torque resisting the rotation caused by the evolving compression in the suspension springs builds up. The momentum may well be sufficient to cause the body to roll beyond the equilibrium position and then rotate back towards it. After some time, the damping in the system will mean that the rotation stops at that displacement for which the torque developed in the suspension is in equilibrium with that produced by the cornering force. The exact nature of the response, the number of overswings, etc. depends on the exact parameter values involved, but the process is the same. Thus, a car will go through a transient response before settling to its final equilibrium.

3.3.2 Weight Transfer From Unsprung Mass

The unsprung mass clearly has no suspension through which to communicate forces, and so, weight transfer that originates this way is transferred much more quickly. This weight transfer still results in compression of the tyres, which are themselves spring dampers, and so, the transfer is not instant, but since damping values here are low compared with the suspension and spring values normally higher, force communication is much faster.

In Fig. 3.11, we can the see that the static load due to each unsprung mass, F_{UZS} is given by

$$F_{UZS} = \frac{m_U g}{2}$$

To determine the unsprung weight transfer, we take moments about the inside contact patch (clockwise +ve) where p is the height of the centre of gravity of the unsprung mass:

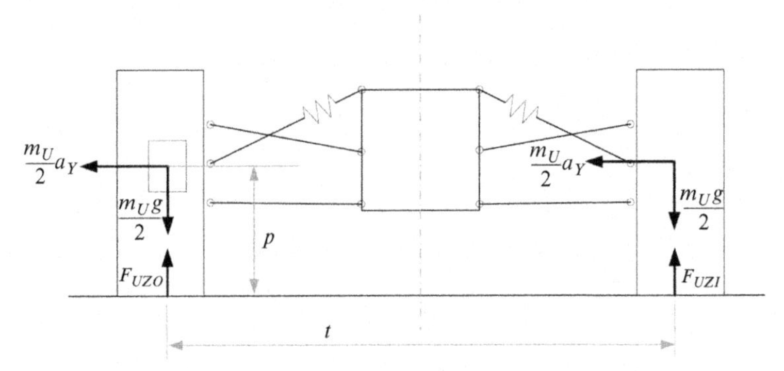

Fig. 3.11 Weight transfer due to the unsprung mass.

$$F_{UZO}t = m_U a_y p + \frac{m_U g}{2}t$$

$$\Rightarrow F_{UZO} = \frac{m_U a_y p}{t} + \frac{m_U g}{2}$$

$$\Rightarrow F_{UZO} = F_{UZS} + \frac{m_U a_y p}{t}$$

Similarly, about the outside contact patch,

$$\Rightarrow F_{UZI} = F_{UZS} - \frac{m_U a_y p}{t}$$

Thus, the amount of weight transfer due to the unsprung mass is given by

$$WT_{UM} = \frac{m_U a_y p}{t}$$

It is useful to see this quantity separately from the weight transfer of the whole vehicle mass as one lump because this weight transfer from the unsprung mass is communicated directly to the tyre instantly following the lateral acceleration. By contrast, the weight transfer from the sprung mass must pass through the suspension, either through the suspension links that being ridged communicate it instantly or through the spring damper, in which case they act after some time.

3.4 WEIGHT TRANSFER FOR MAXIMUM PERFORMANCE

We have seen that weight transfer is an inevitable consequence of acceleration in the horizontal plane, lateral or longitudinal. It has different elements to it; there are the steady-state value and the transient phase, and there are different sources of weight transfer, the sprung and unsprung mass. The obvious question for the vehicle dynamicist is whether it advantageous or not. What makes for the best performance? To answer this, we shall look at the three cases of straight-line acceleration, braking and cornering separately.

3.4.1 Straight-Line Acceleration

In Fig. 3.2, we saw that under forward acceleration, longitudinal weight transfer increases the loads on the rear wheels and decreases it on the front. For a rear-wheel-drive car, this is an advantage because the increased vertical load allows increased tractive effort and so a higher acceleration. It is true that there is less vertical load on the fronts, but if the car is not being steered and has no drive at the front, this doesn't matter. Indeed, it is for this reason that if the designer of a high-performance car is given the choice, rear-wheel drive will be chosen. No dragster was ever made with front-wheel drive!

3.4.2 Braking

Under braking, it is tempting to think that the situation is simply the reverse of straight-line acceleration. However, that is not the case. The difference is that, under braking, all wheels are producing braking effort. With weight transfer acting, there will be more load on the fronts than the rears, but not as much horizontal force will be gained at the front as is lost at the rear, and so, weight transfer is undesirable.

3.4.3 Cornering

Consider a cornering car as shown in Fig. 3.12. The car is in steady-state corner, meaning it has a constant yaw rate and therefore must have no

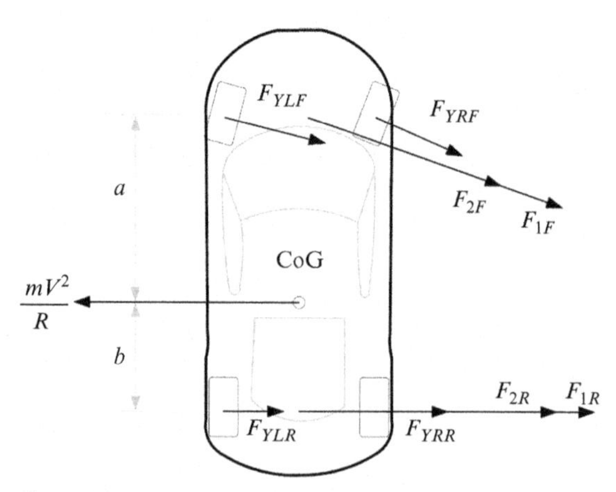

Fig. 3.12 Cornering car.

net torque urging it to accelerate in yaw. We see that lateral forces at the front tyres are given by F_{YLF} and F_{YRF}. The resultant of these is initially taken to be F_{1F}. At the rear, a similar picture pertains, and the resultant force is F_{1R}. The torque acting about the centre of gravity of the force F_{1F} is given by $F_{1F} \times a$ (ignoring the angle, it makes to the direction of mV^2/R, which in practice is small). For the car to be in yaw equilibrium, this must be balanced by the equivalent torque from the rear given by $F_{1R} \times b$.

Consider now the effect of equal weight transfer at the front. Here, F_{YLF} will increase but not by as much as F_{YRF} will decrease, since lateral force is non-linear with vertical load. So, we can see that F_{1F} will get smaller and reduce to F_{2F}. A similar process will apply at the rear, and F_{1R} will get smaller too and become F_{2R}. As a result, mV^2/R will get smaller, and so, the velocity for an equilibrium corner will be decreased as a result of weight transfer. This is undesirable since the car will be slower.

However, there is even more to it than that. Imagine if a *different* amount of weight is transferred at one end, perhaps more at the rear than at the front. Then, the maximum lateral force that can be generated at the rear will be reduced by a larger amount at the front $F_{2R} < F_{2F}$. The driver will then have to reduce the steering angle at the front to reduce the slip angle to keep the moment balance, and a *different* steering wheel position will result in a corner of the *same* radius. This is clearly a much more complex effect, and it certainly doesn't sound desirable. Surely, one would want a given steering wheel position to always produce a corner of the same radius. Understanding this and more importantly learning how, as a vehicle dynamicist, we can control it, is a major topic that we shall solve in Chapter 5 on cornering.

3.5 SUMMARY

In summary, other than in straight-line acceleration, weight transfer is disadvantageous, and it reduces performance. This is the reason that designers pursue reductions in centre of gravity height so assiduously. However, we have seen in this investigation of weight transfer that there are issues that need much more consideration in order to properly understand them. For example,

- Under cornering, if a different amount of weight is transferred at each end, the amount of lateral force at each end will be different. If the driver is to continue on the same path, the steering will have to change to compensate. This is a complicated effect, and when we deal in earnest with cornering in Chapter 5, we shall see this in much more detail.

- As the car enters, a corner weight is transferred and passes through the solid suspension links and also through the suspension spring dampers. This is a complex process. We shall develop a complete understanding of this on both the chapters on cornering and suspension dynamics.
- Under braking, there will be more load on the front axle, so the front brakes will be made to produce more braking effort, but by how much?

So, we see that vehicle dynamics is a highly interrelated set of topics and one topic affects another. It makes perfect sense to have started with the consideration of tyres since these are the elements that produce all the forces that move the car and so must come first. Weight transfer is a logical next step because the horizontal loads that we are so interested in depend directly on the vertical load applied to the tyre. What we have come to see is that the vertical load on each tyre is a highly variable parameter, changing all the time. Now that we have this understanding, we can proceed to the next three topics of straight-line acceleration, cornering and braking.

Finally, we also note that in this section we have moved forward considerably since our first vehicle dynamics models have been made. Admittedly, they are simple models for weight transfer, but they are models never the less. As we shall see in Chapter 9, lap-time simulation and full-vehicle simulation, it is possible to make extremely advanced models of entire vehicles. However, if one is only interested in the weight transfer affecting a particular vehicle, it is so much more practical to make a very simple model that deals only with this case than it is to wade in with full-vehicle simulation. Modelling vehicle dynamics is very much about making plenty of simple small models to understand and optimise parts of a vehicle separately, and some full-vehicle simulation for times when it's expense time and complexity are warranted. In a machine shop, one does not use a seven-axis machining station when a hacksaw will suffice!

3.6 QUESTIONS

1. A car is cornering at 0.75 g with constant speed, weighs 450 kg and has a centre of gravity height of 0.3 m and a track of 1.5 m. Torsional rigidity is such that 0.6 of the weight transfer occurs at the front. The distance '*a*' from the front axle to the centre of gravity is 1.25 m, and the wheelbase is 2.5 m. Calculate the two front wheel loads and comment on the result.
2. A car has a mass 1300 kg, centre of gravity height 0.4 m, a wheelbase of 2.1 m and an '*a*' value of 1.3 m. Determine the static axle loads.

3. For the car in (1) above, what will be the axle loads when decelerating at 0.2 g? What deceleration value will be required to make the rear axle load fall to zero?
4. A vehicle has a centre of gravity height of 0.38 m, an '*a*' value of 0.95, a wheelbase of 2.1 m and a track of 1.38 m. It is travelling around a corner of radius 37 m at 36 mph and braking with 0.25 g. The fraction of weight transfer that occurs at the front is 0.44. Determine the wheel loads for this situation.
5. A vehicle with centre of gravity height 0.5 m, track width of 1.5 m and with downforce of half the weight of the car at the point in consideration. Determine whether it will roll or skid if driven too fast.
6. By making estimates for the dimension involved, use the vehicle in (1) as a starting point and allocate 12% of the vehicle mass to be unsprung. Then, determine the amount of weight transfer that results from spring and unsprung mass.
7. After extending the analysis in the chapter to deal with banked roads, determine the wheel loads for the car in (1) above when travelling over a road banked at 7 degrees and in which the centre of gravity has been place 0.14 m further from the corner centre than the midline of the vehicle.
8. Why is weight transfer always disadvantageous for cornering but can be advantageous for straight-line acceleration?
9. What are the important differences when weight transfer is analysed in the side view compared with the front view?
10. Why is it not the case that increasing spring stiffness on the front suspension will increase weight transfer on the front axle?

3.7 DIRECTED READING

Weight transfer is a fundamental part of vehicle dynamics, and many texts give it a good treatment.

Jazer [10] is a useful reference and deals with two- and four-wheeled vehicles, road inclination, vehicles with more than two axles, vehicle pulling trailers, etc.

Gillespie [11] again provides a good coverage, dealing with grade as well.

Milliken and Milliken [9] in their text have an excellent chapter dealing with all of the above and the asymmetric chassis.

3.8 LEARNING PROJECTS

- Collect data for a range of cars, perhaps road cars, racing cars and even commercial vehicles and gather data for their dimensions, the track, wheelbase and centre of gravity height (centre of gravity can be hard to obtain and may have to be estimated). Produce a chart of this data and use your spreadsheet for weight transfer to estimate the weight transfer for each under normal operating conditions.
- Produce a version of the weight transfer spreadsheet that includes an aerodynamic model. The inputs of the aerodynamics are the forward speed, and you must make a map for the downforce and the migration of the centre of pressure and then include these on top of the basic weight transfer.
- Produce a 'rollover versus skid' comparison spreadsheet. Inputs are made to the spreadsheet, and it then determines whether the vehicle will roll or skid. Analyse a collection of real vehicles to see where they lie with respect to these thresholds.
- Produce an Adams model of a vehicle in the front view and by applying a cornering force determine the roll transient function. Then, slowly increase the length of time taken for the force to build and determine its effects on the roll transient. Examine the effect of the damper value on this relationship and comment on the relevance of this to vehicle dynamics.
- Extend the treatment of weight transfer to include the effect of sloping downhill (front to back) and banking (left to right). To do this, balance moments just as before but starting with an equilibrium diagram in the front and side view.
- Extend the treatment of weight transfer in the front view to account for unequal track front to rear.
- When a car rolls under cornering, weight is transferred as we have seen. It is also transferred by a secondary effect. As the car rolls, the centre of gravity moves physically closer to the outside wheels and becomes asymmetrically placed. Produce a geometrical analysis of this by balancing moments and estimating distances to arrive at your own estimate of the significance of this effect. For a stiff low racing car, you should find that it's not a big consideration, but for a tall soft double-decker bus, it is perhaps more so or indeed this is the case.
- Using data for a car you are familiar with, measure the required dimensions from the vehicle and then estimate the wheel loads for a cornering situation you are familiar with. Obtain estimates for the wheel spring

rates of the car and then estimate the roll angle expected. You can estimate wheel rates by sitting on one corner and measuring the resulting deflection directly. Estimate the real roll angle by getting a friend to film using a digital camera from the inside of the vehicle looking outwards through the windscreen. The tilt of the car from the horizontal can be estimated from the movie afterwards. It is essential to do all this legally and safely!

3.9 INTERNET-BASED RESEARCH AND SEARCH SUGGESTIONS

- It is not easy in practice to determine the centre of gravity height for a vehicle. If one were to hang the vehicle from a rope around the front tyres and then chalk a line extending the line of the rope on the side of the car, one would know that the centre of gravity would be somewhere on that line. Repeating this for the other end would produce another similar line. Where the two cross, is the centre of gravity in the side view. This would indeed work but is scarcely practical. Do some internet-based research to find a method that can be made to work in an ordinary garage or pit making use of corner weight gauges. Look out for sources of error here; some are much more than you might think. Once you have a good understanding of how to do this, produce a spreadsheet into which you can enter the parameters, and it will produce the estimate. Use the spreadsheet to examine how much error in, for example, the linear dimensions used translates into error in the centre of gravity height and then compare this to how much deformation will be experienced in the tyres as this process is carried out. Then, answer the following questions:
  - What would be the best angle to use?
  - Why is the compliance of the tyres such a significant issue?
  - What can be done experimentally to improve accuracy?
- Using data you gain from internet research on the dimensions and performance of dragsters estimate the maximum acceleration before the front wheels lift off.
- Using the internet, gather data for a whole range of vehicles, collecting centre of gravity height, track and wheelbase and produce a plot of track against wheelbase for given centre of gravity heights. Further, refine the library to include power, weight, acceleration times, etc. and produce a database against which new vehicles can be benchmarked.

CHAPTER 4

Straight-Line Acceleration

4.1 INTRODUCTION

In this chapter, we consider one of the simpler aspects of vehicle dynamics that of straight-line acceleration. In this manoeuvre, tractive effort supplied by the engine is communicated to the road by the tyres, and as a result, the car accelerates forwards. This situation is shown in Fig. 4.1.

The dynamics for this could hardly be simpler. The consequences of the tractive force F_X are found by replacing it with an equivalent force system acting at the centre of gravity, CoG. Firstly, this consists of a torque, T, which serves to rotate the car clockwise in the view shown increasing the load on the rear tyres and decreasing it on the front. This is the cause of weight transfer. Secondly, there is the same force F_X that accelerates the car to the left with uniform acceleration, and we may write

$$V = U + at$$

$$V^2 = U^2 + 2as$$

$$S = Ut + \frac{1}{2}at^2$$

where V is the final velocity, U is the initial velocity, t is time for which the linear acceleration applies and a is the acceleration of the car. The force F_X that accelerates the car is related to the tractive torque applied to the wheel by the engine through the gearbox by

$$F_X = \frac{T}{r}$$

where T is the torque at the wheel and r is the radius of the wheel.

4.2 ESTIMATE OF MAXIMUM ACCELERATION VALUE

Note that for maximum acceleration, we shall want to have the weight of the car on the rear wheels, given that the tractive force will be given by

$$F_X = \mu P_Z$$

where P_Z is the vertical load on the driven wheel and that if the car has sufficient tractive effort to put all of the weight force mg on the rear axle, then

Performance Vehicle Dynamics
http://dx.doi.org/10.1016/B978-0-12-812693-6.00004-3

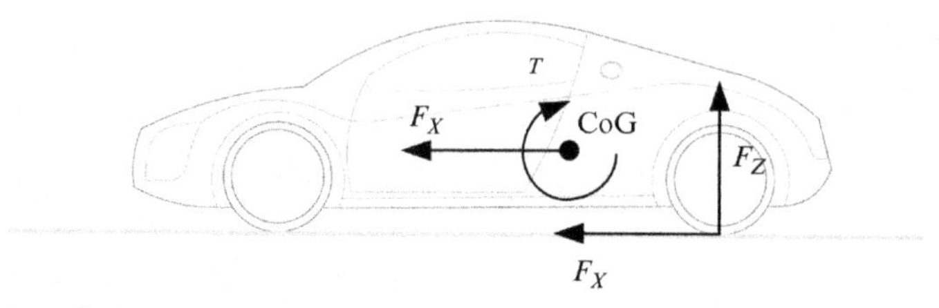

Fig. 4.1 Straight-line acceleration.

$$F_X = \mu mg$$

If we assume that all this force is expended in accelerating the car and so for a maximum longitudinal acceleration of $\hat{a}_X$,

$$m\hat{a}_X = \mu mg$$

or

$$\hat{a}_X = \mu g$$

Thus, we can never do better than μg for longitudinal acceleration, worse possibly, if not all the weight is on the rear, but never better.

In reality, the picture is much more complex than this and the following factors need to be included in the modelling:

- The car experiences rolling resistance from the tyres, from the aerodynamic drag, from friction in the oil of all rotating parts, etc.
- The torque supplied at the wheel is not a constant but depends on engine rpm and the gear selected at the time.
- Not all the tractive force is available for longitudinal acceleration since some is expended in rotational acceleration of the rotating mass.
- The vertical load on the tyre is itself dependent on acceleration. As the car accelerates, load is 'transferred' to the rear, enabling a larger tractive force to be applied.
- The engine may be capable of supplying more torque to the wheel than the tyre is able to transfer to the road. In this situation, the tractive force should be limited to that which the tyre can transfer rather than that available. This is 'grip limited' acceleration (the alternative situation is 'traction limited' meaning the tractive force available limits the acceleration).
- Consideration should be made of the gear shift duration, during which no tractive force is available and the car is coasting.
- The road may be inclined, and the weight force of the car will have a component parallel to the road surface that will either increase or decrease the acceleration value depending on whether the car is accelerating up or down hill.

- As the car pulls away from rest, there is a period when the torque available is determined by the amount of torque being transmitted by the slipping clutch, and this is less than the engine torque.
- The tractive force is not just a coefficient of friction times the vertical load, it is given by a more complex function that depends not just on vertical load but also on the longitudinal slip value that pertains at any given instant, together with other variables such as tyre surface temperature, carcass temperature, inflation pressure and tractive effort history (see Chapter 2 on tyres).
- As the car accelerates, depending on the suspension design, there may be camber change under 'squat' (rear suspension compression caused by weight transfer) that would reduce the tractive force that could be developed.
- As the car speeds up, there may be a reduction in ride height due to aerodynamic downforce again causing camber change and also in increase in the contact patch size.
- Tyre wear will change the radius and rotating inertia of the tyres, thus modifying the acceleration achievable.
- Fuel consumption will affect wheel load and load distribution throughout a race.
- Weather, rain in particular but also temperature, affects the mu value for the tyres.
- Road surface similarly so.
- Altitude and wind direction similarly so.

In the sections that follow, we shall develop progressively more complex models to deal with many of these effects, but firstly, we must develop some important tractive modelling concepts.

4.3 STRAIGHT LINE ACCELERATION MODELLING

We start by introducing some basic concepts.

4.3.1 Torque and Power

Power and torque are related as follows:

$$\text{work done} = \text{force} \times \text{distance}$$

$$\Rightarrow \frac{\text{work done}}{\delta t} = \text{force} \times \frac{\text{distance}}{\delta t}$$

$$\Rightarrow \text{power} = \text{force} \times \text{velocity}$$

For a rotary system,

$$\text{power} = \text{torque} \times \text{angular velocity}$$

where the power is measured in watts, the torque in nm and the angular velocity in rads/s. For a car accelerating forwards,

$$F_X = ma_X$$

However, the tractive force at the wheels is consuming power from the engine:

$$P = F_{YR} \times V_X$$

where P is the engine power and V_X is the current forwards velocity; thus,

$$a_X = \frac{F_X}{m}$$

$$\Rightarrow a_X = \frac{P}{mV_X}$$

Thus, a racing car will have a smaller and smaller forward acceleration as the speed increases, even in the absence of any drag. The acceleration doesn't ever reach zero, it just gets smaller. Consequently, in the absence of drag, the velocity would indeed continue to increase, but not nearly as fast as to begin with. The diagram below shows a vehicles velocity and acceleration determined in this way, together with graphs (Fig. 4.2).

4.3.2 Rotating and Nonrotating Mass

We may improve on this rather simplistic approach by including the effects of the rotating mass. The body of the car merely has to be accelerated forwards, but the wheel and all other rotating masses must be accelerated not only linearly but also rotationally, ignoring any slope on the ground (Fig. 4.3).

The forward force is given by

$$F_X = \frac{T}{r}$$

which is expended firstly against the drag, the remainder then being available to accelerate the car linearly. We shall ignore here any moment generated by F_X or D about the centre of gravity, and note that the portion of the cars mass that rotates is denoted M_R and the rest that only accelerates linearly is denoted M_L, and so, we may write

$$F_X - D = (M_R + M_L)a_X$$

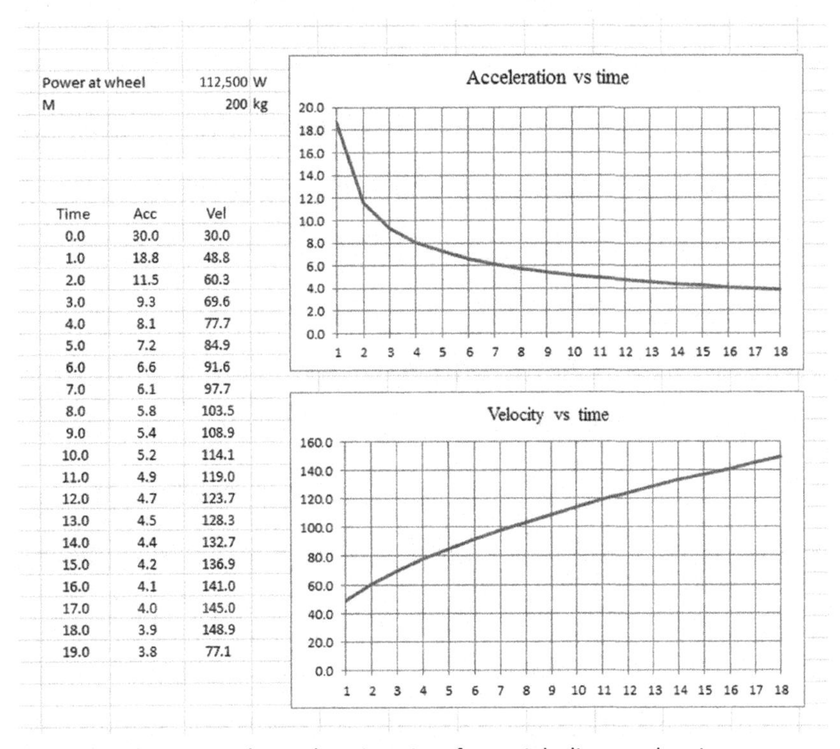

Power at wheel	112,500 W	
M	200 kg	

Time	Acc	Vel
0.0	30.0	30.0
1.0	18.8	48.8
2.0	11.5	60.3
3.0	9.3	69.6
4.0	8.1	77.7
5.0	7.2	84.9
6.0	6.6	91.6
7.0	6.1	97.7
8.0	5.8	103.5
9.0	5.4	108.9
10.0	5.2	114.1
11.0	4.9	119.0
12.0	4.7	123.7
13.0	4.5	128.3
14.0	4.4	132.7
15.0	4.2	136.9
16.0	4.1	141.0
17.0	4.0	145.0
18.0	3.9	148.9
19.0	3.8	77.1

Fig. 4.2 Acceleration and speed against time for straight-line acceleration.

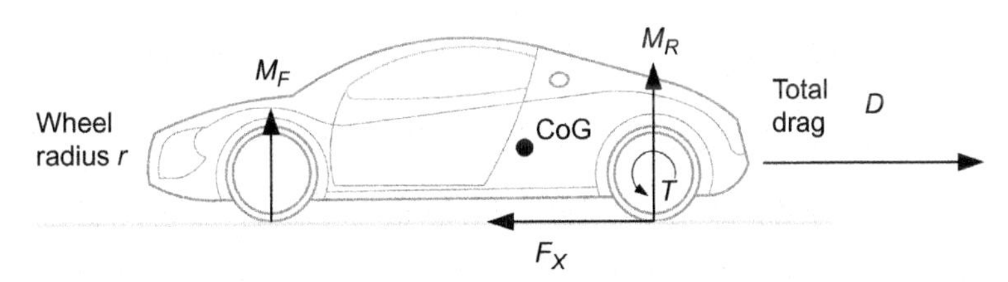

Fig. 4.3 Influences on straight-line acceleration.

Hence,

$$T = r\{(M_R + M_L)a_X + D\}$$

The wheels however must also be accelerated rotationally. If the wheel is rotating with an angular velocity ω (rads/s), then the linear velocity at the edge is

$$V = \omega\, r$$

Thus, the linear acceleration, a_X, at the edge is

$$a_X = \dot{\omega} r$$

The torque necessary to accelerate rotationally a wheel of moment of inertia I is

$$T = I\dot{\omega}$$
$$\Rightarrow a_X = \frac{Tr}{I}$$
$$\Rightarrow T = \frac{a_X I}{r}$$

The total torque is the sum of both the torque taken to accelerate the car linearly and the torque taken to accelerate the wheels rotationally, and noting that we have four wheels,

$$T_{TOT} = r\{(M_R + M_L)a_X + D\} + \frac{4a_X I}{r}$$

which can be rearranged to give the current value of acceleration:

$$\begin{aligned} T_{TOT} &= ra_X(M_R + M_L) + rD + \frac{4a_X I}{r} \\ \Rightarrow T_{TOT} &= a_X\left(r(M_R + M_L) + \frac{4I}{r}\right) + rD \\ \Rightarrow a_X &= \frac{T_{TOT} - rD}{\left(r(M_R + M_L) + \frac{4I}{r}\right)} \end{aligned} \tag{4.1}$$

This is an important step since we can now use the equation to calculate the acceleration from a knowledge of the engine torque and drag characteristics.

4.3.3 Engine Torque and Wheel Torque

A graph of engine torque is shown in Fig. 4.4.

Engines have a limited rpm range over which they can operate, typically between 1000 and 6000 for a road car engine and around 5000–18,000 for an F1 engine. To produce good acceleration in a straight line, it makes sense to have a significant gear reduction between the crankshaft and the wheel;

$$\text{Wheel torque} = \eta_{gear}\,\text{engine torque}$$

where η_{gear} is the gear ratio in which the car is currently being driven. Thus, in principle, we want a very low gear ratio indeed. The lower limit comes when the torque at the wheel is so high that the time spent in the gear before the

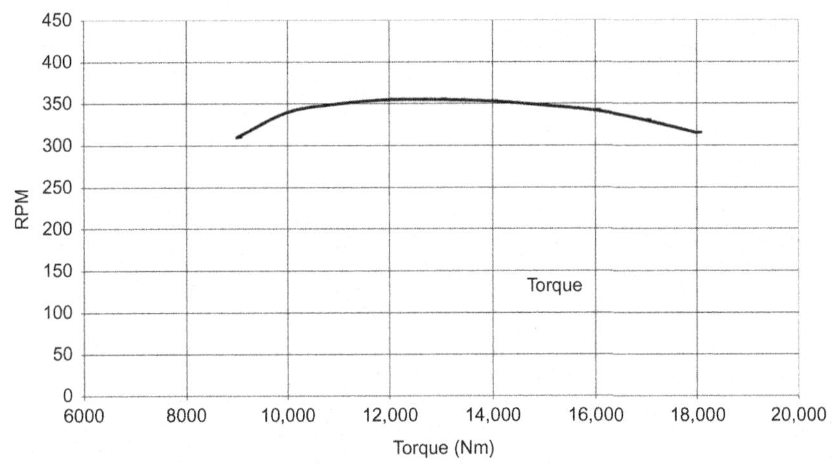

Fig. 4.4 Engine torque versus rpm.

engine reaches its rev limit is impracticably short, and the gear shift is needed after only a few metres after pulling away. Gears are then added until the torque available at the wheels is totally expended in all the sources of drag.

To obtain the straight-line performance, one takes the torque data and scales it for the gear ratios and then plots it against road speed rather than rpm as in the example in Fig. 4.5.

Terminal velocity is reached when then the tractive effort from the wheel torque equates to the drag force from all sources.

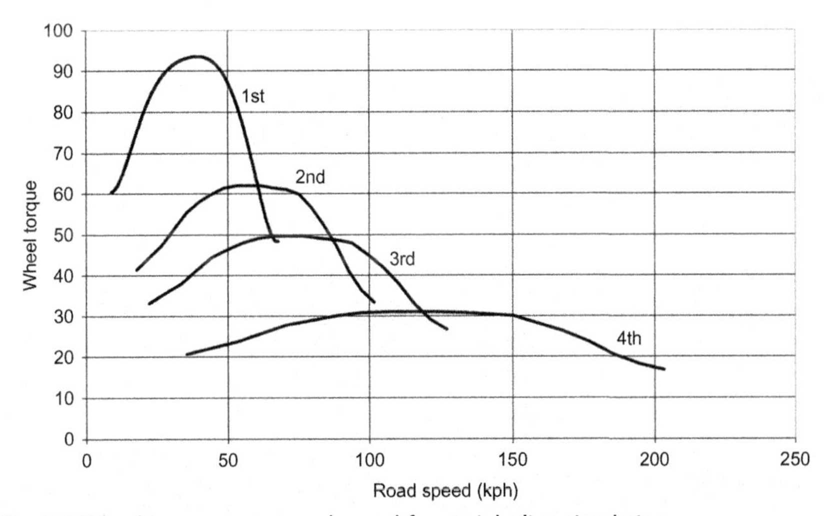

Fig. 4.5 Wheel torque versus road speed for straight-line simulation.

4.3.4 Drag

Before Eq. (4.1) can be used to predict straight-line performance, the drag force 'D' must be determined. This drag force has many sources. Rolling resistance of the tyres, friction acting on rotating components and aerodynamic drag are but a few.

4.3.4.1 Aerodynamic Drag

In high-speed formula or road cars where aerodynamic downforce is a major part of the package, aerodynamic drag is much the largest contributor to the total drag force. In F1, for example, the aerodynamic drag is so large that if a driver lifts off the throttle at top speed, the car will decelerate at around 1 g without even touching the brakes simply due to aerodynamic drag. The aerodynamic drag on a car is given by the formula

$$D_{AERO} = C_D \times \frac{1}{2} \rho A V^2 \tag{4.2}$$

where D_{AERO} is the drag force, C_D is the drag coefficient, ρ is the density of the air, A is the frontal area of the car and V is the velocity of the vehicle. The value for C_D is determined by either wind tunnel testing or computational fluid dynamic simulation. Some approximate values are given below:

Object	C_D
Excellent aerofoil	0.005
Very low drag road car	0.3
Poorly shaped road car	0.5
Hummer	0.56
F1 car	1.2
Flat plate facing the air flow	2.0

The value for an F1 car may seem high here, but it should be remembered that these cars produce huge amounts of downforce many times the weight of the car, and this makes for a lot of drag.

4.3.4.2 Rolling Resistance

A tyre requires a force to roll it along even when it is not cornering or bearing a torque. This is called rolling resistance. There are two sources; firstly, the torque generated by the vertical load acting behind the axle as we saw in Fig. 2.15, and secondly, there are hysteretic losses every time rubber is deformed, and with every rotation, the entire tyre has passed through the deformed contact patch leading to substantial resistance.

A large amount of this rolling resistance comes from the flexing of the rubber as it passes through the deformed shape near the contact patch. The larger the deformation and the more frequently it happens, the larger the rolling resistance, and indeed, rolling resistance increases with both speed and load. A racing tyre at 60 mph will consume several horsepower in rolling resistance.

4.3.5 Coast-Down Test

The best way to obtain a value for drag force to be used in Eq. (4.1) is to measure it. In a coast down test, the car is driven up to near top speed, and then, the driver disengages the vehicle clutch and lets the car slow down to near rest under its own resistance.

In Fig. 4.6, the results of a coast down test are shown. To obtain the drag from the test, simply measure the gradient at several times to provide the deceleration value. From a knowledge of the car weight, this defines the drag force. From this, a map of drag force versus speed can be prepared and then used in Eq. (4.1) to produce a straight-line performance curve. Two gradients are included for use later.

4.3.6 Grip Limited Acceleration

In the above analysis, it has been assumed that the tyres will always grip the road. However, in reality, in a low gear a vehicle may well be capable of

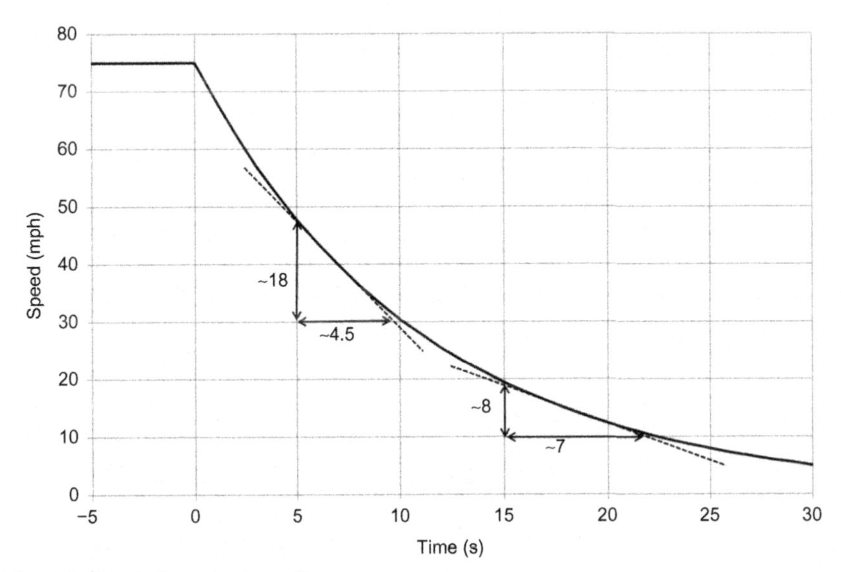

Fig. 4.6 Coast down test results.

providing much more torque than the tyres can transfer to the road, and if the driver does open the throttle beyond this point, the tyres simply spin and much less tractive force is developed than if the tyres are kept on the longitudinal slip peak.

Working from the Fig. 4.3 and using the same approach we used before,

$$\hat{a}_X = \frac{\mu F_X}{M_F + M_R} = \frac{\mu M_R g}{M_F + M_R} = \frac{M_R}{M_{TOT}} \mu g$$

To have the best possible straight-line acceleration, $\hat{a}$, we therefore want μ to be as large as possible. However, if the engine supplies a torque that would result in an acceleration greater than $(M_R/M_{TOT})\mu g$, then this values is not achieved instead the tyres will spin.

The ratio of M_R/M_{TOT} needs some thought. Clearly, the acceleration will be better the larger this ratio is, and so ideally, we want all the weight to be on the rear axle. This would certainly not be a good thing when cornering, but in a straight line it is an advantage. However, tyre μ values decrease with increasing vertical load. In fact, the reduction in μ is small, and even doubling the load only reduces μ by around 0.2, so more is gained than lost by moving the weight to the rear. Whilst increasing the load on the rear tyres in this situation is beneficial, it is not so in others. If all the weight was transferred to the rear by raising the centre of gravity, there is the risk of a 'wheelie' with obvious attendant dangers and compromise. Additionally, very few cars are raced only in a straight line (dragsters being an example), and so, the high centre of gravity that is desirable in a straight line must also be carried round corners where it is certainly not good. Another point is that getting more weight onto the rear is only a benefit when the car is traction limited, and this will only occur at the start of the straight, so high-speed straight-line acceleration is not improved by a high centre of gravity. In a situation where a straight-line race forms part of a series of races, formula student, for example, then its worth doing what the rules allow. At Oxford Brookes University, for example, we often prop the driver up on a pillow so that he or she is raised up as much as is legal.

4.3.7 Determination of Vehicle Acceleration With Weight Transfer

In Section 3.1.1, wheel loads under constant acceleration, we produced an equation for the vertical load on the front and rear wheels of an accelerating car shown in Fig. 4.7.

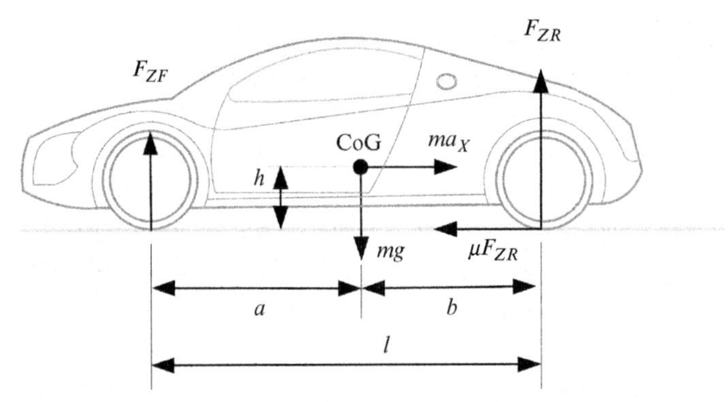

Fig. 4.7 Free body diagram for an accelerating vehicle.

These were

$$\begin{aligned} F_{ZF} &= F_{SZF} - \frac{h}{l} m a_X \\ F_{ZR} &= F_{SZR} + \frac{h}{l} m a_X \end{aligned} \tag{4.3}$$

However, when a car accelerates forwards, the longitudinal force at the driven wheels produces the acceleration:

$$m a_X = \mu F_{ZR} \tag{4.4}$$

There are two approaches we can take to solving for the straight-line acceleration value from here. The first is to solve directly, and the second is to use an iterative approach.

4.3.7.1 Solve Directly

Substituting the expression for F_{ZR} from Eq. (4.3) to Eq. (4.4), we have

$$m a_X = \mu \left(F_{SZR} + \frac{h}{l} m a_X \right)$$

And so,

$$\begin{aligned} m a_X &= \mu F_{SZR} + \mu \frac{h}{l} m a_X \\ &\Rightarrow m a_X - \mu \frac{h}{l} m a_X = \mu F_{SZR} \end{aligned}$$

Thus,

$$a_X = \frac{\mu F_{SZR}}{m \left(1 - \mu \frac{h}{l} \right)}$$

This equation at first sight seems wrong since if μ were to take the value h/l then the denominator would be zero and the acceleration infinite. This corresponds to a car in which the increasing acceleration leads to sufficient weight transfer that the resultant gain in longitudinal acceleration then produces an equal amount of weigh transfer and so on. In reality, long before this situation could be reached, the load on the front axle will have reduced to zero, providing a limit to the process when the car does a 'wheelie'.

4.3.7.2 Iterative Approach

The second method is simply to determine the acceleration that will result with no weight transfer and only the static load on the rear axle F_{SZR}. With this acceleration determined, the weigh transfer can be determined and a new value for the load on the rear axel found. From this, the next weight transfer value can be determined and so on until the increases produced by each iteration are negligible compared with the accuracy desired.

4.4 METHODS FOR THE DETERMINATION OF STRAIGHT-LINE ACCELERATION

In this section, we shall examine three progressively more complete models of straight-line acceleration.

4.4.1 Method One—Basic Calculation

In this approach, we use a simple estimation method, best illustrated by a worked example.

Determine the 0–60 mph time for a small hatchback car with the following data and compare the answer with the measured value:

Mass	1500 kg
Peak torque	120 nm
First gear ratio	4.37
Second gear ratio	3.71
Final drive	3.46
Tyre outer radius	0.32 m
Measured 0–60 mph time	15 s

We shall assume that first gear is used between 0 and 30 mph and second between 30 and 60 mph. Noting that the final drive ratio relates to the ratio drop across the differential and so applies to all gears, we start by calculating the peak torque at the wheels in each gear:

First	$120 \times 4.37 \times 3.46 = 1814.4$ Nm
Second	$120 \times 3.71 \times 3.46 = 1540.39$ Nm

These values of peak torque can be turned into average values that apply over the rev range used for each gear from a knowledge of the torque curve on the engine. In this case, the average torque is roughly 80% of the peak, and so, the torque at the wheel is taken to be

First	$1814.4 \times 0.8 = 1451.5$ Nm
Second	$1540.39 \times 0.8 = 1232.3$ Nm

From these values, we can determine the force exerted at the contact patch since $F_X = T/r$:

First	$1451.5/0.32 = 4535.93$ N
Second	$1232.3/0.32 = 3750.93$ N

Speed loss from coast down test for this car (shown in Fig. 4.6)

At 15 mph,	$8/7$ mph/s $= 1.14 \times 0.44 = 0.5$ m/s^2
At 45 mph,	$18/4.5$ mph/s $= 4 \times 0.44 = 1.76$ m/s^2

and so the drag force is

At 15 mph,	$0.5 \times 1500 = 750$ N
At 45 mph,	$1.76 \times 1500 = 1640$ N

The net accelerating force is

At 15 mph	$4535.93 - 750 = 3785.9$ N
At 45 mph	$3750.93 - 1640 = 2110.93$ N

Net acceleration value is

0–30 mph	$3785.9 - 1500 = 2.52$ m/s^2
30–60 mph	$2110.93 - 1500 = 1.41$ m/s^2

from which the time taken to accelerate through each interval may be determined:

0–30 mph	$13.4 - 2.25 = 5.95$ s
30–60 mph	$13.4 - 1.41 = 9.5$ s

Thus, the total time is

0–60	$5.95 + 9.5 = 15.45$ s

The value compares reasonably well with the quoted figure. Clearly, there are many approximations in the method, and accuracy would be improved if these were reduced. For example, there is no time allowed for the gear change, and no account taken of the real torque curve.

Despite this, the method offers several desirable features. Firstly, it is very simple and one only needs basic data in order to prepare and estimate. Secondly, it is easily amenable to computer solution by spreadsheet for example. Thirdly, it offers obvious methods of refinement, for example, by dividing the speed range into smaller intervals and using accurate torque figures for each interval. In method two below, we shall introduce many of these improvements.

4.4.2 Method Two—Iteration Using Torque Curve and Gear Ratios

We may now improve on the method above by including more factors. The spreadsheet below shows how this may be achieved.

The spreadsheet runs from column 'A' to 'Y', and the process used is similar in many ways to that above. Very simple formulas are applied to determine forces and consequent accelerations. The acceleration event is divided into speed increments; more than just two this time.

The spreadsheet starts at the top left where a table of data is used to provide the engine torque as a function of engine rpm. The table is a 'database' and so Excel can 'look up' in this group of cells the engine torque that pertains to a particular rpm value. In the next region, E2/F17, various input data are provided. This happens again in the box B20/B26 that has gear ratios, and in the box E20/F26, the overall gear ratios including the final drive are presented. The columns H/Y are the main part of the spreadsheet, and in these columns, the spreadsheet starts with the current speed and, working across, determines how long it will take the car to climb through that speed increment. Column X has the time taken for each 10 mph gain, and from this, it is a simple matter of keeping a running total in column Y resulting in a column showing the time taken to achieve any chosen speed. For example, this spreadsheet concludes that the car in question will take about 1.2 s to reach 60 mph, a value that sounds improbably quick (Fig. 4.8).

We now look in more detail at the contents of the cells:

Cells B2/C18—These contain a lookup table listing the torque available from the engine against engine rpm. By using Excel, you can set data like this up as a database so that elsewhere in the sheet, you can look up a value of rpm in the database, and the 'vlookup' command will return the corresponding torque value.

E4/F8 contains data needed generally in the calculation of the spreadsheet. Mu is the Mu value for the tyres, 'torque loss' is a figure put in to

	B	C	E	F	H	I	J	K	L	M	N	O	P	Q	R	S	T	U	V	W	X	Y
2	rpm	Torque							engine rpm						rpm	Engine	gr	Wheel	Long f at	acc	t for	Tot
3		Nm			mph	vel	wheel	wheel	1st.	2nd	3rd	4th	5th	6th	used	Torque		Torque	wheels		10 mph	time
4	6000	240	Mu	1.2		m/s	w	rpm											frm Torq		gain	
5	7000	260	Torq loss Nm	15	0	0	0	0	0	0	0	0	0	0	10000	385	1	6320	19751	26.3	0.17	0.2
6	8000	280	wheel rad m	0.35	10	4	14	133	2190	1698	1421	1083	1022	853	10000	385	1	6320	19751	26.3	0.17	0.3
7	9000	310	aver rad m	0.29	20	9	28	267	4380	3396	2843	2167	2044	1706	10000	385	1	6320	19751	26.3	0.17	0.5
8	10000	340	car mass	750	30	13	42	400	6569	5094	4264	3250	3066	2559	10000	385	1	6320	19751	26.3	0.17	0.7
9	11000	350			40	18	56	534	8759	6792	5686	4333	4088	3411	10000	385	1	6320	19751	26.3	0.17	0.8
10	12000	355			50	22	70	667	10949	8490	7107	5417	5109	4264	10949	385	1	6320	19751	26.3	0.17	1.0
11	13000	355			60	27	84	800	13139	10188	8529	6500	6131	5117	13139	385	1	6320	19751	26.3	0.17	1.2
12	14000	353			70	31	98	934	15328	11886	9950	7583	7153	5970	15328	384	1	6304	19699	26.3	0.17	1.4
13	15000	348			80	36	112	1067	17518	13584	11371	8667	8175	6823	17518	365	1	5992	18725	25.0	0.18	1.5
14	16000	342			90	40	126	1201	19708	15282	12793	9750	9197	7676	19708	335	1	5499	17186	22.9	0.20	1.7
15	17000	330			100	45	140	1334	21898	16980	14214	10834	10219	8529	16980	375	2	4774	14918	19.9	0.22	2.0
16	18000	315			110	49	154	1467	24087	18678	15636	11917	11241	9381	18678	345	2	4392	13724	18.3	0.24	2.2
17	19000	315			120	54	168	1601	26277	20376	17057	13000	12263	10234	17057	365	3	3889	12155	16.2	0.28	2.5
18	20000	310			130	58	182	1734	28467	22074	18478	14084	13285	11087	18478	345	3	3676	11489	15.3	0.29	2.8
19					140	63	196	1867	30657	23772	19900	15167	14306	11940	19899	335	3	3570	11156	14.9	0.30	3.1
20	gear	ratio	final drive	tot	150	67	210	2001	32846	25470	21321	16250	15328	12793	16250	375	4	3046	9518	12.7	0.35	3.4
21	1	2.85	5.76	16.4	160	72	224	2134	35036	27168	22743	17334	16350	13646	17333	365	4	2964	9264	12.4	0.36	3.8
22	2	2.21	5.76	12.7	170	76	237	2268	37226	28866	24164	18417	17372	14499	18417	345	4	2802	8756	11.7	0.38	4.2
23	3	1.85	5.76	10.7	180	80	251	2401	39416	30564	25586	19500	18394	15351	19500	335	4	2721	8502	11.3	0.39	4.6
24	4	1.41	5.76	8.1	190	85	265	2534	41605	32262	27007	20584	19416	16204	19415	335	5	2566	8020	10.7	0.42	5.0
25	5	1.33	5.76	7.7	200	89	279	2668	43795	33960	28428	21667	20438	17057	17057	365	6	2334	7293	9.7		
26	6	1.11	5.76	6.4																		

Fig. 4.8 Spreadsheet determination straight-line performance.

reflect the fact that the data for torque is as measured from a dynamometer; thus, losses due to rotational friction after the engine need to be accounted for. 'Wheel rad' is the wheel radius in metres. 'Average rad' is the average rolling radius. This is the distance between the axle centre line and the ground and with the flat contact patch, which may be very compressed by aerodynamic downforce, this distance will always be less than the wheel rad. Lastly, the car mass in kg is input. Excel allows cells to be 'named' by typing the desired name into the container to the left of the formula bar. Once 'named', a cell can be referenced by its name, making the preparation of formulae much easier to read and quicker to assemble:

B20/C26—These cells contain the gear ratios for each gear and need to be manually entered.

E20/F26—Column E contains the value of the final drive. This is normally the gear ratio that results from the crown gear in the gearbox that drives the differential. The 'overall' gear ratio is obtained by multiplying the individual ratios by the final drive.

Column 'H' contains the increments of mph that will be used for the analysis.

Column 'I' contains the forward velocity of the car at that value of mph in m/s.

Column 'J' contains the wheel angular velocity, omega, in rads/s.

Column 'K' contains the wheel rotational speed in rpm.

Column 'L'—This is an important step. Now that the wheel speed in rpm is known, the engine revs that would correspond if the car was in first gear can be calculated. For example, you can see that at 10 mph, the engine rpm in first gear would be 2190 (a value too low for the engine to maintain and the clutch would need to be slipping at this point). Lower down, you can see that at 60 mph, the engine speed would be 13,139, which is a perfectly acceptable engine speed, and indeed, it would be developing around 400 nm of torque at this point. By the time the wheel speed gets to 2000 rpm, the corresponding engine speed is 32,846, much too high for the engine to manage, and by this point, the driver would need to have selected second gear.

Column 'M' to 'Q'—In a similar way, it is possible to prepare a list of engine rpm for each gear up to sixth gear. Clearly, not all these points are useable, but they represent all points possible so the next task will be to sift out what engine rpm will actually be used.

Column 'R' contains that rpm actually used. One way to prepare this would be to manually look down the column and select the rpm that should be used. For example, in row '5', the rpm is zero, and the engine can't run at zero rpm. We assume here that the slowest the engine can usefully run at is

10,000 rpm. This means the clutch is slipping, and it is running at 10,000 rpm. Again, at 10 mph, the engine rpm is too low to be of use. At 50 mph, we can copy into cell R:10 the value 10,949 because this is the value of rpm that corresponds to 50 mph and is just below our minimum. By the time the car gets to 90 mph, the rpm has risen to 19,708 and is only just within the acceptable range. When 100 mph is reached, the engine rpm that would result from first gear is too large, above 20,000, and the driver has to change to second gear. This process of manually entering the correct value continues until column 'R' is filled. Clearly, it would be much better to write smart Excel coding here that would do all this automatically because otherwise the spreadsheet will have to be remade for each new application.

Column 'S' simply provides the engine torque that results from the current engine rpm and can be automated by looking up the value in column 'R' in the torque database B4/C18.

Column 'T' is present for clarity mainly and shows the current gear. The column can me filled manually or again coded for.

Column 'U'—In the final columns 'U' through 'Y', the knowledge of the torque available is used to determine the acceleration over the current 10 mph increment and from this the total elapsed time into the event. We start with column 'U' in which the torque at the wheel is determined by multiplying the engine torque at its current rpm by the gear ratio.

Column 'V'—From a knowledge of the average radius of the tyre, this torque can be converted into a longitudinal force at the wheels.

Column 'W' calculates the current value of acceleration by dividing the tractive force by the car mass.

Column 'X' contains the time taken to gain 10 mph at this value of acceleration.

Column 'Y' contains the running total of column 'X' to give the total time into the event. For example, we see that the car has reached 60 mph after just 1.2 s.

The method above is clearly an improvement on the method described in Section 4.4.1 but still contains some rather sweeping approximations. Clearly, something is wrong because the 0–60 in 1.2 s sounds rather quick.

4.4.3 Method Three—Straight Line Acceleration With Weight Transfer, Grip Limit and Inertia

The major shortcomings of the model proposed above are as follows:

- No allowance is made for grip or traction limited acceleration. If the tractive effort available at the wheels exceeds that which can be sustained by taking Mu times the vertical load, then the acceleration determined will be unrealistically high.

- No account is taken for the weight transfer during acceleration. As the car accelerates away, there will be more vertical load on the rear tyres, and so, a larger value of acceleration can be sustained.
- The inertia of rotating parts is not accounted for. A part that rotates must be accelerated twice under straight-line acceleration. Firstly to impart linear acceleration and secondly to impart angular acceleration. The material provided above shows how the actual longitudinal acceleration is less than expected due to this effect.

The spreadsheet below shows a straight-line acceleration calculator that takes account of all of these effects. The following comments are of interest:

Cells E2/F17 contain more vehicle dynamic data, for example, the inertia 'I' of the rotating parts. Also, the downforce is modelled as a function that varies with the square of the velocity of the car and supplies 2.2 times the weight of the car as downforce at 100 mph. That is the meaning of 2.2 in cell F2. The frontal area is in cell F9, used to determine total aero drag. The drag force is calculated as

$$Drag = \frac{1}{2} \times C_D \times \rho \times V^2 \times Area$$

where the drag force is in Newtons, the C_D is the drag coefficient for the car being 1.1 in this case. ρ is the density of air, 1.29 kg/m^3; V in the forward velocity in m/s; and $Area$ is the frontal area in m^2. The car weight is 605 kg. Note that CoG height $h=0.3$, wheelbase $=3$ m and Mu value $=2.1$.

rpm	Torque Nm
6000	240
7000	260
8000	280
9000	310
10000	340
11000	350
12000	355
13000	355
14000	353
15000	348
16000	342
17000	330
18000	315
19000	315
20000	310

aero @100mph	2.2
k	1.079
Mu	1.6
Torque loss Nm	15
Cd	1.1
p kg/m^3	1.29
area m^2	1.1
wheel rad m	0.32
wheel mass kg	12
aver rad m	0.29
Ineria kg.m^2	1.000
car mass	500
h	0.3
l	3

gear	ratio	final drive	overall
1	2.9	5.8	16.4
2	2.2	5.8	12.7
3	1.9	5.8	10.7
4	1.4	5.8	8.1
5	1.3	5.8	7.7
6	1.1	5.8	6.4

mph	vel m/s	wheel w	wheel rpm	engine rpm 1st	2nd	3rd	4th	5th	6th	rpm used	Engine Torque	gear	Wheel Torque	Long f at wheels frm Torq	aero rear load /N	Max long f at rear due to Mu	actual long F	load rear
0	0	0	0	0	0	0	0	0	0	10000	325	1	5335	16673	0	3924	3924	2453
10	4	14	133	2190	1698	1421	1083	1022	853	10000	325	1	5335	16673	108	4097	4097	2560
20	9	28	267	4380	3396	2843	2167	2044	1706	10000	325	1	5335	16673	432	4615	4615	2884
30	13	42	400	6569	5094	4264	3250	3066	2559	10000	325	1	5335	16673	971	5478	5478	3424
40	18	56	534	8759	6792	5686	4333	4088	3411	10000	325	1	5335	16673	1727	6686	6686	4179
50	22	70	667	10949	8490	7107	5417	5109	4264	10949	325	1	5335	16673	2698	8240	8240	5150
60	27	84	800	13139	10188	8529	6500	6131	5117	13139	340	1	5581	17442	3885	10140	10140	6337
70	31	98	934	15328	11886	9950	7583	7153	5970	15328	333	1	5467	17083	5288	12384	12384	7740
80	36	112	1067	17518	13584	11371	8667	8175	6823	17518	315	1	5171	16160	6906	14974	14974	9359
90	40	126	1201	19708	15282	12793	9750	9197	7676	19708	300	1	4925	15390	8741	17909	15390	11193
100	45	140	1334	21898	16980	14214	10834	10219	8529	16980	327	2	4163	13008	10791	21190	13008	13244
110	49	154	1467	24087	18678	15636	11917	11241	9381	18678	300	2	3819	11934	13057	24815	11934	15510
120	54	168	1601	26277	20376	17057	13000	12263	10234	17057	315	3	3357	10490	15539	28786	10490	17992
130	58	182	1734	28467	22074	18478	14084	13285	11087	18478	300	3	3197	9990	18237	33103	9990	20689
140	63	196	1867	30657	23772	19900	15167	14306	11940	19899	300	3	3197	9990	21150	37765	9990	23603
150	67	210	2001	32846	25470	21321	16250	15328	12793	16250	327	4	2656	8299	24280	42772	8299	26732
160	72	224	2134	35036	27168	22743	17334	16350	13646	17333	315	4	2558	7995	27625	48124	7995	30077
170	76	237	2268	37226	28866	24164	18417	17372	14499	18417	300	4	2436	7614	31186	53822	7614	33838
180	80	251	2401	39416	30564	25586	19500	18394	15351	19500	300	4	2436	7614	34963	59865	7614	37415
190	85	265	2534	41605	32262	27007	20584	19416	16204	19415	300	5	2298	7182	38956	66253	7182	41408
200	89	279	2668	43795	33960	28428	21667	20438	17057	17057	315	6	2014	6294	43164	72986	6294	45617

drag	Nett F	acce	wt trans	new load rear	new max long F	actual long f	Nett F	acc	accel /g	time fo 10mph gain	Total time	mph	time to..
0	3924	8	392	2845	4552	4552	4552	9	1	0	0.49	0	
16	4081	8	408	2969	4750	4750	4734	9	1	0	0.96	10	
62	4552	9	456	3339	5343	5343	5281	11	1	0	1.39	20	
140	5338	11	534	3957	6332	6332	6192	12	1	0	1.75	30	
250	6437	13	644	4823	7716	7716	7467	15	2	0	2.05	40	
390	7851	16	785	5935	9496	9496	9107	18	2	0	2.29	50	
561	9578	19	958	7295	11672	11672	11111	22	2	0	2.49	60	2.5
764	11620	23	1162	8902	14243	14243	13479	27	3	0	2.66	70	
998	13976	28	1398	10756	17210	16160	15161	30	3	0	2.81	80	
1263	14127	28	1413	12606	20169	15390	14127	28	3	0	2.96	90	
1559	11449	23	1145	14388	23021	13008	11449	23	2	0	3.16	100	
1887	10047	20	1005	16514	26423	11934	10047	20	2	0	3.38	110	
2246	8244	16	824	18816	30105	10490	8244	16	2	0	3.65	120	1.2
2635	7355	15	735	21425	34280	9990	7355	15	1	0	3.96	130	

Ttot =F*r	Ttot-rD	accounting for wheel I r(mr+ml)	4I/r	ay	time for 10mph gain	Tot time	mph	
1457	1457	160	13	8	0.53	0.53	0	
1520	1515	160	13	9	0.51	1.04	10	
1710	1690	160	13	10	0.46	1.49	20	
2026	1981	160	13	11	0.39	1.88	30	
2469	2389	160	13	14	0.32	2.21	40	
3039	2914	160	13	17	0.26	2.47	50	
3735	3556	160	13	21	0.22	2.69	60	2.6
4558	4313	160	13	25	0.18	2.87	70	
5171	4852	160	13	28	0.16	3.03	80	
4925	4521	160	13	26	0.17	3.20	90	
4163	3664	160	13	21	0.21	3.41	100	
3819	3215	160	13	19	0.24	3.65	110	
3357	2638	160	13	15	0.29	3.94	120	1.3
3197	2353	160	13	14	0.33	4.27	130	

Fig. 4.9 Solution including weigh transfer and wheel inertia.

The spreadsheet works in a similar way to before, but columns 'V' and 'X' contain the grip limited and torque limited longitudinal forces. Once an acceleration value is determined in column 'D' 31 downwards, this is used to produce an estimate of the weight transfer and so the new maximum longitudinal force that can be developed. Only one iteration is used here. After this, cells R28-Y51 are used to produce a refined acceleration time accounting for the inertia of rotating parts (Fig. 4.9).

4.5 QUESTIONS

1. A car accelerates at 0.4 g for 3 s and then coasts for 5 s. more. Determine the distance travelled.
2. A road car has 150 hp available at the wheels. Determine the acceleration achievable at 30 mph.
3. Explain the difference between 'grip limited' acceleration and 'torque limited' acceleration.
4. A vehicle has mass of 1050 kg, a wheelbase of 2.6 m and a centre of gravity height of 0.4 m. The torque available at the wheels is 180 nm. Determine the speed at which the acceleration becomes torque limited. You may assume your own gearbox ratios.
5. A vehicle has a mass of 900 kg, peak torque 350 nm, first gear ratio 4:1 and second 2.5:1, the final drive is 3:1 and tyre outer radius 0.35 m. Determine the 0–60 mph time.

4.6 DIRECTED READING

A detailed chapter on acceleration by Genta [12] is included in this very analytical book.

Gillespie [11] again provides a good coverage including drive torque reactions.

Jazer [10] deals with two- and four-wheeled vehicles, road inclination, vehicles with more than two axles, vehicle pulling trailers, etc.

4.7 LEARNING PROJECTS

- Produce a spreadsheet analysis for a car of your choice that analyses the acceleration and velocity as a function of time.
- For a vehicle, you have access to and under safe legal conditions, perform a coast down test. In the absence of telemetry, a colleague can video the car during the test from which a graph of the kind shown in Fig. 4.6

should be prepared. Use this to produce an estimate of straight-line performance of the car using the three techniques explained in the chapter.

- For a range of vehicles, obtain pictures of the front view and estimate the frontal area. From this, estimate the drag force for a range of speeds over which the vehicle is used. Compare this with the engine torque to produce a simple estimate of the terminal velocity.
- Refine the spreadsheet that includes rotating mass and estimate the error that originates in the numerical analysis by reducing the step length by a factor of 5 and 10 and comparing the results.
- Extend the spreadsheet to account for parameters not included, thinking of as many as you can. Once you have an estimate of the effect, rank them in order. Some starting topics are
  - change in tyre radius as a result of wear,
  - change in tyre mass as a result of wear,
  - fuel load empty and full,
  - increase in driver mass after lunch,
  - reduction in aerodynamic drag of 1%,
  - effect of going against a wind of 15 mph and with the same wind.
- Extend the spreadsheets to consider road inclination.

4.8 INTERNET-BASED RESEARCH AND SEARCH SUGGESTIONS

- Research the fastest accelerating vehicles you can find. Dragsters are well up this league though not at the top. From research workout, the peak acceleration achieved by dragsters. Given that longitudinal acceleration is governed by Mu, and rubber does get much above 1.5 for steady-state value, find out what is done to the track to allow such high acceleration values?
- Research a collection of vehicles you are interested in and produce a single graph of Mu against 0–60 mph time with a point in the chart for each vehicle.
- A 'rolling road' is a device used to determine the torque and power available from the wheels of a vehicle. Research what these machines are like and how they are used.
- Produce a simple sentence to accurately describe when a driver wanting the best straight-line performance should change gear. Does this necessarily happen at peak torque?
- Research the best rapid gear shift mechanisms available. How short is the shortest shift?

CHAPTER 5

Cornering

5.1 INTRODUCTION

In this chapter, we examine the dynamics of cornering. The consideration of straight-line acceleration and braking we have already met are not too taxing from a dynamic point of view, but cornering, by contrast, is an altogether much more complex process. Cornering results from a blend of forwards velocity and rotation about a vertical axis through the vehicle, called yaw. This yaw rotation is caused by the lateral force at the front yawing the vehicle one way and that at the rear yawing it the other. Many people start a consideration of cornering by thinking that a car goes around a corner in much the same way that a train goes along a railway track. It certainly can feel like that, when one is inside a vehicle, however, the process is much more complex than that. It is best to think of the car as a *free body*; it accelerates forwards and backwards in response to throttle and brake control inputs, and it accelerates laterally in response to steering inputs. The driver's job is to operate the controls in such a way that the forces generated by the tyres accelerate the vehicle as required that the resulting displacement is the desired path and that it is covered at the desired speed. Its nothing like a train, if anything, it's much more like a plane trying to follow a largely two-dimensional path in the sky. Once one knows just how complex the cornering process is, it can come as a surprise that vehicles ever do make it around corners.

5.2 YAW AND YAW RATE

The term *yaw rate* was taken originally from aeronautics and refers to rotation about an axis normal to the plane of the road. Imagine yourself in a helicopter immediately above the centre of gravity of a vehicle in motion, looking downwards onto the car. The helicopter always remains directly above the cars' centre of gravity and always points in the same direction. In this coordinate system, the longitudinal motion of the car cannot be seen. However, the yaw response is made very clear. When a car enters a corner, there will be

Performance Vehicle Dynamics
http://dx.doi.org/10.1016/B978-0-12-812693-6.00005-5

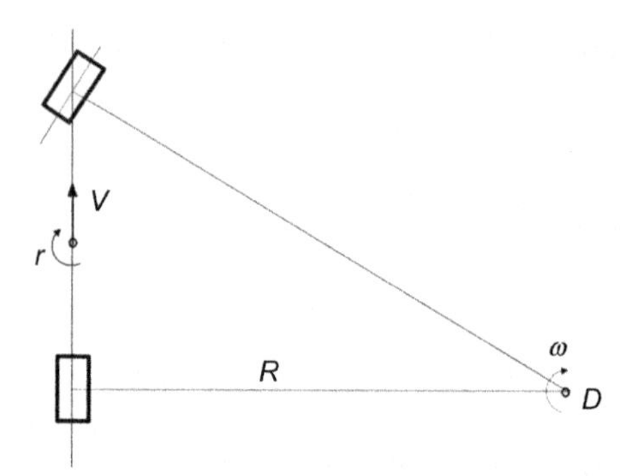

Fig. 5.1 Yawing of a vehicle.

acceleration in yaw. When it is in a steady-state corner, the yaw rate will be constant, and we shall see the car rotating at constant angular velocity.

We see this graphically in Fig. 5.1. The yaw rate of a vehicle is usually given the symbol r that initially seems strange given that it is a *rate of change* of angle and might be expected to be called $\dot{r}$. (It isn't really any more odd than labelling velocity 'V' when it might more properly be $\dot{x}$.) In Fig. 5.1, we see that the vehicle has a yaw rate r about its centre but the whole vehicle is rotating about the corner centre 'D' with an angular velocity ω. A positive yaw rate is defined as clockwise when viewed from above. Elementary consideration of this shows that $r=\omega$ since the vehicle must complete one rotation about its own centre in the same time that it completes one rotation about the corner centre. The forwards speed, 'V', of the vehicle is given by

$$V=R\omega$$

and so,

$$r=\frac{V}{R} \tag{5.1}$$

In Eq. (5.1), one must be attentive to the difference between r, for yaw rate, and R, for corner radius. In Table 5.1, we see the yaw rate r for various

Table 5.1 Yaw rate for various manoeuvres

Manoeuvre	***V* (m/s)**	***R* (m)**	***r* (rads/s)**	**a_Y (g)**
Car at 23 mph on a roundabout	10.3	16	0.64	0.67
130R—Suzuka at 155 mph	69	130	0.58	3.7
The moose test at 35 mph	15.6	24	~0.65	1.0

vehicles taking a corner of radius R at velocity V together with the consequent lateral acceleration a_Y experienced.

The ordinary car going around a roundabout is travelling quickly and exerts 0.67 g; most passengers would find this very uncomfortable. The Grand Prix car, negotiating 130R at Suzuka, has the benefit of downforce, and so, the maximum acceleration it can sustain is much more. The *moose test* is a test performed to characterise road cars and features a slalom event, as might be employed to avoid an obstacle such as a moose in the road. It's an extreme test, as the lateral acceleration a_Y shows, meant really for accident avoidance, rather than normal handling.

5.3 FOUR REGIMES OF CORNERING

We shall now examine four different regimes of dynamics that apply to a cornering vehicle.

5.3.1 Quasistatic Cornering

Quasistatic cornering means travelling sufficiently slowly that the slip angles at the front and rear are negligibly small, not just in the linear region of the tyre, but negligible. The situation is shown in Fig. 5.2.

In the diagram, we start at the rear. The slip angle here is negligible, and we approximate that the centre of the corner must lie on a line at right angles to the tyre pointing direction. A similar line drawn at the front tyre is made at right angles to the front tyre pointing direction, and where these two lines cross, at point D, we have the corner centre. In the insert, we see the diagram drawn more to scale from which we can see that to a good approximation:

$$l = \partial R$$

Rearranging and using degrees

$$R = \frac{l}{\delta} \times 57.3$$

In this situation, we see that the radius of the corner, R, is simply given by the ratio of the wheelbase to steering input angle ∂. In particular, it does not vary with vehicle speed. This would indeed be a very desirable thing. It would mean that, regardless of speed, for any given steering input, the vehicle would always reply with the same corner radius. A car that offers this arrangement is described as having *Ackermann geometry*, after Ackermann to whom it is attributed, though in fact it has its origins much earlier than

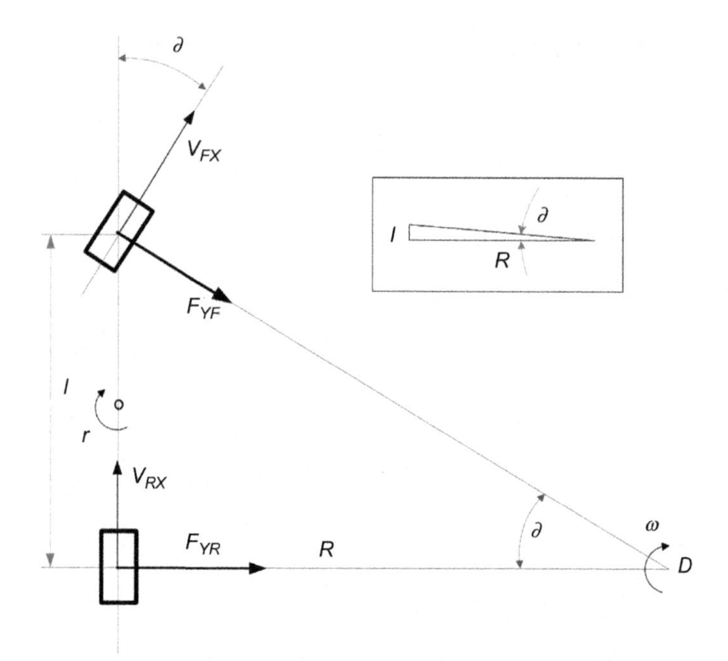

Fig. 5.2 Quasistatic cornering of the bicycle model.

Ackerman, in the design of horse-drawn carriages where it was used on luxury models to prevent gravel drives from being disturbed any more than necessary.

5.3.2 Low Speed Cornering

We define low-speed cornering to mean that slip angles have evolved front and rear but that these are still low in the linear range. This situation is shown in Fig. 5.3.

The centre of mass is in the middle of the wheelbase and we assume that the cornering stiffnesses are equal front and rear. In this situation, we again start at the rear. Given that the vehicle has some modest forwards speed, there must be a lateral force being developed at both the front and the rear tyre. This in turn requires that the rear tyre has a slip angle. This is α_R. With this in place, we can again draw a line perpendicular to the direction of travel, and where it meets a similarly constructed line from the front tyre, we find the centre about which the vehicle is cornering to be at D. This centre has moved forwards and is well ahead of the rear tyre centre line, though not as far as being level with the centre of gravity.

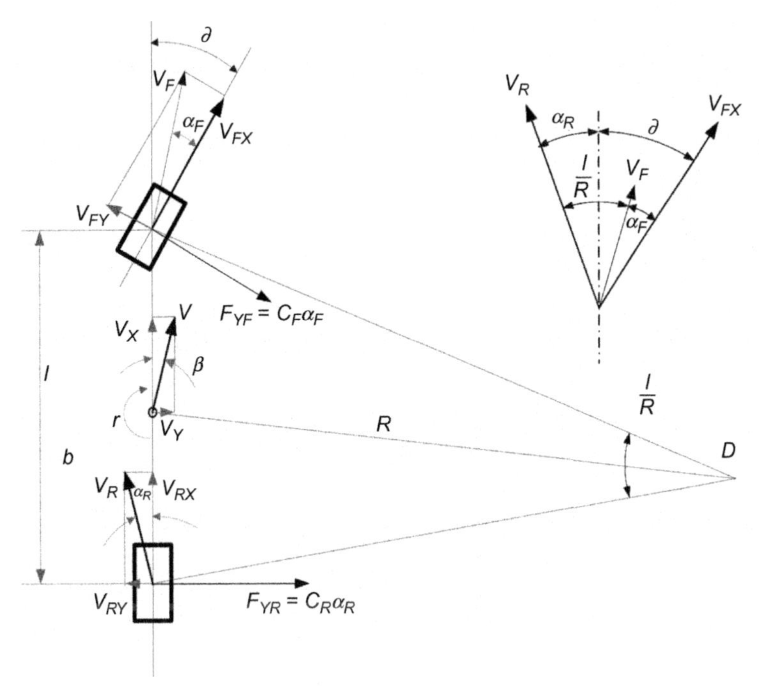

Fig. 5.3 Low-speed cornering.

Notice that the rear tyre velocity V_R is at right angles to the line from the rear tyre to point D but the lateral force generated by the rear tyre is at right angles to the pointing direction of the tyre. At the vehicle centre, the vehicle velocity, 'V', is again at right angles to a line from the vehicle centre to point D. This velocity has components V_X and V_Y. If the point D was moved further forwards and was level with the vehicle centre, then we could see that the V_Y component would be zero and the vehicle centre velocity would be tangential to a line from the vehicle centre to the corner centre. This speed is called the *tangent speed* for this reason.

The insert in the diagram shows the important angles involved, and by inspection, we may write

$$\alpha_R + \delta = \frac{l}{R} + \alpha_F$$

or

$$\delta = \frac{l}{R} + \alpha_F - \alpha_R \tag{5.2}$$

Thus, there is simple relationship between corner radius, the wheelbase and the slip angles that determines how much steered angle will be required at the front wheel in order to negotiate the corner. This important but simple equation leads to a lot of relevant points, and we shall return to it later in a lot more detail.

5.3.3 Medium Speed Cornering

In a medium-speed corner, the tyre lateral forces are still a linear multiple of the slip angle, but we are at the limit of this region. From Fig. 5.4, it is clear that the corner centre about which the car is rotating has moved further forwards. Fig. 5.5 shows how the centre of the corner migrates as the slip angle front and rear increase. Position 1 corresponds to the quasistatic case. Position 2 is the *tangent speed*, at which the corner centre is in line with the geometric centre of the vehicle. This occurs when the slip angles have reached half the steered angle ∂.

When the slip angles reach the full value of the steered angle ∂, the corner centre is level with the front axle. From here, it continues to move forwards,

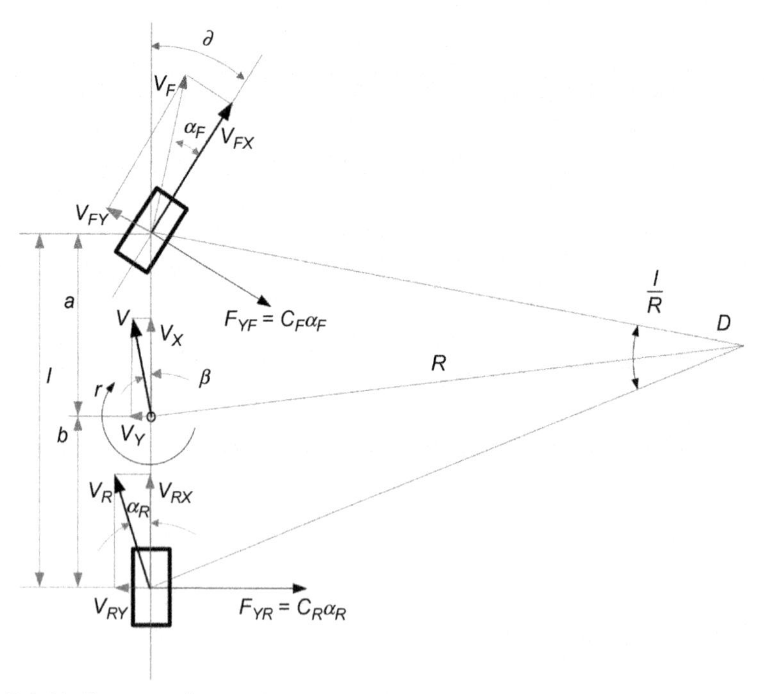

Fig. 5.4 Medium-speed cornering.

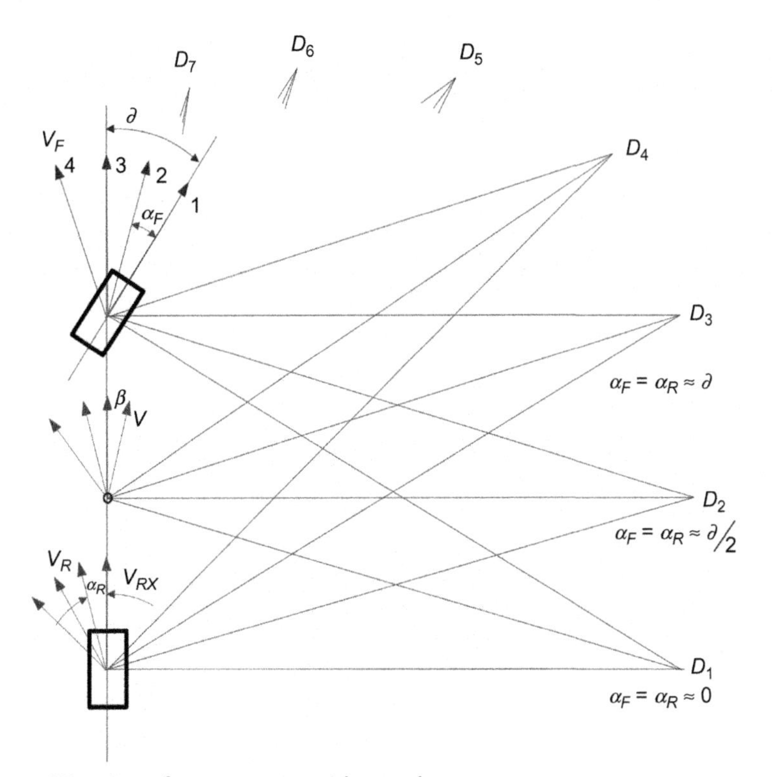

Fig. 5.5 Migration of corner centre with speed.

and further points in positions D_5, D_6 and D_7 in the sequence are shown. In practice, such extreme positions can never be achieved because, before they are reached, the slip angles will exceed the values at which the peak lateral force is developed. Control is lost, and the car is no longer cornering but spinning out instead.

Consideration of the diagram with practical steering inputs and at real scale shows that, even with zero steered angle '∂', the position of the corner centre at the limit of about 8 degrees of slip angle would not have moved much in advance of the front axle and, further, that its movement forwards will be along an approximately straight line and constant distance from the car. Thus, approximating the corner radius to R, as shown in Fig. 5.2, is acceptable. We might conclude from this that the migration of the corner centre is unimportant. However, in order to understand how a neutral steering car can keep the same corner radius whilst the slip angles increase and to understand why the vehicle has a slip angle, β, it is essential to first grasp this dynamic.

5.3.4 High Speed Cornering and Limit Behaviour

Moving now to a high-speed corner as shown in Fig. 5.6, the situation is different from that above in two important ways. Firstly, the speed has become sufficiently high that the tyre forces have moved out of the linear range and indeed are now very close to the limit as is shown in the insert graph. The second difference is that to illustrate limit performance more clearly the car the centre of gravity has been moved backwards and is much nearer the rear axle than before, when it was in the geometric centre.

If we consider the forces at the contact patches front and rear, it is clear that in order for the car to be in moment equilibrium about the centre of gravity we require

$$F_{YR}b = F_{YF}a$$

Given that a is now larger than b, it follows that $F_{YR} > F_{YF}$ and, if this is true, then $\alpha_R > \alpha_F$. This is shown in the insert graph, and we can see that, if the speed increases much from here, the rear tyre will not be able to generate the force necessary for equilibrium and the car will no longer be able to

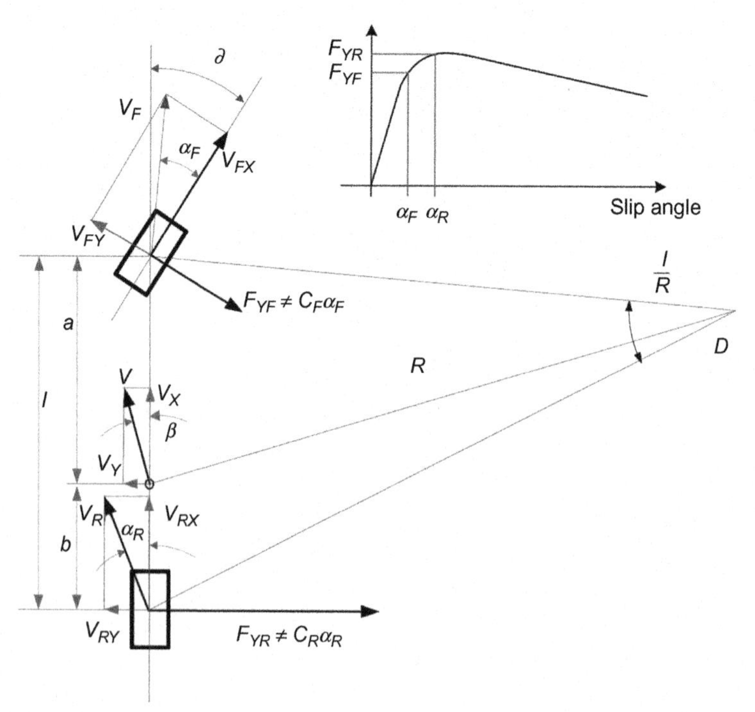

Fig. 5.6 High-speed cornering.

maintain its path around the corner. It is the tyre curve that sets the limit and, in the case shown, it will be the rear that loses control first eventually sliding outwards. The path taken by such a car as it goes over the limit is shown in Fig. 5.7.

We follow the dynamics by starting in position 1. The car is travelling forwards, and both tyres are capable of producing the moment balance necessary to keep it in equilibrium. The car is cornering. If the speed is then increased between position 1 and 2, the force required for equilibrium at the rear can no longer be produced by the rear tyre, the slip angle increases, and the lateral force starts to decrease. This in turn means that the rear end starts to slid outwards, increasing the slip angle, which further reduces the lateral force. The rear then swings wildly outwards, and control is lost. This proceeds until the vehicle swaps ends, and depending on the initial

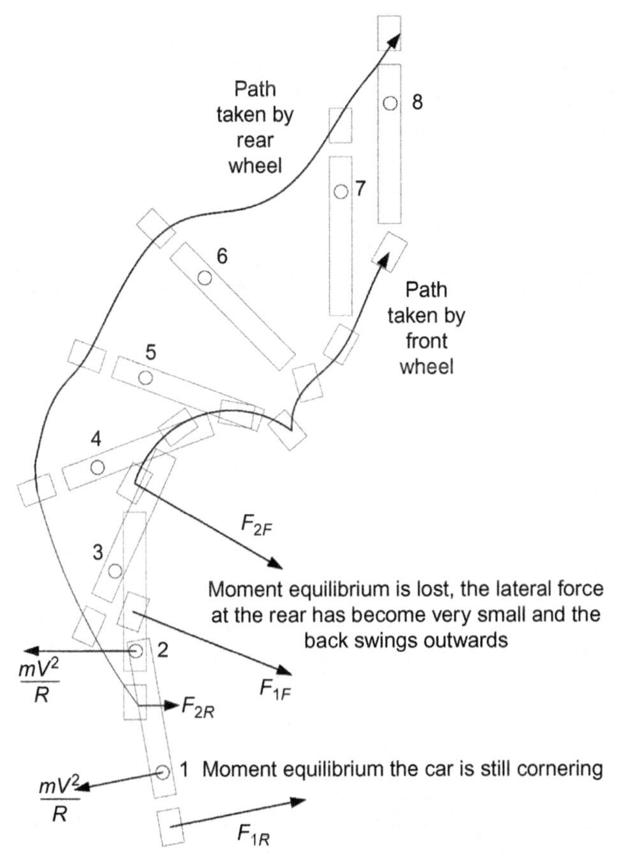

Fig. 5.7 Limit path taken by the oversteering car.

conditions and the driver's response, the car can regain control and slide backwards approximately as shown. The vernacular term for this dynamic is *getting the back end out* for fairly obvious, if rather irreverent, reasons.

It is simple enough to see that if the centre of gravity were moved forwards and $b > a$ then it would be the front lateral force that would reach the limit first. In this case, the behaviour is as shown in Fig. 5.8.

In this case, we start in position 1, and the car is cornering. The lateral force at the front and rear produces moment equilibrium and balances the centripetal force mV^2/R. This centripetal force continues to act on the vehicle from here on but is not shown in the diagram for clarity. At position 3, the increasing car speed has meant that the front tyre has exceeded its maximum lateral force generating capacity and the force F_{3F} is suddenly reducing. This means that the moment balance is lost and the front starts to slide outwards. The car consequently rotates (anticlockwise in the figure), and this in turn serves to reduce the slip angle at the rear. By position 5, the car is sliding outwards from the original corner. Drivers sometimes refer to

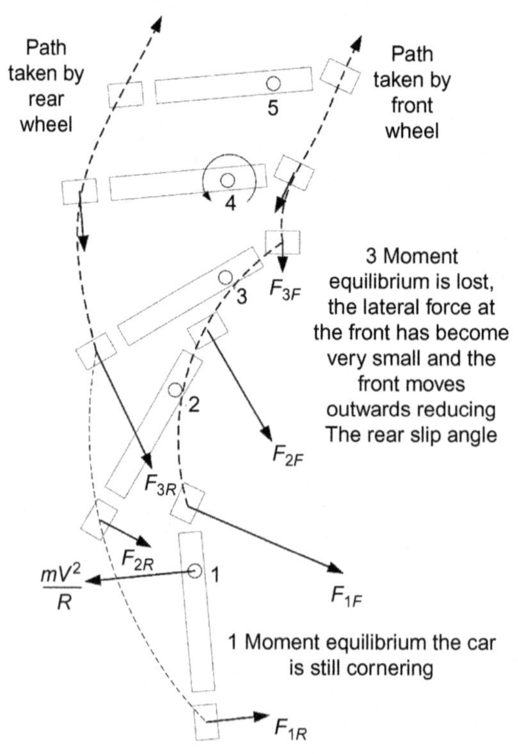

Fig. 5.8 Limit path taken by the understeering car.

this dynamic as *pushing*. The manoeuvre is less dramatic than for the over-steering car, the front moves outwards and the car remains facing roughly forwards. However, the corner radius is not maintained, and control is still lost. The website for the book contains videos generated in the Adams multibody package that demonstrate these behaviours vividly.

5.4 OPTIMISED STEERING AND CORNERING PERFORMANCE

Given the material above, we are now in a position to begin to understand how the ideal car would behave, firstly in terms of producing a given corner radius for a given steering input and secondly in terms of ensuring the car is as fast as possible through a corner.

5.4.1 Optimised Corner Radius in Response to Steering Input

Consider the steering response first. We have seen that the fundamental equation of cornering is given by

$$\delta = \frac{l}{R} + \alpha_F - \alpha_R$$

If we imagine a car in which the slip angles front and rear evolve at the same rate and therefore take the same value at any point in time, then $\alpha_F = \alpha_R$, and the equation of cornering reduces to

$$\delta = \frac{l}{R}$$

This is a highly desirable result because it means that as the speed increases (and so to do the slip angles) the radius of the corner remains unchanged for any given steering input. Such a car will always reply to a given steering input with exactly the same corner radius regardless of speed or indeed any other parameters. This is a highly desirable result. We can see how this desired outcome translates into vehicle design by considering Fig. 5.9.

We start with the simplifying assumptions for the time being that the cornering stiffnesses are the same front and rear, the centre of gravity is in the geometric centre and, given the true proportions of the diagram, the lateral force at the front acts parallel to that at the rear. As a consequence, the slip angles are also the same front and rear, and so are the lateral forces; therefore,

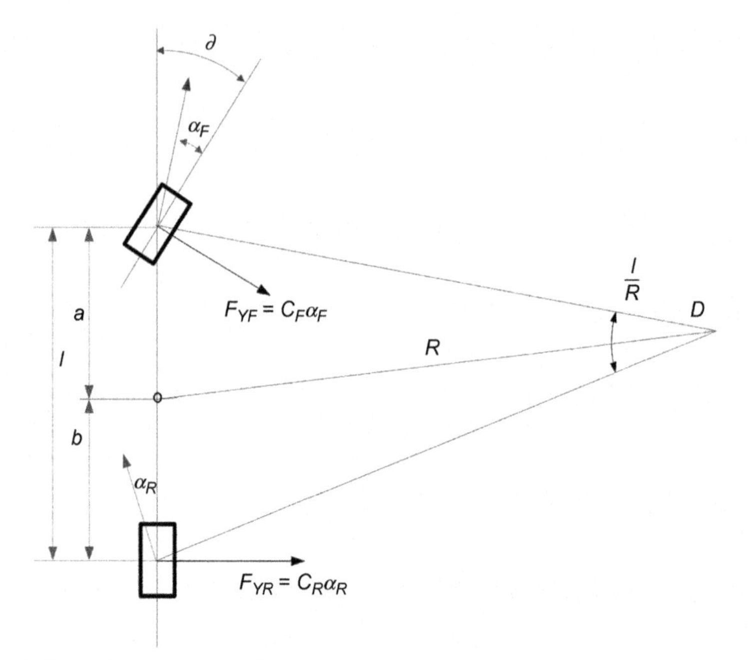

Fig. 5.9 Cornering moment balance.

$F_{YF} = F_{YR}$. If the car is in equilibrium in a corner, then the torque generated by $a \times F_{YF}$ must equal the torque generated by $b \times F_{YR}$.

Such a car is called *balanced* to reflect the ideal moment balance or alternatively *neutral steering* to reflect that fact that the corner radius is *neutral* with respect to speed, a highly desirable result and one that we strive to achieve all the time.

We now consider a car with unequal cornering stiffnesses front and rear. This might occur because different tyres are used as is common in racing where the rears are larger or because the vertical loads are different but the tyres the same, as is common with a road car. We still want $\alpha_F = \alpha_R$ and that the moment equilibrium is maintained, so,

$$aF_{YF} = bF_{YR}$$
$$\Rightarrow aC_F\alpha_F = bC_R\alpha_R$$

and so,

$$\frac{a}{b} = \frac{C_R}{C_F}$$

Thus, it is easy to keep any car neutral; all that has to happen is that the location of the centre of mass must be placed along the car in proportion to the ratio of cornering stiffnesses.

5.4.2 Optimised Cornering Performance

A vehicle with optimised cornering performance means one that is as fast as possible through the corners. The centripetal force required to keep an object on a circular path is given by

$$F = \frac{mV^2}{R}$$

Thus, for V to be as large as possible, we require F to be as large as possible. Referring to Fig. 5.9, we can see that, in order for the total lateral force, given by the sum of F_{YF} and F_{YR}, we require that $\alpha_F = \alpha_R$ because in this way both the front and rear axle will reach the peak of the lateral force curve at the same time. If this is not true, then one or other of them will lag behind and not reach its maximum at the time when the car loses control and the speed at that time will be lower than the case when $\alpha_F = \alpha_R$.

This is exactly the same car developed above for the optimised steering response, and so, the neutral steering car not only gives the ideal steering response but also is the fastest through the corners. The same comments developed above apply to a car with unequal cornering stiffnesses front and rear, and a neutral steering car also provides the largest force at the front and rear simultaneously. These two pressures together make achieving the neutral steering car all the more important.

It is worth noting here that even a neutral steering car will be very difficult to control when driven at the limit. Imagine a car entering a corner and the driver managing to get it to reach the peak of the lateral force generating capacity at the apex of the corner. When the driver comes to exit the corner, all will be well, because as the steering input is reduced, the lateral force at the front will decrease, and the car will start to yaw out of the corner. If, however, the driver has made even a tiny misjudgement and more lateral force is needed to tighten the corner, this will not be possible because the tyres are already saturated. The driver will have to slow down to reduce the lateral force required and bring it within the range of the tyres. Slowing down is not helpful in a racing situation. Furthermore, in this situation, it is actually difficult to do; the tyre capacity to generate a horizontal force in any direction is completely consumed in lateral force generation, and to brake, all one can do is lift off the throttle and wait for drag to slow the car and then start to gently apply braking effort. If this is not possible in the space available, then spinning out is inevitable.

5.5 THE THREE REGIMES OF STEERING RESPONSE

We have seen that the ideal car will be arranged to have $\alpha_F = \alpha_R$ and that such a car is called *neutral steering*. We have also introduced the idea that the centre of gravity need not be in the geometric centre of the chassis. If it isn't, then the end that it is nearest will need to develop more lateral force to keep the car in yaw equilibrium. In practice, it might not be possible to do this, and this leads to the existence of three steering response possibilities for a car to have:

- **(i)** CoG in front of the geometric centre, meaning that α_F will be greater than α_R
- **(ii)** CoG at the geometric centre, meaning that α_F will be equal to α_R
- **(iii)** CoG behind the geometric centre, meaning that α_F will be less than α_R

Each of these is shown below.

In Fig. 5.10, we see these three cars, each in yaw equilibrium. Thus, the lateral force at the rear of the car on left is small because the centre of gravity is significantly forwards of the geometric centre. The lateral force at the front

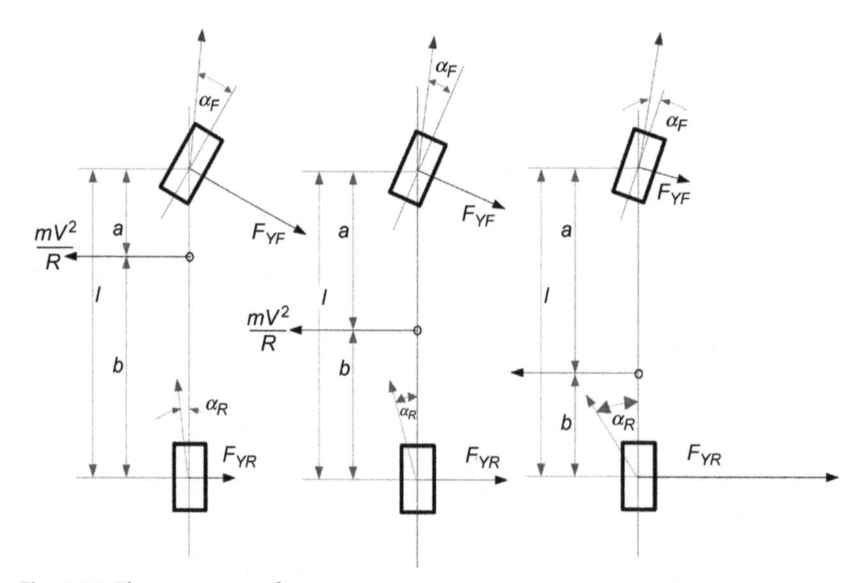

Fig. 5.10 Three regimes of steering response.

is much larger. The car in the centre has the centre of gravity in the geometric centre, and the forces front and rear are equal. Moving to the car on the right, this situation is reversed. The movement of the centre of gravity causes the slip angles front and rear to change. Looking at the fundamental equation of cornering, we can see that, if the difference between α_F and α_R is positive, a larger steering input will be needed to achieve a given corner than for a car with a central centre of gravity, and if it is negative, a smaller one will be needed:

$$\delta = \frac{l}{R} + \alpha_F - \alpha_R$$

These three possibilities are now examined in more detail, starting with the neutral steering car because it is the simplest.

5.5.1 Neutral Steering Car

We start by examining the simplest situation, the car in which the centre of gravity is in the middle and the cornering stiffness of the front and rear tyres are the same. In this case, the magnitude of the front and rear forces will be the same. As the car speed increases, the magnitude of these forces will increase too; the slip angles front and rear will be the same, and the radius that results from the steering demanded by the constant steering wheel input will not change as the speed increases. Such a car is called a *neutral* steering car.

In Fig. 5.11, we see again the diagram we first developed in Fig. 5.3, but in this case, the angles are rearranged so that α_R is next to α_F. Given that this is a neutral steering car, any increase in speed will require an equal increase in lateral forces front and rear and therefore α_R will increase by the same amount as α_F. Thus, δ and $1/R$ will remain unchanged.

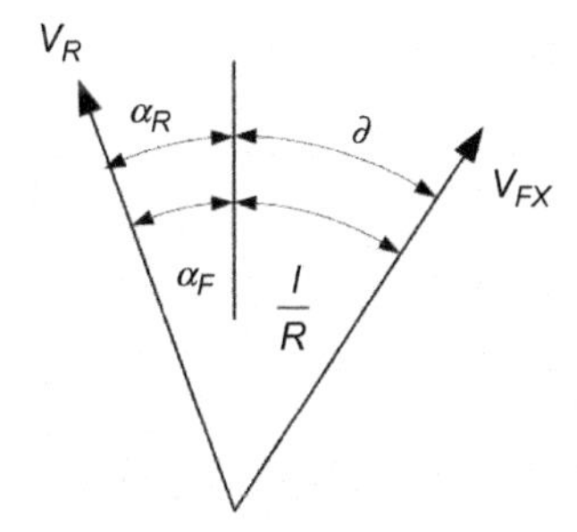

Fig. 5.11 Cornering angles, neutral steering car.

This result is equally clear when looking at the fundamental equation of cornering developed above:

$$\delta = \frac{l}{R} + \alpha_F - \alpha_R$$

The centre about which the car is rotating will move forwards as we have seen, but the corner radius will not change.

5.5.2 Understeering Car

With the centre of gravity moved forwards, the force at the front will have to be larger than the rear in order to maintain moment equilibrium. Fig. 5.12 shows the situation graphically. Both α_R and α_F together add to the same amount, but α_F is larger than α_R. As a result, the steering angle δ must be larger. Once this is applied, the car will produce the same radius corner as before, but it will need the larger steering input to do so. Such a car is termed *understeering* because it produces *less corner*, meaning a larger radius and wider path than the neutral steering car. We should expect this effect to be speed-dependent because as the vehicle speed is increased, so a larger lateral force will be needed from both the front and rear. However, α_F is larger than α_R, so, an equal increase in both will result in a larger difference between them. Then, since the angle l/R remains unchanged with the same corner radius, the steering angle δ will have to be larger to make up. Thus, moving the centre of gravity forwards causes an understeering car, and the amount of understeer will increase with speed.

Again, we can see this result from the fundamental equation of cornering. For the understeering car, α_F is larger than α_R:

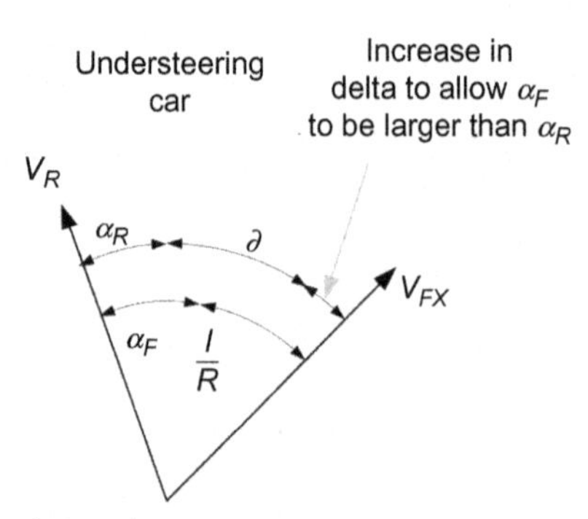

Fig. 5.12 Cornering angles understeering car.

$$\delta = \frac{l}{R} + \alpha_F - \alpha_R$$

Looking at the equation, the term $\alpha_F - \alpha_R$ equates to positive number; thus, the steered angle δ must be larger for the same wheelbase and corner ratio.

5.5.3 Oversteering Car

By inspection, the opposite is true if we move the centre of gravity rearwards and an oversteering car is produced. In this case, the force at the rear wheel must be larger to maintain moment equilibrium, so too must be the rear slip angle. Since the total lateral force equilibrium must be maintained, if there is more force at the rear, there must be less at the front and so a smaller steered angle as shown in Fig. 5.13. Such a car is called *oversteering* because it produces *more corner* resulting in a smaller radius and narrower path than the neutral steering car. As before, the effect will be speed-dependent because with an increase in speed comes more lateral force if both slip angles increase, but the rear maintained larger than the front; the resulting steered angle δ will be smaller for a corner of the same radius.

Again, we can get the same result from the fundamental equation of cornering:

$$\delta = \frac{l}{R} + \alpha_F - \alpha_R$$

Looking at the equation and with α_R greater α_F, the term $\alpha_F - \alpha_R$ equates to negative number; thus, the steered angle δ must be smaller for the same wheelbase to corner ratio. After considering the oversteering and understeering case, the name 'neutral steering' seems to make more

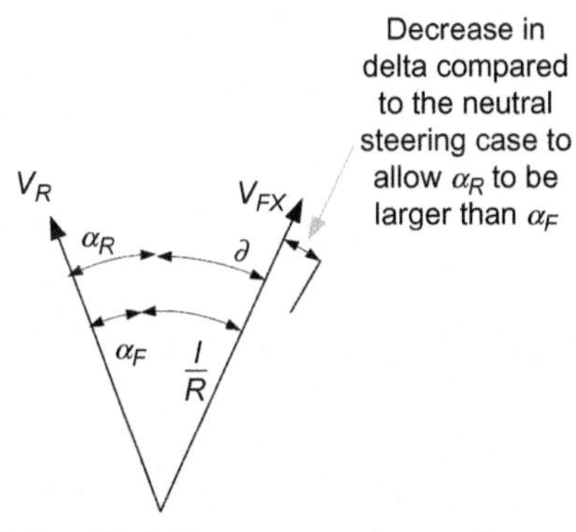

Fig. 5.13 Graphical relationship between cornering angles oversteering car.

sense in that it is the configuration that avoids these problems and is 'neutral to' or unaffected by speed.

5.5.4 Path Response for the Oversteering and Understeering Cars

Fig. 5.14 shows the same dynamic as above but in a different way. In considering the corner above, the driver responded to the differing cars and applied the steering input required to achieve the desired corner. We could, however, examine how the differing cars respond to the same steering input, and this is shown below. In many ways, this is a better way to consider the problem since it makes it much more like a control problem in which the corner radius is the 'output' and the steering wheel displacement is the 'input'.

In the figure, the steering wheel is supplied with the same angular step input on arrival at the corner, but three-path responses are different for the oversteering, neutral and understeering car. As can be seen, the oversteering car produces a smaller radius corner for the same steering input compared with the understeering car.

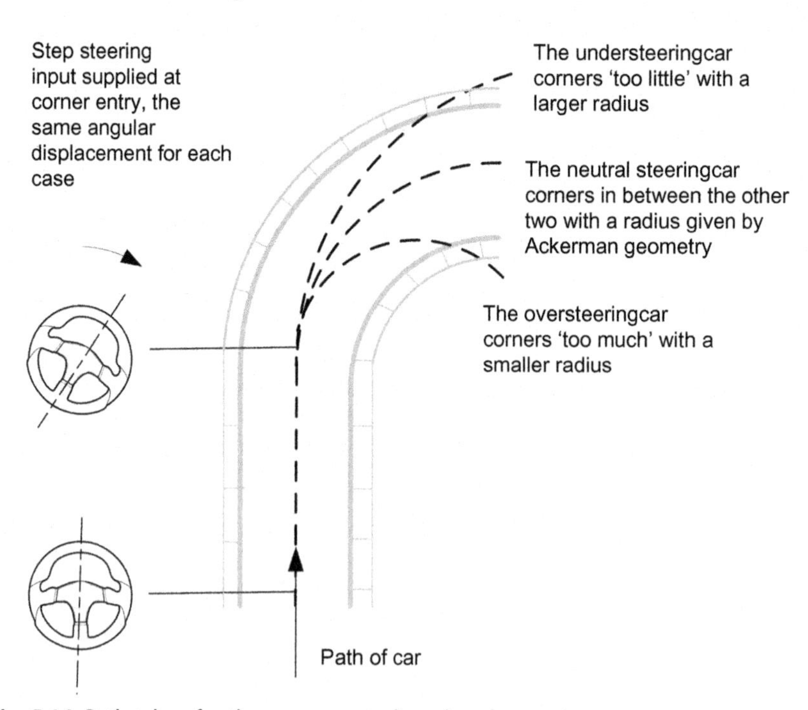

Fig. 5.14 Path taken for the over-, neutral- and understeering car.

5.6 QUANTIFICATION OF THE UNDERSTEER GRADIENT

We have seen that a car may be understeering, neutral steering or oversteering and that the response to steering input depends on the kind of car and how severely it displays the tendency it has. What is needed now is the ability to quantify the amount of steering error. Looking again at the fundamental equation of cornering,

$$\delta = \frac{l}{R} + \alpha_F - \alpha_R$$

Clearly, if a car is not neutral steering, then as the speed increases, the slip angle at one end will increase more than the other, and the steered angle ∂ will change with speed. It is common to see this dependence as relating to lateral acceleration rather than forwards speed, and we can represent the equation in the form

$$\delta = \frac{l}{R} + Ka_Y$$

In this equation, we have a number, 'K', which could be used to quantify the amount by which the steering response deviates from Ackermann where $\delta = l/R$. If K is negative, then the car would require less steering input for a given radius corner than a car with Ackermann steering. Alternatively, if the same steering input is maintained, then the car will corner with a smaller radius. In Fig. 5.14, the path taken by a car using a smaller radius is shown. One can think of this car as providing *more corner* than demanded, and this is why it's known as an oversteering car. If K is positive, then the car would require more steering input than the Ackermann car to achieve the same radius. Alternatively, if the same steering input is maintained, then the car will corner with a larger radius. In Fig. 5.14, the path taken by a car using a larger radius is shown. One can think of this car as providing *less corner* than demanded, and this is why it's known as an understeering car. In between, when K is zero, the car is of course Ackermann.

To derive an expression for the understeer gradient, we start by working from the figure and for lateral equilibrium:

$$F_{YR} + F_{YB} = \frac{MV^2}{R}$$

Moment equilibrium requires

$$F_{YF}\,a = F_{YR}\,b$$

So,

$$\frac{MV^2}{R} = F_{YR} + F_{YR}\frac{b}{a}$$
$$= F_{YR}\left(1 + \frac{b}{a}\right)$$
$$= F_{YR}\left(\frac{a+b}{a}\right)$$
$$= F_{YR}\frac{l}{a}$$

And so,

$$F_{YR} = \frac{MV^2}{R}\frac{a}{l}$$

Similarly,

$$F_{YF} = \frac{MV^2}{R}\frac{b}{l}$$

Fig. 5.15 shows the vertical equilibrium:
Working from Fig. 5.15 and taking moments about the front tyre,

$$mg \times a = F_{ZR}l$$
$$\Rightarrow F_{ZR} = mg\frac{a}{l}$$
$$\Rightarrow \frac{F_{ZR}}{g} = \frac{ma}{l}$$

Similarly,

$$\frac{F_{ZF}}{g} = \frac{mb}{l}$$

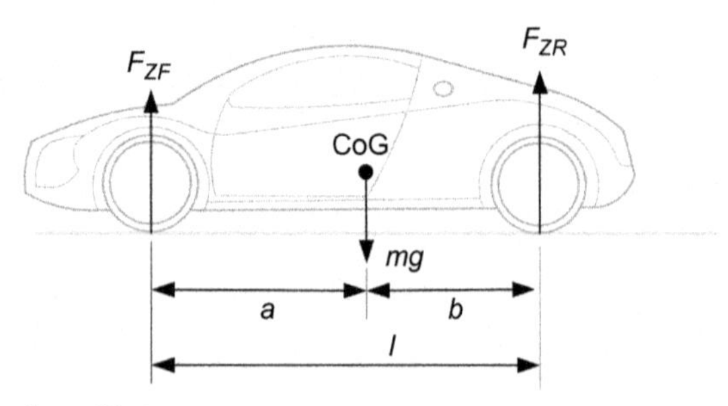

Fig. 5.15 Vertical equilibrium.

However,

$$F_{YR} = \frac{mV^2}{R}\frac{a}{l} = \frac{V^2}{R} \times \frac{ma}{l}$$
$$= \frac{V^2}{R}\frac{F_{ZR}}{g}$$

Now, since

$$F_{YR} = C_R \alpha_R$$
$$\Rightarrow \alpha_R = \frac{F_{YR}}{C_R}$$

So,

$$\alpha_R = \frac{V^2}{C_R R}\frac{F_{ZR}}{g}$$

By a similar method,

$$\alpha_F = \frac{V^2}{C_F R}\frac{F_{ZF}}{g}$$

Substituting into the fundamental equation of cornering,

$$\delta = 57.3\frac{L}{R} + \alpha_F - \alpha_R$$
$$\Rightarrow \delta = 57.3\frac{L}{R} + \frac{V^2}{C_F R}\frac{F_{ZF}}{g} - \frac{V^2}{C_R R}\frac{F_{ZR}}{g}$$

Thus,

$$\delta = 57.3\frac{L}{R} + \left(\frac{F_{ZF}}{C_F} - \frac{F_{ZR}}{C_R}\right)\frac{V^2}{gR} \quad (5.3)$$

Comparison with

$$\delta = \frac{l}{R} + Ka_Y$$

From above shows that

$$K = \left(\frac{F_{ZF}}{C_F} - \frac{F_{ZR}}{C_R}\right)$$

Inspection of this equation shows that K can be either positive or negative. Consider a car with equal cornering stiffness front and rear. If F_{ZR} is larger than F_{ZF}, as will be the case if the centre of gravity is towards the rear, then K will be negative. If F_{ZR} and F_{ZF} are the other way around, as for a car with the centre of gravity towards the front, then K will be positive.

Giving thought now to the magnitude of K, we see that if the car is travelling with a velocity such that V^2/R equates to one g, then the value of K gives the steering correction from Ackermann steering that applies at one g. Thus, in summary, a positive value of K indicates an understeering car, a negative one indicates an oversteering car and the value gives the number of degrees per unit a_Y, of steering deviation from the neutral or Ackermann steering geometry that apply.

Worked Example

Consider a small road car with the following dynamic data:

Wheelbase l:	3 m
Front tyre –CoG distance a:	1.2 m
Mass m:	950 kg
Velocity V:	18 m/s
Corner radius:	25 m
Cornering stiffness front:	−1050 N/degree
Cornering stiffness rear:	−850 N/degree

Determine the angle of steer required at the front wheels for this car and how much this value changes if the centre of gravity is moved forwards 0.3 m.

We can first find the lateral force at the rear by taking moments about the front:

$$850\alpha_R \times 3 = \frac{950 \times 18^2}{25} \times 1.2$$
$$\Rightarrow \alpha_R = 5.8 \text{ degrees}$$

Similarly, taking moments about the rear,

$$\frac{950 \times 18^2}{25} \times (3 - 1.2) = 1050\alpha_F \times 3$$
$$\Rightarrow \alpha_F = 7.0 \text{ degrees}$$

The amount of steered angle can now be determined:

$$\begin{aligned}\delta &= \frac{l}{R} + \alpha_F - \alpha_R \\ &= \frac{3}{25} + 7.0 - 5.8 \\ &= 1.3^0\end{aligned}$$

5.6.1 Steering Angle vs Speed for a Range of Understeer Gradients

It is a simple enough matter to produce a spreadsheet that can calculate the steered angle necessary to maintain a given corner radius using the material above and in particular Eq. (5.3). In the spreadsheet below, the input data are supplied top left. From this, the b value and vertical loads on the front and rear axle are determined. Finally, an output table is calculated, giving the steered angle and velocity for these input parameters, and this can be seen in the block-headed 'output'. The table shown has data for a vehicle with an understeer gradient of -1.025 degrees/g. This spreadsheet was then used to prepare data for the graph shown. The data in the table were adjusted with different centre of gravity locations, and five examples were prepared that offer understeer gradient values of, 1.0, 0.5, 0, -0.5 and -1.0. The data from each evaluation were stored and the graph produced with them all on the one pair of axes. It is seen that, for the understeering car, with K values that are positive, the amount of steering angle needed to maintain the given corner radius increases with speed, whilst, for the oversteering car, the amount needed decreases. Interestingly, the amount needed for the oversteering car actually reaches zero at just over 90 mph.

The speed at which the oversteering car requires zero steering input to maintain the corner of given radius is known as the *critical speed* and can be estimated using Eq. (5.3):

$$\delta = 57.3\frac{L}{R} + \left(\frac{F_{ZF}}{C_F} - \frac{F_{ZR}}{C_R}\right)\frac{V^2}{gR}$$

Using the data from the spreadsheet, we have

$$0 = 57.3\frac{3.05}{25} + \left(\frac{4294}{-1200} - \frac{3064}{-1200}\right)\frac{V_{CRIT}^2}{9.81 \times 25}$$

$$\Rightarrow V_{CRIT}^2 = \frac{-6.99}{-0.0042}$$

$$\Rightarrow V_{CRIT} = 40.8\,\text{m/s} \cong 89.7\,\text{mph}$$

A figure that agrees well with the value read off from the graph for a car of understeer gradient −1.0. It may seem strange that a car can corner with no steering input at all, but one must remember that, for an oversteering car, the rear is being urged outwards by the rearward centre of gravity, increasing the effect of the steering input at the front, and as the speed increases, less steering input is needed, none eventually. A similar thing happens the other way around for the understeering car, and the forward-biased centre of gravity urges the front outwards, decreasing the effect of the steering input. Notice that the critical speed is not dependent on the corner radius R. This is because it appears on the bottom row of two terms that sum to zero and if rearranged they would cancel. However, this begs the question, what would the corner radius actually be for the oversteering car at the critical speed? According to the above, *any* radius is possible. We shall revisit this ambiguity later when dealing with the derivative analysis.

Before moving on, we must revisit the assumption made at the start of this section where the lateral force generated at the tyre was taken to be a linear function of slip angle. This is important now because, if we consider Fig. 5.16, it is clear that the calculated value of steered angle required at low speed for a given corner radius will be reasonably accurate; after all, the tyre lateral force is a linear function for a small slip angle. However, it must become a poor approximation once the tyre force leaves the linear region,

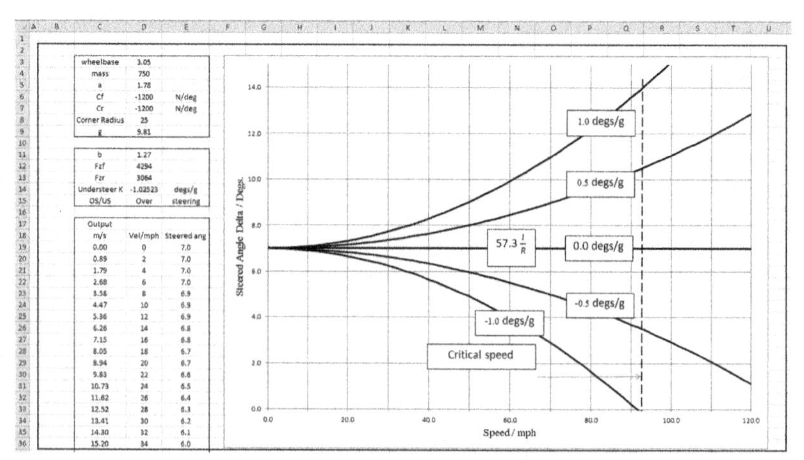

Fig. 5.16 Spreadsheet analysis of steered angle vs speed for a range of understeer values.

and eventually, it must be plain wrong once the actual tyre function has reached its maximum and started to decrease, by this time the approximation is nothing like reality.

5.6.2 Other Contributions to the Understeer Gradient

So far, we have identified that the forwards and aft location of the centre of gravity of a vehicle and the speed at which it is travelling will have an effect on the steering response. However, we have only considered the effect of this limited set of parameters. Is there anything else to worry about, and how can we know?

In principal, the answer to this question is easy. Consider again the car in plan view shown in Fig. 5.10. With the lateral force at the front and rear in moment balance with each other, it is clear that this moment balance is the cause of the understeer and oversteer effects we have dealt with. If, however, there is *anything* else that affects the size of the lateral force generated at the front and rear, then this too will affect the moment balance and so has the capacity to affect whether the car is understeering or oversteering. For example, imagine a performance passenger car fitted with a double wishbone suspension at the rear and MacPherson struts at the front. When such a car corners, the body will roll, and the front tyre will follow this roll as dictated by the front suspension. Both front tyres will be under positive camber and lose lateral force generation. Both rears will remain unchanged. The net effect is therefore an understeering steering effect because to compensate for the reduced front lateral force the driver will have to rotate the steering wheel into the turn more to compensate. Alternatively, a car fitted with an antiroll bar facilitating different weight transfer front to rear will again be influenced. For example, if the antiroll bar at the rear is stiffened, weight transfer at the rear will increase (and at the front decrease). This will result in reduced lateral force generation at the rear and an increase at the front. This is an understeering effect because to offset the effect the driver will have to reduce the steering input for the same corner radius. Indeed, effects remote from the suspension can affect the amount of understeer; for example, a vehicle in which the centre of aerodynamic pressure in the side view is towards the rear of the car will experience oversteer when a gust of wind is presented from the side and will be understeering because the driver will need less steering input, some having been provided by the gust. Thus, a whole collection of contributors affect the amount of understeer.

As a consequence of this, we need to recognise that the amount of understeer is not just related to the location of the centre of gravity, and we need a more general way of dealing with the amount of understeer. We do this by making use of an alternative from of the cornering equation:

$$\delta = \frac{l}{R} + Ka_Y \tag{5.4}$$

In this version, we have the steered angle, wheelbase and corner radius as before, but now, the quantity K is introduced. This is called the *understeer gradient* and is a summation of all the different effects that contribute to the amount of understeer. It is called the understeer gradient because as we can see it determines whether the graph of steered angle against forwards speed has a slope that *increases* or *decreases* as acceleration increases and so it controls the *gradient* of this line.

If we compare Eq. (5.3) above with Eq. (5.4), then it is clear by inspection that

$$K = \left(\frac{F_{ZF}}{C_F} - \frac{F_{ZR}}{C_R}\right)$$

This may easily be rearranged by balancing moments to give

$$K = \frac{mg}{l}\left(\frac{a}{C_R} - \frac{b}{C_F}\right) \tag{5.5}$$

But if we were wanting to consider other sources of understeer, then we should need to determine other terms for K to be added to that above:

$$K = \underbrace{\left(\frac{F_{ZF}}{C_F} - \frac{F_{ZR}}{C_R}\right)}_{\text{CoG effect}} + \underbrace{?}_{\text{Roll steer}} + \underbrace{?}_{\text{Camber gain}}$$

Sometimes, vehicle dynamicists refer to the 'understeer budget' meaning the sum of all contributions with an amount worked out for each. This can then be compared with the measured amount to provide a secure and analytical measure of performance. A consideration and quantification of many other such sources is beyond the scope of this text, but we can outline a few.

5.6.2.1 Roll Steer

When a vehicle rolls as a result of lateral acceleration, it is entirely possible for the kinematic design of the suspension to result in toe changes at one end of the vehicle then the other. This is inevitable in cases where, for example, a vehicle has a MacPherson strut at the front and a wishbone at the rear. The toe change being different at each will result in slip angle changes front to rear and with them lateral force changes. A relationship can be developed between roll angle and the difference in lateral forces and also a relationship between understeer gradient and roll angle.

5.6.2.2 Camber Gain

A difference in camber gain under roll between the front and rear will result in a difference in lateral force generation between front and rear, and again, a relationship between understeer gradient and camber gain can be developed.

5.6.2.3 Steering Compliance

Imagine a chassis in which the steering rack is very rigid and securely mounted to the chassis but with wishbone mounts that are by comparison more compliant and deflect more than the rack mounts under lateral load. This movement of one with respect to the other will result in a steering input that depends on lateral acceleration and originates from a difference in chassis compliance between one mount and the other. This is a contribution to understeer gradient that comes from chassis compliance. Its modelling would require extensive FEA modelling, and its quantification is no small undertaking. This possibility should alert the chassis designer that the anchorages are at least equally compliant as a starting point.

5.6.2.4 Migration of Centre of Aerodynamic Pressure

In a car with significant aerodynamic downforce, it is normal for the centre of pressure to migrate with speed. A significant cause of this is that the front wing is a very efficient part of the package and remains good up to high speed, where other elements operate in disturbed air and don't increase as much. Consequently, the centre of pressure normally moves forwards with speed. This in turn affects the vertical load on the front wheels, and so, again, the lateral forces generated are changed. In such a car, the forwards speed of the car affects the understeer gradient in a way that originates from the aerodynamic design of the car.

5.6.2.5 Differential

The differential is a device that permits two driven wheels to rotate at different speeds and yet still have torque transmitted to both. The simplest differential is one in which no resistance to the difference in angular velocity between the two driven wheels is provided; such a differential is called an *open* differential. This is not problematic for modest weight transfer and good grip. However, if a car is cornering at the limit, there will be virtually no load on the inside wheel, and it may even lift off. Without restriction to the difference in angular velocity, this wheel will spin up, and since it can rotate freely, no drive will be supplied to the outer wheel in good contact with the road. There will be a loss of speed. The same thing will happen under the conditions of *split mu* meaning one tyre experiencing good grip whilst the other has very little, for example, a road car with one wheel in a puddle or on icy ground near the road edge.

To prevent this, one could simply use a *solid* rear axle, meaning that both wheels are connected to the same rotationally rigid axle. This does indeed solve the problem at conditions of full weight transfer or split mu but does not solve the problem that at modest speeds a difference in rotational speed is needed to allow the outside wheel to complete a larger distance, being on a great radius, than the inside wheel. In addition to this, however, a solid rear axel is very undesirable from a handling point of view since the solid rear axle will mean that the inside wheel is force to rotate faster than the radius of the corner upon which it moves demands and this will induce longitudinal slip producing a greater longitudinal force in the forwards direction than is generated at the outside wheel. The resulting torque will urge the car to turn out of the corner and continue straight ahead. Thus, a solid rear axle is a strongly understeering effect.

To solve this, we make use of a *limited-slip* differential. Such a device works by offering very little resistance to a difference in angular velocities, whilst rotational speeds are low (allowing the car to drive both wheels in modest cornering without causing understeer), but offering considerable resistance when the difference is high. This behaviour means that when split mu or heavy cornering is in progress the wheel able to turn freely will initially spin up but very soon the difference in angular velocities will cause the differential to resist, and even though the difference exists, drive is still transferred to the wheel with poor grip. The internet-based research guide at the end of the chapter provides many sources to visit to understand the operation of these differentials. Here, we are concerned with the effect on handling, and this is best illustrated by the following treatment where we consider a car with an open diff and the effect of this when the axle is solid. These two extremes provide limits within which all differentials will lie.

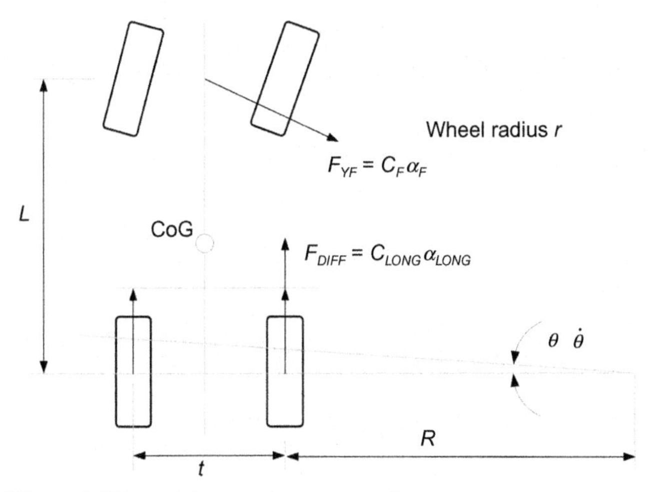

Fig. 5.17 Effect of differential on understeer gradient.

In Fig. 5.17, the vehicle is going around a right-hand corner. Weight is transferred to the outside. The outer rear wheel is gripping the road well, and because the rear axle is solid, the inner rear wheel is rotating faster than it should, given its linear velocity. Thus, it has a longitudinal slip and produces a longitudinal force, F_{DIFF}, in addition to the force it is producing because of the torque it is bearing.

Forwards velocity of the inner wheel, $R\dot{\theta}$
Forwards velocity of outer wheel, $(R+t)\dot{\theta}$
Angular velocity of outer wheel, $\frac{V}{r}=\frac{(R+t)\dot{\theta}}{r}$

Since the inner wheel is forced to rotate with the same angular velocity as the outer one and then remembering that the longitudinal slip ratio is given by $\left(1-\frac{r\omega}{V}\right)\times 100$, we have

$$\text{Longitudinal slip ratio} \quad =1-\frac{\frac{(R+t)\dot{\theta}}{r}}{R\dot{\theta}}\times 100$$

$$=\frac{t}{R}\times 100$$

We can now balance moments to determine how much lateral force generated at the front axle will be needed to balance the yawing torque caused by F_{DIFF}:

$$C_\alpha \alpha_F L = F_{DIFF}\frac{t}{2}$$

$$\Rightarrow \alpha_F = \frac{1}{C_\alpha L} C_{LONG}\alpha_{LONG}\frac{t}{2}$$

$$\Rightarrow \alpha_f = \frac{C_{LONG}}{C_\alpha L}\frac{t}{R}\frac{t}{2} \times 100$$

$$= \frac{C_{LONG}t^2}{2C_\alpha LR} \times 100$$

Physically, this means that a slip angle α_f is needed simply to offset the understeer contribution made by the solid rear axel. Clearly, even more will be needed to make the car go around the corner. We can estimate this for the following conditions appropriate for a formula student car on a 20 m skid pan:

C_{LONG}: 1100/%
C_{lat}: 1200 N/degree
R: 20 m
T: 1.5 m
L: 1.6 m

$$\alpha_f = \frac{C_{LONG}}{2C_\alpha L}\frac{t^2}{R} \times 100 = \frac{1100}{2 \times 1200 \times 1.6}\frac{1.5^2}{20} \times 100$$

$$= 3.2 \text{ degrees}$$

The size of this steered angle is the cause for concern. Most racing tyres develop their peak lateral force at around 8–10 degrees of slip. This means that a car without a differential would already be using up 2.6 degrees of slip in countering the effect of the differential and only around 6.4 will be left for cornering. If the car were fitted with a limited-slip differential, this would be much improved.

From this, we conclude that for this car a solid rear axle will result in a loss of around one-third of the lateral force generating capacity whilst an open diff would result in none. However, the solid rear axle will continue to provide all the torque supplied by the engine to the outside wheel even after the inside one has lifted off. In between these two extremes will be an operating point at which the losses due to one effect balance the gains from the other, and this should be the design goal.

To achieve this is not simple. Firstly, a cornering analysis based on the derivative analysis presented below in this chapter should be prepared but modified to incorporate a loss of cornering force at the front wheel dependent on the force F_{DIFF} above. The characteristic for F_{DIFF} is best determined from a differential dynamometer, an experimental piece of test equipment that will measure the force F_{DIFF} as a function of the difference in angular speed of the two driven half-shafts. This can then be applied to the model. Alternatively, an analysis can be completed to determine the optimal differential function for the case in hand, and this function can be used to design the differential. Further to these design possibilities is the *active differential*. In this approach, the differential has the ability to control actively the difference in speed between the half-shafts. Sometimes, this is done by having each half-shaft flitted with a brake under computer control. If the control system senses that the car is not yawing sufficiently for the combination of steering input, yaw rate and forwards speed then the inside wheel brakes until it matches. If it is too much, the outside wheel is braked until a match is reached. Such systems make possible the design of a car that is neutral steering all the time even though the basic chassis is not. This approach should however be used with care; it is better to cure the illness than treat the symptom. A naturally neutral steering car will always be potentially faster since longitudinal torque is not wasted in correcting the understeer gradient. In a road car, this may be of much less important, and there is more to gain by correcting a car whose understeer gradient cannot be made neutral for good design reasons, and more is gained than lost be introducing a limited-slip differential. To simulate these kinds of effects accurately requires the full-vehicle approach to modelling presented in Chapter 9.

5.6.3 Experimental Determination of the Understeer Gradient

The form of Eq. (5.4) is eminently practical for experimental determination of the understeer gradient. On the car, one only needs to measure the lateral acceleration at the centre of gravity (or work it out from accelerometers at each axle if need be) and the steered angle. After this, the corner radius must be known, and following a known circular track is simple enough. The wheelbase can be measured in the garage. The vehicle is then driven at constant speed around the known radius circle and data recorded. To parametrise the effect, it is usual to do this for a range of speeds and corner radii and obtain a map of the understeering performance. In a car fitted with telemetry, the understeer gradient currently be experienced by the car can be derived from these channels using GPS data for the telemetry to have a knowledge of current corner radius. Once good data are being recorded, the comparison of the real car to computer simulations can take place, and rational comparison of measured data can begin. It is, of course, possible to test a vehicle and gain a qualitative description of it from experience, but until numerical data and simulation are compared, it is all rather subjective and lacks any rigour.

5.7 EQUATIONS OF MOTION FOR A CORNERING RACING CAR

In this section, we shall develop a pair of equations that describe the response of a car to a step-steering input. Sometimes, it's helpful to have a sneak preview of the answer before embarking on a complex analysis, and this is just such a case. It's also very informative to have thorough qualitative understanding of the effects at work before beginning such a derivation. For these reasons, we shall first spend some time considering exactly how it is that a car negotiates a corner.

Fig. 5.18 illustrates cornering dynamics by making use of the bicycle model. Initially, the car is as shown in position '0.0 s'. The car is travelling in a straight line, and there is no steering input. The 'trail line' running backwards from the centre of gravity shows where the car has just been. It's the line that would be made if there were a pot of paint with a hole in the bottom inside the car pouring slowly through a hole in the chassis and onto the

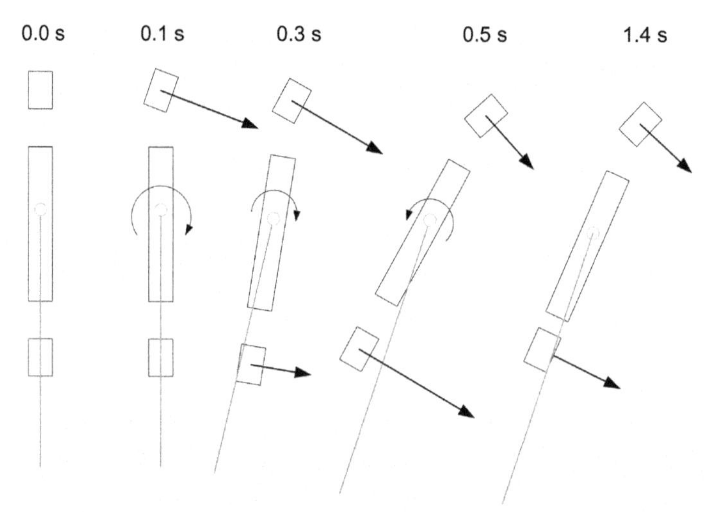

Fig. 5.18 The evolution of side-slip, yaw velocity and yaw displacement.

tarmac below. At position '0.1 s' the driver supplies a rapid step input to the steering. Since the front wheel is now operating at a small slip angle, it generates a lateral force. Because this force has not been applied for very long, the car has not yet yawed much. The car has a small yaw velocity as shown by the rotational arrow. Slightly later, the car is in the position '0.3 s'. The lateral force acting at the front of car has caused the whole car to yaw; this in turn means that the slip angle has evolved at the rear wheel. The radius of the corner that the car is currently travelling in is still very large, meaning that the lateral force necessary to keep the car in circular motion is very small. Importantly, it is smaller than the lateral force generated by the front and rear tyres together. The lateral force not expended in keeping the car in circular motion therefore accelerates the car to the right in the diagram. The trail line clearly shows this; however, there is still the yawing moment acting on the car and as time goes by the radius of the corner gets tighter and tighter and the yaw velocity builds up. In position '0.5 s', the car has yawed further and actually overshot the equilibrium position. The yaw moment serves to accelerate the car in the other direction coming as it does from a reduced lateral force at the front and increased lateral force at the rear. In position '1.4 s', the car has settled down to the equilibrium. The moment produced by the lateral forces at the front and rear is zero, the yaw velocity is constant, and the side-slip velocity is constant.

The graph in Fig. 5.19 shows the yaw velocity (continuous line) and yaw displacement (dotted line). In the graph, the time along the bottom axis corresponds to the times given in the dynamic above. The yaw velocity can clearly be seen to overshoot the equilibrium value and then settled down to it in much the same way that a resonant single-degree-of-freedom system would. The yaw displacement is basically a straight line of positive gradient as is to be expected for a car that is in constant circular motion. However,

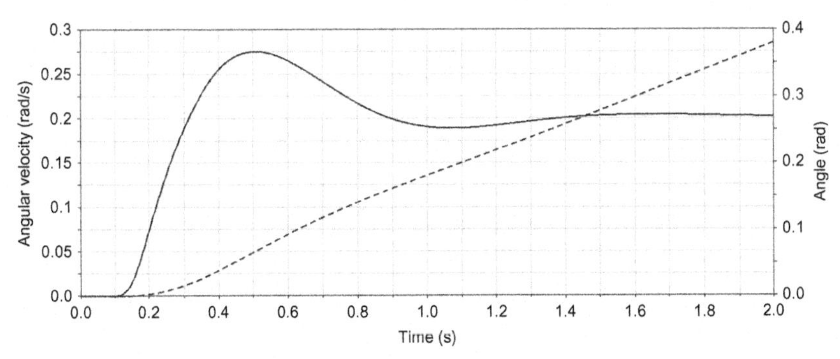

Fig. 5.19 Yaw rate and yaw angle response to step steer.

the resonant behaviour visible in the yaw velocity graph produces undulations in the yaw displacement curve. Close scrutiny of the curve shows that it is not a straight line. The peak of the yaw velocity curve coincides with the point of maximum gradient on the yaw displacement curve, as we should expect.

Earlier, we have studied the behaviour of a racing car under steady-state cornering. We have developed the concept of the understeer gradient and even seen how a racing car will behave at the limit when either over- or understeering. This analysis is very useful and certainly allows us to understand vehicle dynamics in a way that helps us design cars. It is abundantly clear, for example, that we need a neutral steering car if we are to design the best performance into the car. However, these were very much steady-state considerations. The dynamic above is clearly a *transient* one, meaning that it is present only for a short time, after which it has decayed away and we are left only with the steady state. Being able to model this transient dynamic is going to be very beneficial. For example, we shall see that there is a minimum time within which any given vehicle can complete this transient process and this cannot be improved upon. Imagine being a race engineer for a frustrated driver who complains bitterly that the turn-in response of the car is too slow. If one then diligently sets about making chassis set-up changes designed to improve it, that effort might be completely wasted. Instead, one should perform the derivative analysis introduced below and determine how close to the theoretical best time the car actually is. If it is returning this time, the driver is doing the best that can be done with the car and the car design itself must be improved. If not, then there is no point improving the car until the driver has improved. Unpalatable as this case may be for some drivers, the truth is always the best platform from which to make decisions.

We need now to set about an analytical mathematic analysis of the cornering car, one that produces the time-based response to steering input.

We shall start by examining the free-body diagram we met earlier but consider it in more detail now in Fig. 5.20. There are a lot of parameters to consider, and we start by noticing the car is travelling around a corner of radius, R. It has a velocity, V, which is tangential to a line from the centre of gravity to the corner centre. This velocity can be resolved into two components in the coordinate system of the vehicle. Firstly, the lateral velocity V_Y and then a forwards velocity V_X, which together produce a resultant velocity for the car. Since the car in a corner, V, is not parallel to the vehicle axis, the whole vehicle is at a slip angle of β. In addition to these velocities,

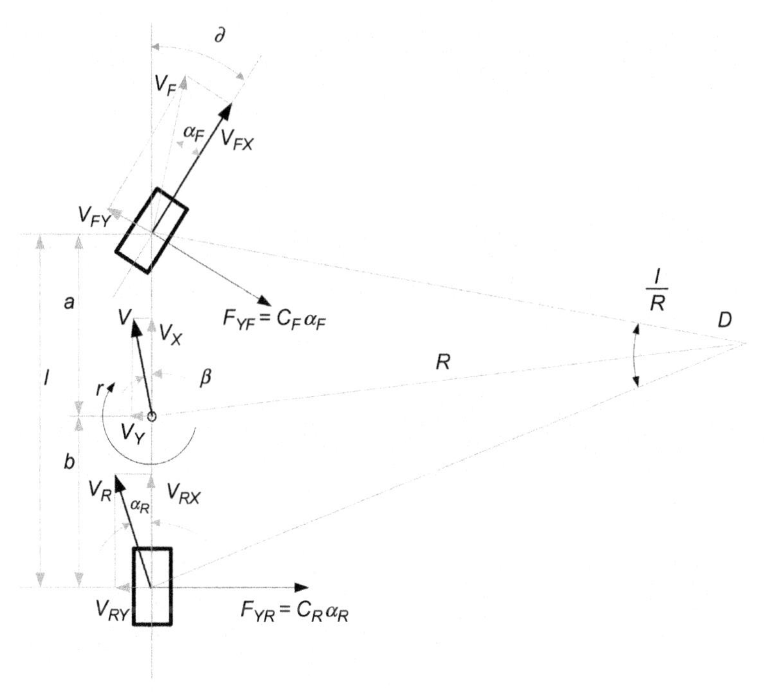

Fig. 5.20 Free-body diagram for a cornering vehicle.

the car has a yaw velocity. The rate of yaw is r and occurs about the centre of gravity. At the front, a slip angle α_F applies and results in a force at right angle to the tyre. α_R applies in a similar way at the rear. We shall denote the total yawing moment, N, and it will result from summing up each lateral force multiplied by its perpendicular distance from the centre of gravity. We shall denote the total lateral force Y.

5.7.1 Lateral Force Equilibrium

The lateral acceleration of the car comes from two sources. Firstly, the car is rotating around the corner centre, and secondly, there is a lateral acceleration:

$$a_Y = \frac{V^2}{R}$$

Remembering that the angular velocity, ω, of the car about the corner centre is equal to the yaw velocity r, we can write

$$V = \omega R \;\Rightarrow V = rR$$

$$\frac{V}{R} = r$$

Thus, the lateral acceleration that results from rotation about the corner centre is

$$a_Y = \frac{V^2}{R} = \frac{V}{R}V$$
$$\Rightarrow a_Y = rV$$

However, the car also has a horizontal side-slip velocity, V_Y, and if this is changing that it clearly will during the evolution of a corner, then we have a component of horizontal acceleration:

$$a_Y = \dot{V}_Y$$

Thus, the total lateral acceleration is the sum of these two components and

$$a_Y = Vr + \dot{V}_Y$$

Now, since the angles involved are small and the velocity V is constant,

$$V_Y = V\sin\beta$$
$$\Rightarrow V_Y \approx V\beta$$
$$\Rightarrow \dot{V}_Y = \frac{d}{dt}(V\beta) = V\dot{\beta}$$

So, the total lateral acceleration is given by

$$a_Y = V\left(r + \dot{\beta}\right) \tag{5.6}$$

We look now at the rear of the vehicle to determine the rear slip angle α_R. The rear tyre is sliding to the right because the whole vehicle is doing so with velocity V_Y. In addition, it is sliding to the left because the whole car is yawing about the centre of gravity. The magnitude of this component is $r \times b$. The total lateral velocity is therefore

$$V_{YR} = V_Y - rb$$

By definition, the slip angle at the rear is given by

$$\alpha_R = \tan^{-1}\left(\frac{V_{YR}}{V_X}\right) \approx \frac{V_{YR}}{V_X}$$

Thus,

$$\alpha_R = \frac{V_Y - rb}{V_X}$$

Approximately,

$$\beta = \tan^{-1}(V_Y/V_X)$$
$$\approx V_Y/V_X$$

And since the velocity in the x direction is approximately the vehicle velocity, $V_X \approx V$, so

$$\alpha_R = \beta - \frac{rb}{V}$$

At the front, a similar analysis is applied. Here, however, the yaw velocity is moving from the front to the right not the left, and the steered angle δ must be included. Thus, the velocity of the front tyre V_{YF} is

$$V_{YF} = V_Y + ra$$

Since the projection of V_{YF} in the y-axis is approximately, V_{YF}, and remembering that α_F is negative, the angle between the front-wheel velocity, V_F, and the x-axis, $(\delta - -\alpha_F)$, can be approximated as

$$\partial + \alpha_F = \tan^{-1}\left(\frac{V_{YF}}{V_X}\right)$$
$$\approx \frac{V_{YF}}{V_X}$$
$$\Rightarrow \alpha_F = \frac{V_{YF}}{V_X} - \delta$$

Substituting for V_{YF}

$$\alpha_F = \frac{V_Y + ra}{V_X} - \delta$$

that simplifies to

$$\alpha_F = \beta + \frac{ra}{V} - \delta$$

where α_F is in the sense shown in Fig. 5.20, since we now know the front and rear slip angles, the front and rear lateral forces are simply

$$F_{YF} = C_F\left(\beta + \frac{ar}{V} - \delta\right)$$
$$F_{YR} = C_R\left(\beta - \frac{rb}{V}\right)$$

The total lateral force is therefore

$$F_{TOT} = C_F\left(\beta + \frac{ar}{V} - \delta\right) + C_R\left(\beta - \frac{rb}{V}\right)$$

$$\Rightarrow F_{TOT} = (C_F + C_R)\beta + C_F\frac{ar}{V} - C_R\frac{rb}{V} - C_F\delta$$

Since the total lateral acceleration is given by

$$a_Y = V\left(r + \dot{\beta}\right)$$

So, we may write

$$\Rightarrow mV\left(r + \dot{\beta}\right) = (C_F + C_R)\beta + \frac{r}{V}(C_F a - C_R b) - C_F\delta \tag{5.7}$$

5.7.2 Moment Equilibrium

We now consider the moments acting on the body of the car that make it yaw. The total moment causing the car to yaw is called the yawing moment. When the car is in a straight line or in a steady-state corner, the yawing moment is zero because the car is not accelerating in yaw. On corner entry and exit, the yawing moment is large:

$$N = N_F + N_R$$

$$\Rightarrow N = F_{YF}a - F_{YR}b$$

$$\Rightarrow N = C_F a\beta + C_F\frac{a^2 r}{V} - C_F a\delta - C_R b\beta + C_R\frac{rb^2}{V}$$

The total yawing moment accelerates the car in yaw according to $T = I\ddot{\theta}$. Here, r is the yaw rate, and so, the yaw acceleration is given by $\dot{r}$:

$$\Rightarrow I\dot{r} = (C_F a - C_R b)\beta + \frac{r}{V}\left(C_F a^2 + C_R b^2\right) - C_F a\delta \tag{5.8}$$

where I is the polar moment of inertia about the vertical axis of the car through the centre of gravity.

5.7.3 Equations of Motion Derivation Summary

With a lengthy derivation like the one above, it is easy to lose one's way, so here is a reduced version showing all the processes at once (Fig. 5.21).

Understanding the lateral acceleration in this derivation can be difficult. Imagine the moment instantly after a significant step steer has been supplied to the front wheel (position '0.1 s' in Fig. 5.18). At this time, the car will clearly yaw since a torque has been applied. However, it will also accelerate

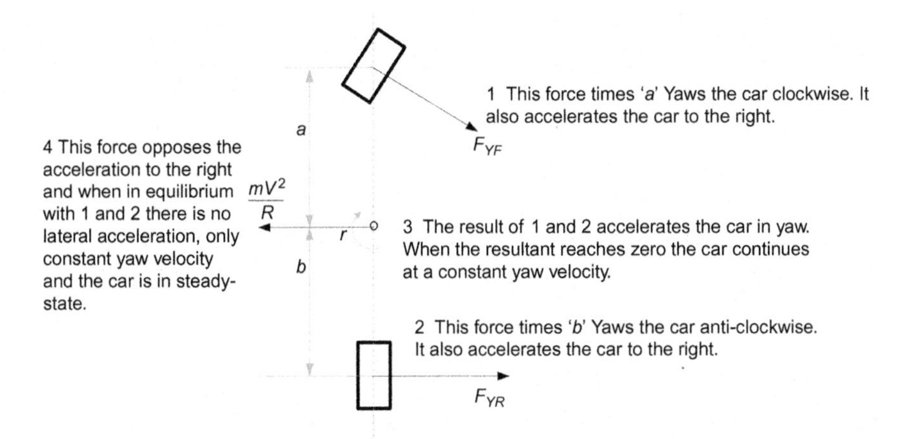

Fig. 5.21 Summary of the derivation of the equations of motion.

to the right since the centripetal force, mV^2/R, has not evolved yet, the car is still in a straight line, and R is infinite. Indeed, scrutiny of Fig. 5.5 shows that the side-slip velocity can only be zero when the vehicle is at the *tangent speed*.

5.7.4 Equations of Motion

Eqs (5.7), (5.8) are the lateral force equilibrium and moment equilibrium equations for a cornering car and are shown again below. They are extremely interesting equations for a vehicle dynamicist. They relate the cornering response of a general vehicle to the steering input and when solved will yield a time-based response. The parameters contained must mean that a solution can be obtained that will relate corner radius to steering input and reveal its dependence on speed and indeed all this can be done. Consider again

$$\Rightarrow mV\left(r+\dot{\beta}\right) = (C_F + C_R)\beta + \frac{r}{V}(C_F a - C_R b) - C_F\delta$$

$$\Rightarrow I\dot{r} = (C_F a - C_R b)\beta + \frac{r}{V}\left(C_F a^2 + C_R b^2\right) - C_F b\delta$$

The equations are differential and link together the yaw rate r, the steering input δ and the vehicle side-slip angle β. If we were to solve them, we should be able to produce relationships between these quantities. Since both equations contain β, we could produce one equation containing only δ and r. We could then solve this for the yaw angle. This would clearly be advantageous; it would allow us, for example, to work out how long it would take a car to adopt a new heading in response to a given steering input. In designing a racing car, we would seek to make this be very short because this would allow rapid negotiation of corners. One might argue that an infinitely short response time would be best; however, this can never be obtained since the car is bound to have some value for yaw inertia I, so it will always take some time to respond to an applied torque no matter how large the moment applied may be. In addition, there is little point in having response times that are vastly better than the driver can make use of because of the limits of human reaction times.

Our problem is therefore how to solve these two equations. There are two approaches; the first is to use a modelling approach and use a computer to do it. The second is to develop an analytical approach. Both are important. When a computer simulation is used, solution speeds are high, nonlinear effects that can't be solved analytically, can be dealt with easily, and very good agreement between model and practice can be developed. However, the computer simulations only ever tell us *what* will happen; they don't explain *why*. If we want to make improvements to the performance, we shall be left simply making a change to something and see if it makes things better or worse. There are so many variables that this approach is impractical, and even if it were, there is another more serious problem. Suppose we found some local region of best performance where any change to a parameter made the car worse. There might be another minima elsewhere in the parameter space that is even better. The problem is the same as trying to find the lowest spot in a huge area of high hills. You don't just need to know that you have found a local lowest point; you need to know you have found the lowest one of all. To do this, you need the map; you need to understand *why* the car is behaving as it is. For this reason, we need not only the ability to solve the equation by computer but also the ability to solve them analytically so that we know *why* things happen as they do and so make informed suggestions for improvement, not just guesses.

We start by looking at the numerical solution of the equations, and then, in Section 5.7.6, we shall develop the analytical understanding.

5.7.5 The Numerical Solution of the Bicycle Model

Using multibody code such as Adams, it is possible to produce a simulation of the bicycle model complete with splines to store tyre characteristics, input steering functions as desired and even deal with varying speed through a corner (Fig. 5.22).

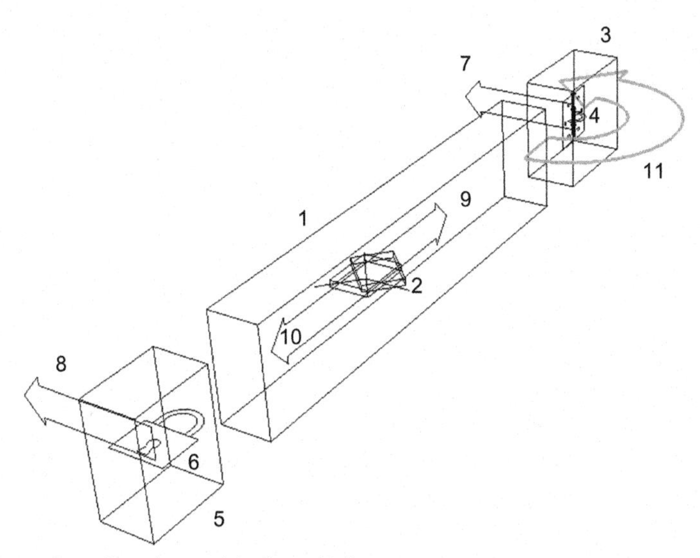

Fig. 5.22 Adams bicycle model of a vehicle.

On the website for this book, there are downloadable files for the bicycle model together with short movies of its operation and instructions on its creation. At this point, it is important to consider models prepared in this way because they offer advantages over the analytical solution that follow. The best thing to do here is to make your own working bicycle model and use it to demonstrate first-hand the dynamics we have covered because its only in this way that one can become truly adapt at vehicle dynamics. The section at the end of the chapter contains some ideal projects to undertake. The model consists of the following features.

The central body (1) is free to move left-right, forward-aft and to yaw as controlled by the planar joint (2) between it and the ground. The front wheel (3) is constrained to rotate about a hinge joint (4) between it and the 'body'. The rear wheel (5) is fixed to the body by the padlock joint (6). Lateral forces (7) and (8) are applied to the model at the front and rear wheels. These move with the body and so always provide a force at right angles to the pointing direction of the 'tyre'.

Internally, Adams must be programmed, making use of the state variables it offers, to determine the slip angles that pertain at any instant to each tyre. The slip angle is then used to determine the lateral force that would apply by looking the slip angle up in a spline entered at the start to model the tyre characteristic. The model is set in motion by applying a forwards constant force to accelerate forwards (9) and a rearward drag force (10), proportional to the velocity squared. These two forces keep the vehicle at nearly constant speed even during cornering when the rearward pointing vector of the front tyre lateral force serves to slow the vehicle down. Finally, any desired input steering function (11) is supplied through a second spline over time.

The advantage of such a model is that it makes use of real tyre splines, real steering inputs and being a numerical solution contains none of the approximations that we shall have to introduce in the next section to produce an analytical solution to Eqs (5.7), (5.8). The model can be used to demonstrate the response of the oversteering, neutral and understeering car and accurately predict the performance that a given car will offer. When developing a sound knowledge of vehicle dynamics, it is best to fully understand the analytical methods first and then build numerical models such as this to validate the theory. Finally, one needs to obtain data from a real car and validate that against both theoretical and numerically prepared data.

5.7.6 The Analytical Derivatives Approach to Vehicle Dynamics

We have seen the two differential equations that govern the motion of a car:

$$\Rightarrow mV\left(r+\dot{\beta}\right) = (C_F + C_R)\beta + \frac{r}{V}(C_F a - C_R b) - C_F\delta$$

$$\Rightarrow I\dot{r} = (C_F a - C_R b)\beta + \frac{r}{V}\left(C_F a^2 + C_R b^2\right) - C_F a\delta$$

These equations are too unwieldy to be solved directly; instead, we shall borrow an approach first used in aeronautics to model the dynamic behaviour of aeroplanes. Examination of these equations shows that the total lateral force and yawing moment, the left-hand side of the above equations, are functions of three variables and we can summarise the equations as

$$mV\left(r+\dot{\beta}\right) = f(\beta, r, \delta)$$
$$I\dot{r} = f(\beta, r, \delta)$$

The derivative approach involves taking the partial derivatives of each independent variable, adding them together and equating this to the change in total lateral force and yawing moment that would result from a change in β, r or δ at a given particular operating point:

$$\Delta\left(mV\left(r+\dot{\beta}\right)\right) = \left(\frac{\partial Y}{\partial \beta}\right)\beta + \left(\frac{\partial Y}{\partial r}\right)r + \left(\frac{\partial Y}{\partial \delta}\right)\delta$$

$$\Delta I\dot{r} = \left(\frac{\partial N}{\partial \beta}\right)\beta + \left(\frac{\partial N}{\partial r}\right)r + \left(\frac{\partial N}{\partial \delta}\right)\delta$$

Most people find this step confusing when they first meet it. In essence however it is very simple. What we are saying is that the equations are too difficult to solve directly so instead of doing that we shall approximate the solution to a linear response locally. Imagine that the solution curve is represented by the surface of a steep hill upon which we are walking. We could approximate the surface locally to being a flat surface that has the same gradient in the north-south and east-west directions as the hill itself has at this location. As long as we don't go too far from where we are, the flat sheet will give a good approximation to the real hill surface.

Most texts abbreviate the partial derivative terms to make the equations more wieldy. In addition, the Δ is often omitted. Strictly, the Δ should be included since, having taken partial derivatives, we have lost the absolute values and are obtaining the change in value by adding together the local slope multiplied by the distance travelled in each of the three dimensions of the equation. However, since the whole derivative approach is one of perturbation from a given point, it is acceptable to drop the Δ as long as we are clear about why we have done so. Thus, most scholars on the subject would write

$$mV\left(r+\dot{\beta}\right) = Y_\beta\beta + Y_R r + Y_\delta\delta$$
$$I\dot{r} = N_\beta\beta + N_R r + N_\delta\delta$$

We obtain the equations for each of these derivative terms by the direct comparison of the above with the equations of equilibrium reproduced below:

$$mV\left(r+\dot{\beta}\right) = (C_F + C_R)\beta + \frac{r}{V}(C_F a - C_R b) - C_F\delta$$

$$I\dot{r} = (C_F a - C_R b)\beta + \frac{r}{V}\left(C_F a^2 + C_R b^2\right) - C_F a\delta$$

Thus, the derivative terms are

$$\begin{aligned}
Y_\beta &= C_F + C_R \\
Y_R &= \frac{1}{V}(C_F a - C_R b) \\
Y_\delta &= -C_F \\
N_\beta &= (C_F a - C_R b) \\
N_R &= \frac{1}{V}\left(C_F a^2 + C_R b^2\right) \\
N_\partial &= -C_F a
\end{aligned} \tag{5.9}$$

5.7.7 The Physical Significance of the Derivative Terms

Each of these six-derivative terms, $Y_\beta, Y_R, Y_\delta, N_\beta, N_R, N_\delta$, has an understandable physical roll to play in the dynamics of the vehicle. They are described in turn below:

$$Y_\beta = C_F + C_R$$

Relates how much lateral acceleration results from each degree of whole vehicle side-slip. Looking at Fig. 5.20 and remembering that the force that produces this lateral acceleration comes from the slip angles at the tyres, we can see that the force always acts in the opposite direction to which the vehicle is travelling in side-slip. Thus, Y_β produces a force, proportional to a velocity, acting to reduce that velocity. This is exactly like the force in a damper unit, and Y_β is often referred to as side-slip-damping. Imagine a neutral steering car travelling forwards in a straight line; if one were magically able to fly alongside and give it a heavy sideways kick at the centre of gravity, the car would recoil, rapidly settling to a new parallel new path alongside the old one. It is the side-slip damping that brings this lateral displacement to rest:

$$Y_R = (1/V)(C_F a - C_R b)$$

Relates how much lateral acceleration results from each unit of yaw velocity. It may seem odd that lateral acceleration could result from yaw velocity, but if one wheel is much more distant from the centre of gravity than the other and both are moving laterally, one left and one right, because of a yaw velocity on the car, then the total lateral force will not be in balance, and a lateral acceleration will result. If the car were at rest and being yawed, there would be a certain lateral acceleration that would result; if the car is then moving forward, around the corner, the fact that it is moving in a circle serves to reduce the slip angle front and rear by an amount proportional to the velocity. Thus, Y_β is inversely proportional to forwards velocity:

$$Y_\delta = -C_F$$

Relates how much lateral acceleration results from each unit of steering input angle the driver demands. The whole car slip angle evolves once a steering demand is initiated at the front wheels. Once the whole car has adopted its side-slip angle β, the lateral acceleration value is determined by Y_β. However, the parameter that determines how large this force grows to is Y_δ. Clearly, this depends on the cornering stiffness at the front (steered) wheels and the angle through which they are steered. The distance between the centre of gravity and the front wheels affects how large the yawing moment is, which in turn affects how quickly the car responds to the steering input, not the value to which the yaw displacement rises. Since the car is going around a corner, an increase in Y_δ will cause an increase in lateral acceleration since the car will now be travelling around a tighter corner with the same speed:

$$N_\beta = (C_F a - C_R b)$$

Relates how much yaw acceleration results from each degree of whole vehicle side-slip. As with Y_R, it may seem odd that side-slip results in yaw. However, by thinking of a car with equal cornering stiffnesses front and rear but with the rear axle much closer to the centre of gravity than the front, then a sideways shove at the centre of gravity will result in a greater moment being generated by the front axle than the rear, and so, yaw acceleration will indeed result. This derivative is very important in determining vehicle stability. If N_β is negative, then a little side-slip produces a yawing moment that turns the car more at the front than the rear. This in turn alters the slip angles front and rear with a greater slip angle at the front than the rear. Once this has reached equilibrium, the yaw moment is in balance again. Thus, a unit of side-slip input will result in a given amount of yaw angle once the car has settled down from the transient effects of its sudden application. N_β is therefore the oversteer-understeer derivative; if it is zero, no yaw acceleration results from side-slip, and the car is neutral. Once understood, this in fact seems obvious from the shape of the equation for N_β that is one moment minus another; if they equate, there is no yaw:

$$N_R = (1/V)\left(C_F a^2 + C_R b^2\right)$$

Relates how much yaw acceleration results from each unit of yaw velocity. The yaw acceleration that results from a given value of yaw velocity is always in the sense that opposes the yaw velocity. Thus, if we try to yaw a car, N_R acts to oppose that yaw input, and it is therefore a source of yaw damping. If an input pulse of yaw acceleration is provided by some external means, the yaw damping will serve to dissipate it and make the yaw displacement tend asymptotically to a limit value. Since it depends on the square of the separation of the axles from the centre of gravity, it is a maximum when a and b are equal. As with Y_R, the fact that the vehicle is moving around a circle serves to reduce the slip angles and so N_R is inversely proportional to forwards velocity:

$$N_\delta = -C_F a$$

Relates how much yaw acceleration results from each unit of steering input. Clearly, the yawing moment of the front axle depends on the angle through with it is steered, the cornering stiffness of the tyre and the distance from the centre of gravity to the front axle. Increases in any of these bring about a larger yaw moment.

Having completed this understanding of the equations of motion and the concept of solution by means of the derivative analysis, we are now able to proceed to solve them and determine a relationship between the steering input provided and the yaw response of the vehicle.

5.8 TRANSIENT RESPONSE TO CONTROL INPUT

5.8.1 Qualitative Response to Steering Input

Referring back to Figs 5.18 and 5.19, we saw the transient behaviour that a car must pass through before settling into its steady-state equilibrium. It is clear that understanding this would assist the design of a performance car. For example, a car capable of reaching its steady-state position more quickly than another will provide the driver with a more rapid response. Additionally, a car without any overswing in the initial corner transient can more quickly and easily be taken to the limit of handling than one with overswing that would require a more complex control input to master with a good driver having to provide a compensatory steering input that would in some way negate the overswing inherent in the car.

Using the derivative analysis developed above, it is possible to simplify, rearrange and produce a single differential equation of motion from the two equations of motion developed above. In fact, we simply eliminate β from the equations of cornering directly and produce a second-order differential equation that is identical in form to that used to describe the motion of a single-degree-of-freedom mass, spring and damper system. Once this is done, we can examine the system response in yaw velocity to step-steering inputs and so determine how quickly the vehicle will adopt the yaw velocity demanded. Since this is a second-order system, we can expect a single resonant frequency, and all the dynamics that apply to a single-degree-of-freedom system with a mass, spring and damper (as presented in Chapter 8 suspension dynamics) will also apply to a car experiencing a yaw velocity as a result of a yaw demand.

5.8.2 Derivatives Equations

Starting with derivative equations of motion

$$mV(r+\dot{\beta}) = Y_\beta \beta + Y_R r + Y_\delta \delta$$

and

$$I\dot{r} = N_\beta \beta + N_R r + N_\delta \delta$$

we can rearrange the second equation to find β:

$$\Rightarrow \beta = \frac{1}{N_\beta}\{I\dot{r} - N_R r - N_\delta \delta\}$$

$$\Rightarrow \dot{\beta} = \frac{1}{N_\beta}\{I\ddot{r} - N_R \dot{r} - N_\delta \dot{\delta}\}$$

So, substituting into the first equation of motion

$$mV\left(r+\left(\frac{1}{N_\beta}\{I\ddot{r} - N_R\dot{r} - N_\delta\dot{\delta}\}\right)\right) = Y_\beta\left(\frac{1}{N_\beta}\{I\dot{r} - N_R r - N_\delta \delta\}\right) + Y_R r + Y_\delta \delta$$

$$\Rightarrow mVr + \frac{mVI}{N_\beta}\ddot{r} - \frac{mVN_R}{N_\beta}\dot{r} - \frac{mVN_\delta}{N_\beta}\dot{\delta} = \frac{IY_\beta}{N_\beta}\dot{r} - \frac{Y_\beta N_R}{N_\beta}r - \frac{Y_\beta N_\delta}{N_\beta}\delta + Y_R r + Y_\delta \delta$$

And so,

$$\Rightarrow \frac{mVI}{N_\beta}\ddot{r} - \left(\frac{mVN_R}{N_\beta} + \frac{IY_\beta}{N_\beta}\right)\dot{r} + \left(mV + \frac{Y_\beta N_R}{N_\beta} - Y_R\right)r$$

$$= \frac{mVN_\delta}{N_\beta}\dot{\delta} + \left(Y_\delta - \frac{Y_\beta N_\delta}{N_\beta}\right)\delta$$

Thus,

$$\Rightarrow I\ddot{r} - \left(N_R + \frac{IY_\beta}{mV}\right)\dot{r} + \left(N_\beta + \frac{Y_\beta N_R}{mV} - \frac{N_\beta Y_R}{mV}\right)r$$

$$= N_\delta\dot{\delta} + \left(\frac{Y_\delta N_\beta}{mV} - \frac{Y_\beta N_\delta}{mV}\right)\delta \tag{5.10}$$

Scrutiny of this equation shows that it is of the form

$$\Rightarrow I\ddot{r} - c\dot{r} + kr = C_1\dot{\delta} + C_2\delta \tag{5.11}$$

where

$$I = I$$

$$c = -\left(N_R + \frac{IY_\beta}{mV}\right)$$

$$k = \left(N_\beta + \frac{Y_\beta N_R}{mV} - \frac{N_\beta Y_R}{mV}\right)$$

$$C_1 = N_\delta$$

and

$$C_2 = \left(\frac{Y_\delta N_\beta}{mV} - \frac{Y_\beta N_\delta}{mV}\right)$$

Eq. (5.11) is the same differential equation that governs the time response of a freely suspended mass, linked to the ground by a spring and therefore relates the yaw rate r and its differentials to the steering input displacement and its rate of change with time. It can be used to solve for the yaw rate function provided that the input steering function is known. This might be a sinusoidal input or even a general input obtained from telemetry and then solved using numerical techniques. We shall now consider the simpler case of the response to a step steer.

5.8.3 Transient Response to a Step-Steer Input

The response of a car to a step-steer input is an important response for several reasons. Firstly, a good response to a step steer will make for a car with good handling, and so, understanding how to optimise a step-steer response is essential. In addition, the experimental measurement of a step-steer response is more easily obtained than other responses, for example, sinusoidal or ramp inputs. Indeed, all inputs may be thought of as starting with a step of some kind.

A step steer is obtained by noting that the steering input rate $\dot{\partial}$ is zero reducing Eq. (5.11) to become

$$\Rightarrow I\ddot{r} - c\dot{r} + kr = C_2\delta \tag{5.12}$$

where δ is the step-steering input value that is constant with time. The solution of this equation is well known in engineering and is made up of the sum two parts. The first, the particular integral, is obtained by solving the equation as it is, and second, the complimentary function, is obtained by setting the right-hand side to zero. Proof of this is not trivial and involves the demonstration of the uniqueness of solution and the *Wronskian determinant* among other mathematical techniques. Boyce and DiPrima [13] offer a good treatment.

The particular integral, corresponding to the steady-state solution value of r denoted r_∞, can be obtained by noticing that in the steady state, $\ddot{r} = \dot{r} = 0$. Physically, r_∞ is the yaw rate that the car will settle to once the transient response to control input has faded away and is obtained by setting the right hand side to zero. Therefore, putting in $\ddot{r} = \dot{r} = 0$, we have

$$kr_\infty = C_2\delta$$

$$\Rightarrow r_\infty = \frac{C_2\delta}{k}$$

And so, the final yaw rate, to which the system will settle once the transient has passed, is easily determined. The complimentary function is obtained by setting the right-hand side *to zero*, and we have

$$\Rightarrow I\ddot{r} - c\dot{r} + kr = 0$$

We first assume a solution of the form

$$r = e^{st}$$

where s is a constant to be determined. Substituting, we have

$$\left(s^2 + \frac{c}{I}s + \frac{k}{I}\right)e^{st} = 0$$

that can only be true if the bracketed term is zero for all t meaning that

$$\left(s^2 + \frac{c}{I}s + \frac{k}{I}\right) = 0$$

Thus, using the equation for the determination of the roots of a quadratic,

$$\begin{aligned} s_{1,2} &= \frac{-\frac{c}{I} \pm \sqrt{\left(\frac{c}{I}\right)^2 - 4\frac{k}{I}}}{2} \\ &= -\frac{c}{2I} \pm \sqrt{\left(\frac{c}{2I}\right)^2 - \frac{k}{I}} \end{aligned} \quad (5.13)$$

and we can now substitute to obtain a general solution by adding the C.F. and the P.I:

$$r = Ae^{S_1 t} + Be^{S_2 t} + r_\infty \quad (5.14)$$

It is important to be clear on the normalisation. If this equation were not divided by r_∞, then its solution would produce the yaw rate as a function of time for the vehicle being analysed. If we were dealing with, for example, a Hummer IV, then no doubt the response would be very different to an F3 car. Thus, the problem comes when we want to compare one car with another. Certainly, the yaw rate for the Hummer would be worse than the F3 but we expect that. By dividing the response by its final, steady-state value, we shall have a graph that settles to unity in both cases. Then, by comparing the responses, we can more meaningfully compare the shape of the response and say whether the Hummer is doing the best it can or the F3 the best it can. The unnormalised response is clearly the version that must be used to obtain actual values of yaw rate against time for the vehicle in hand.

The constants A and B depend upon initial conditions. The constants S_1 and S_2 dictate the nature of the solution. If they happen to evaluate to positive numbers, then there will be two values s_1 and s_2 that will be real numbers, and upon substitution into Eq. (5.14), the yaw rate will exponentially decay to its steady-state value. If the square root in Eq. (5.13) evaluates to a negative number, there will again be two values, but this time, they will be complex numbers, and the solution will be periodic. The value of the damping coefficient that marks the border between these behaviours is the one that pertains when the square root evaluates to zero. In this case, there will only be one common value for S_1 and S_2. The solution will still be exponential, but it will mark the boundary between the periodic and harmonic solutions. This value of damping is called the critical damping value, C_c, and is given by setting the term inside the square root in Eq. (5.13) to zero, and so,

$$\left(\frac{C_c}{2I}\right)^2 = \frac{k}{I}$$

$$\Rightarrow C_c = 2I\sqrt{\frac{k}{I}}$$

The quantity $\sqrt{k/I}$ has special significance. If we consider an undamped, free system, we may write

$$\Rightarrow I\ddot{r} + kr = 0$$

Rearranging and introducing

$$\omega_n^2 = k/I$$

$$\frac{1}{\omega_n^2}\ddot{r} + r = 0$$

By inspection, this equation has a harmonic solution of the form $r = A\sin\omega_n t + B\cos\omega_n t$; hence, a periodic solution for a system with zero damping will be given by

$$\omega_n = \sqrt{\frac{k}{I}}$$

Thus also,

$$C_c = 2I\omega_n$$

The ratio of system damping to critical damping is called the damping ratio, zeta, ς, and is defined as

$$\varsigma = \frac{c}{C_c}$$

Noting that

$$\frac{c}{2I} = \frac{\varsigma C_C}{2I} = \frac{2I\omega_n\varsigma}{2I} = \omega_n\varsigma$$

We can now re-express in terms of these system parameters, since

$$s_{1,2} = -\varsigma\omega_n \pm \sqrt{(\varsigma\omega_n)^2 - \omega_n^2}$$

$$\Rightarrow s_{1,2} = \omega_n\left(-\varsigma \pm \sqrt{(\varsigma^2 - 1)}\right)$$

$$\Rightarrow s_{1,2} = \omega_n\left(-\varsigma \pm i\sqrt{(1 - \varsigma^2)}\right)$$

We can now substitute these for s_1 and s_2 into the general solution and produce an equation of motion for the three cases: underdamped ($\varsigma < 1$), critically damped ($\varsigma = 1$) and overdamped ($\varsigma > 1$).

5.8.4 Underdamped, Critically Damped and Overdamped Steering

5.8.4.1 Case 1—Underdamped With Spreadsheet Solution

We start by considering the unnormalised response, and so,

$$r = Ae^{\left(-\omega_n\varsigma + i\sqrt{1-\varsigma^2}\omega_n\right)t} + Be^{\left(-\omega_n\varsigma - i\sqrt{1-\varsigma^2}\omega_n\right)t}$$

$$\Rightarrow r = e^{-\varsigma\omega_n t}\left(Ae^{i\sqrt{1-\varsigma^2}\omega_n t} + Be^{-i\sqrt{1-\varsigma^2}\omega_n t}\right)$$

Since

$$e^{i\theta} = \cos\theta + i\sin\theta$$

the bracketed term on the right-hand sign therefore equates to real and imaginary components. These may be simplified to a vector at a phase angle whose magnitude and phase angle are determined from initial conditions. Thus, we have

$$r = Xe^{-\varsigma\omega_n t}\sin\left(\sqrt{1-\varsigma^2}\omega_n t + \phi\right)$$

This is an exponentially decaying, oscillatory sinusoidal motion offset from the input motion by a phase angle ϕ. The frequency of this motion is called the damped natural frequency, ω_d, and we can see that it is given by

$$\omega_d = \sqrt{\left(1-\varsigma^2\right)} \times \omega_n$$

The total solution is given by the sum of the particular integral and the complimentary function, and so, the yaw rate response to a step steer for the underdamped vehicle is given by

$$r = Xe^{-\varsigma\omega_n t}\sin\left(\omega_d t + \phi\right) + \frac{c_2}{K}\delta_0 \tag{5.15}$$

where X and ϕ are the constants determined from initial conditions. In this case, the vehicle is responding to a step-steer input; thus, the initial yaw velocity will be zero, so we have $t=0 \Rightarrow r=0$. The initial yaw acceleration can be determined from the initial yaw moment. At the instant the steering input is applied, this will come from the lateral force at the front contact patch multiplied by the distance a. Thus,

$$T_0 = C_f \partial a$$

This initial torque, at $t=0$, results in a positive initial acceleration which remembering that in our coordinate system cornering stiffnesses are negative, is given by:

$$-\dot{r}_0 = \frac{C_f \partial a}{I}$$

We may now substitute for these initial conditions and find the constants X and ϕ. Substituting $t=0 \Rightarrow r=0$,

$$\Rightarrow 0 = X\sin\left(\phi\right) + \frac{c_2}{K}\delta_0$$

$$X = \frac{-C_2\partial}{K\sin\phi}$$

To determine ϕ, we start by differentiating Eq. (5.15) above to give

$$\dot{r} = -\varsigma\omega_n X e^{-\varsigma\omega_n t}\sin(\omega_d t + \phi) + X\omega_d e^{-\varsigma\omega_n t}\cos(\omega_d t + \phi)$$

Then, substituting at $t=0$,

$$\Rightarrow -\frac{C_f\partial a}{I} = -\varsigma\omega_n X\sin\phi + X\omega_d\cos\phi$$

$$\Rightarrow -\frac{C_f\partial a}{I} = \frac{\varsigma\omega_n C_2\partial\sin\phi}{K\sin\phi} + \frac{-C_2\partial\omega_d\cos\phi}{K\sin\phi}$$

$$-\frac{C_f\partial a}{I} = \frac{\varsigma\omega_n C_2\partial}{K} + \frac{-C_2\partial\omega_d}{K\tan\phi}$$

$$\Rightarrow \frac{C_2\partial\omega_d}{K\tan\phi} = \frac{C_f\partial a}{I} + \frac{\varsigma\omega_n C_2\partial}{K}$$

So, $\Rightarrow \tan\phi = \dfrac{\left\{\dfrac{C_2\partial\omega_d}{K}\right\}}{\left\{\dfrac{C_f\partial a}{I} + \dfrac{\varsigma\omega_n C_2\partial}{K}\right\}}$

Multiplying top and bottom by KI and cancelling δ,

$$\tan\phi = \frac{C_2\omega_d I}{\{C_f aK + \varsigma\omega_n C_2 I\}}$$

Thus, the derivative analysis models the yaw rate response to a step-steer input for the underdamped steering system with the following equation:

$$r = Xe^{-\varsigma\omega_n t}\sin(\omega_d t + \phi) + \frac{c_2}{K}\delta_0$$

where

$$\tan\phi = \frac{C_2\omega_d I}{\{C_f aK + \varsigma\omega_n C_2 I\}}$$

$$X = \frac{-C_2\partial}{K\sin\phi}$$

$$k = \left(N_\beta + \frac{Y_\beta N_R}{mV} - \frac{N_\beta Y_R}{mV}\right) \tag{5.16}$$

$$\phi = \tan^{-1}\left(\frac{\omega_d}{\varsigma\omega_n}\right)$$

$$c_c = 2I\omega_n \tag{5.17}$$

$$c = -\left(N_R + \frac{IY_\beta}{mV}\right) \tag{5.18}$$

$$\varsigma = \frac{c}{c_c} \tag{5.19}$$

$$\omega_n = \sqrt{\frac{k}{I}} \tag{5.20}$$

$$\omega_d = \sqrt{\left(1-\varsigma^2\right)} \times \omega_n \tag{5.21}$$

A spreadsheet to determine the yaw rate response (for an underdamped steering response) to a step-steer input can easily be prepared using the above analysis, and an example is reproduced in Fig. 5.23.

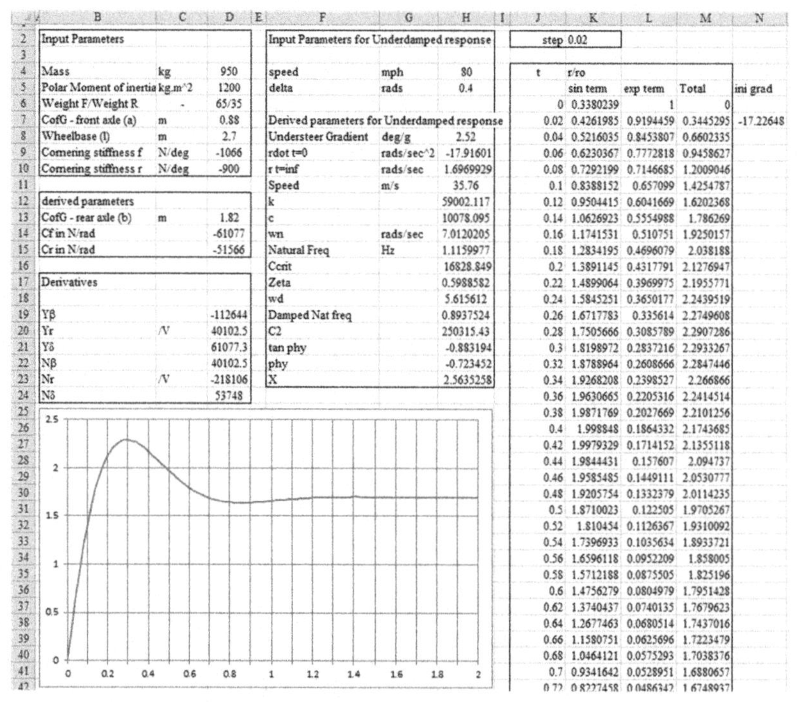

	B	C	D	E	F	G	H	I	J	K	L	M	N
2	Input Parameters				Input Parameters for Underdamped response				step	0.02			
3													
4	Mass	kg	950		speed	mph	80		t	r/ro			
5	Polar Moment of inertia	kg.m^2	1200		delta	rads	0.4			sin term	exp term	Total	ini grad
6	Weight F/Weight R	-	65/35						0	0.3380239	1	0	
7	CofG - front axle (a)	m	0.88		Derived parameters for Underdamped response				0.02	0.4261985	0.9194459	0.3445295	-17.22648
8	Wheelbase (l)	m	2.7		Understeer Gradient	deg/g	2.52		0.04	0.5216035	0.8453807	0.6602335	
9	Cornering stiffness f	N/deg	-1066		rdot t=0	rads/sec^2	-17.91601		0.06	0.6230367	0.7772818	0.9458627	
10	Cornering stiffness r	N/deg	-900		r t=inf	rads/sec	1.6969929		0.08	0.7292199	0.7146685	1.2009046	
11					Speed	m/s	35.76		0.1	0.8388152	0.657099	1.4254787	
12	derived parameters				k		59002.117		0.12	0.9504415	0.6041669	1.6202368	
13	CofG - rear axle (b)	m	1.82		c		10078.095		0.14	1.0626923	0.5554988	1.786269	
14	Cf in N/rad		-61077		wn	rads/sec	7.0120205		0.16	1.1741531	0.510751	1.9250157	
15	Cr in N/rad		-51566		Natural Freq	Hz	1.1159977		0.18	1.2834195	0.4696079	2.038188	
16					Ccrit		16828.849		0.2	1.3891145	0.4317791	2.1276947	
17	Derivatives				Zeta		0.5988582		0.22	1.4899064	0.3969975	2.1955771	
18					wd		5.615612		0.24	1.5845251	0.3650177	2.2439519	
19	Yβ		-112644		Damped Nat freq		0.8937524		0.26	1.6717783	0.335614	2.2749608	
20	Yr	/V	40102.5		C2		250315.43		0.28	1.7505666	0.3085789	2.2907286	
21	Yδ		61077.3		tan phy		-0.883194		0.3	1.8198972	0.2837216	2.2933267	
22	Nβ		40102.5		phy		-0.723452		0.32	1.8788964	0.2608666	2.2847446	
23	Nr	/V	-218106		X		2.5635258		0.34	1.9268208	0.2398527	2.266866	
24	Nδ		53748						0.36	1.9630665	0.2205316	2.2414514	
25									0.38	1.9871769	0.2027669	2.2101256	
26									0.4	1.998848	0.1864332	2.1743685	
27									0.42	1.9979329	0.1714152	2.1355118	
28									0.44	1.9844431	0.157607	2.094737	
29									0.46	1.9585485	0.1449111	2.0530777	
30									0.48	1.9205754	0.1332379	2.0114235	
31									0.5	1.8710023	0.122505	1.9705267	
32									0.52	1.810454	0.1126367	1.9310092	
33									0.54	1.7396933	0.1035634	1.8933721	
34									0.56	1.6596118	0.0952209	1.858005	
35									0.58	1.5712188	0.0875505	1.825196	
36									0.6	1.4756279	0.0804979	1.7951428	
37									0.62	1.3740437	0.0740135	1.7679623	
38									0.64	1.2677463	0.0680514	1.7437016	
39									0.66	1.1580751	0.0625696	1.7223479	
40									0.68	1.0464121	0.0575293	1.7038376	
41									0.7	0.9341642	0.0528951	1.6880657	
42									0.72	0.8227458	0.0486342	1.6748937	

Fig. 5.23 Underdamped xls derivative analysis for response to step steer.

One can produce a working version of the spreadsheet above using the following points for guidance:

(1) The cells B2:D10 contain the input data. In this case, a vehicle similar to a small hatchback is being modelled.

(2) B12:D15 have the derived parameters, b is obvious and the cornering stiffnesses are converted into N/rad as must be used in the formula.

(3) B17:D24 have the derivative terms worked out. You can use the values in the spreadsheet to check that your version is working correctly. Once you have it working, you can make adjustments to the input parameters and observe the effect on the derivative terms. For example, N_β is itself a measure of the understeer gradient, and you should find that as the parameter a is varied and used to make the understeer gradient in H8 pass through zero, so too does N_β. Two of the derivative terms, Y_R and N_R, are divided by the velocity V. In the box B17:D24, these terms have not yet been divided by velocity since the cells here contain the coefficients themselves and have not yet been applied to a particular velocity. It must be remembered that, when these two terms are used, they must indeed be divided by the velocity pertaining. The other four derivative terms do not have velocity dependence and are simply coefficients. It's very easy to forget to do that.

(4) F2:H23 have a whole range of the parameters that need to be determined before the actual yaw rate function can be determined.

(5) H4 has the speed at which the spreadsheet is doing the model.

(6) H5 has the steering input.

(7) H8 has the understeer gradient $(m/l)(a/C_R - b/C_F)$ in degree/m/s^2.

(8) H9 has a quantity labelled 'rdot $t=0$'. This is the value of initial yaw acceleration. It is used here as a check. The column M contains the yaw rate at each time t listed in column J; this is the output answer we are looking for. N7 contains a value *in grad*. This is the initial yaw acceleration as determined by taking the yaw velocity in cell M7 and dividing it by the step length. Thus, it is the initial yaw acceleration as determined by the spreadsheet. The value here compares well with the value in cell H9, suggesting that the spreadsheet is working.

(9) Cell H10 calculates $r\,t = inf$ and is the yaw rate value to which the output should eventually settle. As can be seen, the value here agrees with the final value in the graph, suggesting again that the spreadsheet is working correctly.

(10) H11 has speed that is just the value in SI units for the calculation instead of the everyday units of miles per hour.

(11) H12 through H23 contain evaluations of all the parameters needed to calculate the expression for yaw rate, and by going through the material above, you should be able to get the spreadsheet to work. Clearly, the values in the figure can be used to test your own spreadsheet, and on the website for the book, electronic copies are also available.

(12) Column J has time; it is in increments of *step* as entered manually into cell K2, so if you want to change the step length, it is an easy matter.

(13) Columns K and L contain the exponential and sinusoid terms of the equation

$$r = Xe^{-\varsigma\omega_n t}\sin(\omega_d t + \phi) + \frac{C_2}{K}\delta_0$$

used to calculate the yaw rate response. One could put it all in one cell if you desired, but it makes for easy debugging if terms are kept manageably short.

(14) Finally, column M has the total solution, and it is this column that is graphed in the screenshot.

5.8.4.2 Case 2—Critically Damped With Spreadsheet Solution

With the damping ratio $\varsigma = 1$, the general solution is

$$s_1 = s_2 = -\omega_n$$

$$\Rightarrow r = Ae^{-\omega_n t} + Be^{-\omega_n t}$$

$$r = Ce^{-\omega_n t}$$

where $C = A + B$. This is exponentially decaying motion. However, the solution in this form does not contain enough constants to account for two initial conditions, and an alternate solution must be used:

$$r = (A + Bt)e^{-\omega_N t}$$

The total solution to Eq. (5.12) for the case when $\varsigma = 1$ is again given by the sum of the particular integral and the complimentary function:

$$r = (A + Bt)e^{-\omega_N t} + \frac{C_2\delta}{k}$$

Using the initial conditions for a step steer as before, $t = 0 \Rightarrow r = 0$, we have

$$0=A+\frac{C_2\delta}{k}$$

$$\Rightarrow A=-\frac{C_2\delta}{k}$$

Also, $t=0$ implies

$$\dot{r}_0=\frac{-C_f\partial a}{I}$$

$$r=Ae^{-\omega_N t}+Bte^{-\omega_N t}$$

$$\Rightarrow \dot{r}=-\omega_n Ae^{-\omega_N t}+Be^{-\omega_N t}-\omega_n Bte^{-\omega_N t}$$

$$\Rightarrow \frac{-C_f\partial a}{I}=-\omega_n A+B$$

$$\Rightarrow B=-\frac{C_f\partial a}{I}-\omega_n\frac{C_2\partial}{k}$$

Thus,

$$r=\left(-\frac{C_2\partial}{k}+\left\{-\frac{C_F\partial a}{I}-\omega_N\frac{C_2\partial}{k}\right\}t\right)e^{-\omega_N t}+\frac{C_2\partial}{k}$$

This equation can be normalised again by dividing through by r_∞; hence,

$$\frac{r}{r_\infty}=\left(-1+\left\{-\frac{C_F\partial a}{Ir_\infty}-\omega_N\right\}t\right)e^{-\omega_N t}+1$$

This therefore relates the yaw rate response of a vehicle with a critically damped steering response to a step input. It can be seen that the response tends to zero when t is large because the exponential outside the large bracket tends to zero. When $t=0$, the normalised yaw rate is zero as expected since the function simplifies to $-1+1$. The constants used in the expression are

$$r_\infty=\frac{C_2\partial}{k}$$

$$k=\left(N_\beta+\frac{Y_\beta N_R}{mV}-\frac{N_\beta Y_R}{mV}\right)$$

$$\omega_n=\sqrt{\frac{k}{I}}$$

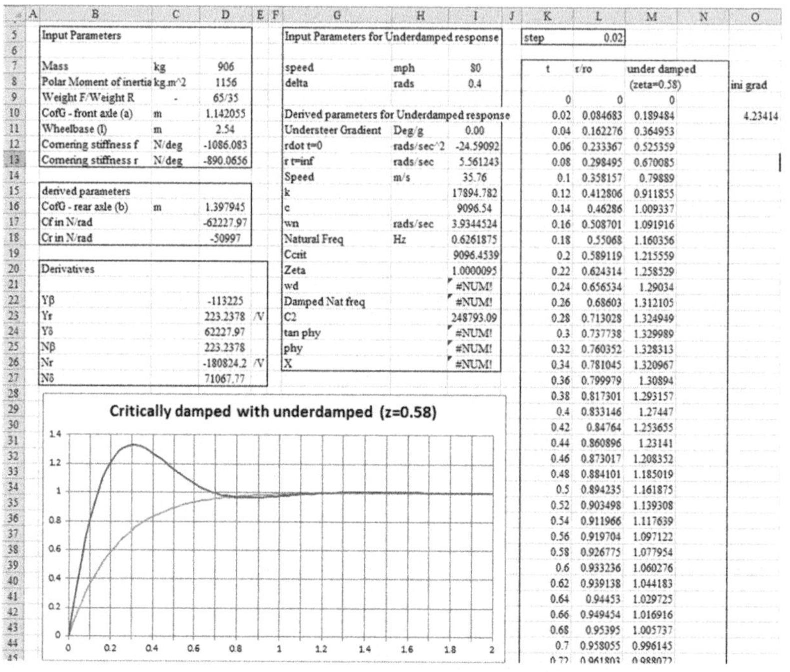

Input Parameters		
Mass	kg	906
Polar Moment of inertia	kg.m^2	1156
Weight F/Weight R	-	65/35
CofG - front axle (a)	m	1.142055
Wheelbase (l)	m	2.54
Cornering stiffness f	N/deg	-1086.083
Cornering stiffness r	N/deg	-890.0656

derived parameters		
CofG - rear axle (b)	m	1.397945
Cf in N/rad		-62227.97
Cr in N/rad		-50997

Derivatives		
Yβ	-113225	
Yr	223.2378	/V
Yδ	62227.97	
Nβ	223.2378	
Nr	-180824.2	/V
Nδ	71067.77	

Input Parameters for Underdamped response		
speed	mph	80
delta	rads	0.4
Derived parameters for Underdamped response		
Understeer Gradient	Deg/g	0.00
rdot t=0	rads/sec^2	-24.59092
r t=inf	rads/sec	5.561243
Speed	m/s	35.76
k		17894.782
c		9096.54
wn	rads/sec	3.9344524
Natural Freq	Hz	0.6261875
Ccrit		9096.4539
Zeta		1.0000095
wd		#NUM!
Damped Nat freq		#NUM!
C2		248793.09
tan phy		#NUM!
phy		#NUM!
X		#NUM!

step	0.02

t	r/ro	under damped (zeta=0.58)	ini grad
0	0	0	
0.02	0.084683	0.189484	4.23414
0.04	0.162276	0.364953	
0.06	0.233367	0.525359	
0.08	0.298495	0.670085	
0.1	0.358157	0.79889	
0.12	0.412806	0.911855	
0.14	0.46286	1.009337	
0.16	0.508701	1.091916	
0.18	0.55068	1.160356	
0.2	0.589119	1.215559	
0.22	0.624314	1.258529	
0.24	0.656534	1.29034	
0.26	0.68603	1.312105	
0.28	0.713028	1.324949	
0.3	0.737738	1.329989	
0.32	0.760352	1.328313	
0.34	0.781045	1.320967	
0.36	0.799979	1.30894	
0.38	0.817301	1.293157	
0.4	0.833146	1.27447	
0.42	0.84764	1.253655	
0.44	0.860896	1.23141	
0.46	0.873017	1.208352	
0.48	0.884101	1.185019	
0.5	0.894235	1.161875	
0.52	0.903498	1.139308	
0.54	0.911966	1.117639	
0.56	0.919704	1.097122	
0.58	0.926775	1.077954	
0.6	0.933236	1.060276	
0.62	0.939138	1.044183	
0.64	0.94453	1.029725	
0.66	0.949454	1.016916	
0.68	0.95395	1.005737	
0.7	0.958055	0.996145	
0.72	0.961803	0.988072	

Fig. 5.24 Critically damped analysis for response to step steer.

In Fig. 5.24, a new version of the spreadsheet developed above but one that makes use of the equations that pertain to the critically damped case. Some parameters, for example, tan(phy), are still present in the spreadsheet but only have meaning for an underdamped response, and so, the spreadsheet returns an error, '#NUM!' when trying to evaluate the terms. The response of the critically damped car is shown in comparison with the underdamped response developed above. It can be seen that the critically damped car offers a very different response to the underdamped case. Whilst the car takes around the same length of time to settle, it does so in a much smoother fashion and in particular exhibits no overswing. This kind of response is much more desirable, and in general, vehicle dynamists design racing cars to have a critically damped steering response.

5.8.4.3 Case 3 – Over-damped With Spreadsheet Solution

In the case of the over-damped car, the solution of Equation 5-12 consists of the sum of the homogeneous solution and particular integral. In this case, since the term $\varsigma^2 - 1$ is positive, its square root is a real number and the solution is simply:

$$r = Ae^{\left(-\varsigma + \sqrt{\varsigma^2 - 1}\right)\omega_n t} + Be^{\left(-\varsigma - \sqrt{\varsigma^2 - 1}\right)\omega_n t} + r_\infty$$

Using the initial conditions of our step input

$$t = 0 \Rightarrow r = 0 \quad \text{and} \quad t = 0 \Rightarrow \dot{r}_0 = \frac{-C_F \partial a}{I}$$

The minus sign is introduced in front of C_F because in our coordinate system a clockwise rotation is +ve. However, cornering stiffness are negative, see Fig 2.28. This is a consequence of the SAE coordinate system used. Looking at figure 5.20, where the equilibrium analysis used for deriving the equations of motion starts it is clear that the car will yaw clockwise (+Ve) and thus a minus sign, negating the negative cornering stiffness, is correct. (This was done on page 146 for the undamped case.)

The constants A and B can be determined. Setting $t = 0$,

$$0 = A + B + r_\infty$$
$$\Rightarrow B = -(A + r_\infty)$$

and introducing f and g where $f = \left(-\varsigma - \sqrt{\varsigma^2 - 1}\right)\omega_n$ and $g = \left(-\varsigma + \sqrt{\varsigma^2 - 1}\right)\omega_n$.

$$\dot{r} = Age^{gw_n t} + Bfe^{fw_n t}$$
$$\dot{r}_0 = Ag + Bf$$
$$\Rightarrow \dot{r}_0 = A(g - f) - r_\infty f$$
$$A = \frac{1}{g - f}\left\{\frac{-C_f \partial a}{I} + fr_{\infty r}\right\}$$

and we have:

$$r = Ae^{gt} + Be^{ft} + r_\infty$$

where

$$r_\infty = \frac{C_2 \partial}{k} \quad \dot{r}_\infty = \frac{-c_f \partial a}{I}$$

$$f = \left(-\varsigma - \sqrt{\varsigma^2 - 1}\right)\omega_n \quad g = \left(-\varsigma + \sqrt{\varsigma^2 - 1}\right)\omega_n$$

$$\omega_n = \sqrt{\frac{k}{I}} \quad A = \frac{1}{g - f}\left\{\frac{-C_F \partial a}{I} + fr_\infty\right\} \quad B = -(A + r_\infty)$$

Which is an exponentially decaying yaw rate value as a function of time. Figure 5-25 below, shows the total solution of the vehicle as developed above for a vehicle with an overdamped steering response. It can be seen that the over damped case provides for a car with a longer, and so less desirable settling time than the critically damped or under-damped car. Thus we confirm that a steering response with critical damping is the best choice and will result in a car that settles to its final yaw rate as quickly as possible. The settling time for this car corresponds to around the time taken for one period at the natural frequency and nothing can be done to improve on this.

The vertical axis of the graph on the left shows the normalised yaw rate and the horizontal axis shows time in seconds for a car with the dynamic data shown. The un-normalised response is shown in the graph on the right and also shown is the same curve obtained from Adams bicycle model with the same vehicle parameters.

The graph shows that, for the overdamped car, the higher the damping ratio the longer the car takes to reach the yaw rate demanded and that the motion is exponentially decaying. For the underdamped car, it is still true that reducing the damping ratio reduces the time taken to first reach the yaw rate demanded; however, the motion is always oscillatory for the underdamped system, and the system therefore

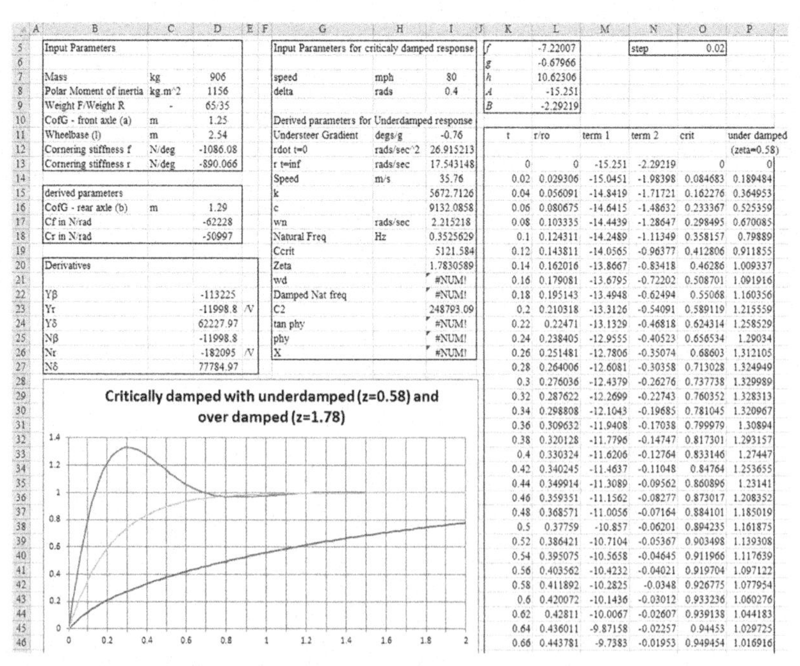

Input Parameters		
Mass	kg	906
Polar Moment of inertia	kg.m^2	1156
Weight F/Weight R	-	65/35
CofG - front axle (a)	m	1.25
Wheelbase (l)	m	2.54
Cornering stiffness f	N/deg	-1086.08
Cornering stiffness r	N/deg	-890.066

derived parameters		
CofG - rear axle (b)	m	1.29
Cf in N/rad		-62228
Cr in N/rad		-50997

Derivatives		
Yβ	-113225	
Yr	-11998.8	/V
Yδ	62227.97	
Nβ	-11998.8	
Nr	-182095	/V
Nδ	77784.97	

Input Parameters for criticaly damped response		
speed	mph	80
delta	rads	0.4
Derived parameters for Underdamped response		
Understeer Gradient	degs/g	-0.76
rdot t=0	rads/sec^2	26.915213
r t=inf	rads/sec	17.543148
Speed	m/s	35.76
k		5672.7126
c		9132.0858
wn	rads/sec	2.215218
Natural Freq	Hz	0.3525629
Ccrit		5121.584
Zeta		1.7830589
wd		#NUM!
Damped Nat freq		#NUM!
C2		248793.09
tan phy		#NUM!
phy		#NUM!
X		#NUM!

f	-7.22007
g	-0.67966
h	10.62306
A	-15.251
B	-2.29219

step	0.02

t	r/ro	term 1	term 2	crit	under damped (zeta=0.58)
0	0	-15.251	-2.29219	0	0
0.02	0.029306	-15.0451	-1.98398	0.084683	0.189484
0.04	0.056091	-14.8419	-1.71721	0.162276	0.364953
0.06	0.080675	-14.6415	-1.48632	0.233367	0.525359
0.08	0.103335	-14.4439	-1.28647	0.298495	0.670085
0.1	0.124311	-14.2489	-1.11349	0.358157	0.79889
0.12	0.143811	-14.0565	-0.96377	0.412806	0.911855
0.14	0.162016	-13.8667	-0.83418	0.46286	1.009337
0.16	0.179081	-13.6795	-0.72202	0.508701	1.091916
0.18	0.195143	-13.4948	-0.62494	0.55068	1.160356
0.2	0.210318	-13.3126	-0.54091	0.589119	1.215559
0.22	0.22471	-13.1329	-0.46818	0.624314	1.258529
0.24	0.238405	-12.9555	-0.40523	0.656534	1.29034
0.26	0.251481	-12.7806	-0.35074	0.68603	1.312105
0.28	0.264006	-12.6081	-0.30358	0.713028	1.324949
0.3	0.276036	-12.4379	-0.26276	0.737738	1.329989
0.32	0.287622	-12.2699	-0.22743	0.760352	1.328313
0.34	0.298808	-12.1043	-0.19685	0.781045	1.320967
0.36	0.309632	-11.9408	-0.17038	0.799979	1.30894
0.38	0.320128	-11.7796	-0.14747	0.817301	1.293157
0.4	0.330324	-11.6206	-0.12764	0.833146	1.27447
0.42	0.340245	-11.4637	-0.11048	0.84764	1.253655
0.44	0.349914	-11.3089	-0.09562	0.860896	1.23141
0.46	0.359351	-11.1562	-0.08277	0.873017	1.208352
0.48	0.368571	-11.0056	-0.07164	0.884101	1.185019
0.5	0.37759	-10.857	-0.06201	0.894235	1.161875
0.52	0.386421	-10.7104	-0.05367	0.903498	1.139308
0.54	0.395075	-10.5658	-0.04645	0.911966	1.117639
0.56	0.403562	-10.4232	-0.04021	0.919704	1.097122
0.58	0.411892	-10.2825	-0.0348	0.926775	1.077954
0.6	0.420072	-10.1436	-0.03012	0.933236	1.060276
0.62	0.42811	-10.0067	-0.02607	0.939138	1.044183
0.64	0.436011	-9.87158	-0.02257	0.94453	1.029725
0.66	0.443781	-9.7383	-0.01953	0.949454	1.016916

Fig. 5.25 Overdamped analysis for response to step steer.

overshoots the value demanded. As the damping is reduced, more and more overshoot is presented. A system with zero damping would remain in oscillation forever.

5.9 MULTIBODY CODE SOLUTION FOR STEERING RESPONSES USING ADAMS

The bicycle model developed in Section 5.7.5 can be used to examine the behaviour of a vehicle over a range of designs by simply running the model with different parameters and comparing the results. For example, by moving the centre of gravity marker in the model fore and aft, one can make the car oversteering, neutral or understeering, a simple matter in simulation but rather more problematic in real life!

The website for the book contains short videos prepared in this way that make very clear the oversteering, neutral steering and understeering vehicle behaviour. There are also guidance notes on the preparation of such models in Adams. These are available firstly to illustrate the effects and secondly to enable the reader to prepare your own such models and use them to simulate cars for yourself.

Clearly, these two methods (firstly that of solving the derivative equations analytically using a spreadsheet and secondly by using a multibody code such as Adams) are different in their approach, and so, we must expect differences in the results. For example, if we use a tyre characteristic modelled as a spline within Adams, then the limit behaviour will be correctly modelled since when the slip angle at either axle exceeds the value for which a maximum lateral force is returned then the force at that end will quickly decrease and the path no longer maintained. By contrast, the spreadsheet solution provided above has linearised functions for the cornering stiffness, and the lateral force produce simply increases linearly with slip angle, and there is no limit behaviour. (This would make for exceptional performance if indeed it could be achieved!)

Below, some observations are made about the differences in modelling techniques and the consequent differences in output that we should expect:

- As the front wheel is steered, the lateral force has a component acting rearwards on the vehicle. This component serves to slow the vehicle down. In real life, a cornering car with steered input does indeed slow down rather faster than in a straight line for exactly this reason. In the bicycle model, this decrease in speed is modelled and could be validated if a rolling resistance drag force is included in the model.

- The derivative analysis assumes constant speed, and errors will occur if this is not maintained. In the case of the bicycle model, there is no such constraint, and so, if we want to model a corner taken at nonconstant speed, the model will incur errors as a result.
- In Adams, a smooth but very short 'step' function must be used, for example, the haversine step. In the analytical solution, an infinitesimally short step is used.

5.9.1 Comparison of the Models Utility

In the case of the derivative analysis, we have successfully solved the equations of motion for a cornering car from first principals. The analysis has the very great advantage that it is easy and quick. It doesn't seem easy and quick when learnt for the first time, but once learnt, it may simply be reapplied and rapidly provide an excellent summary of vehicle dynamic performance. However, extending the analysis to deal with more complex situations will be problematic. The maths will get intractable vey quickly if we try to accommodated, nonlinear acceleration, real tyre characteristics or general steering inputs, for example, let alone weight transfer, downforce or chassis compliance. Perhaps, its greatest strength is that the derivative analysis explains *why* a vehicle will behave as it does. The presence of a critical speed, for example, or the idea of a minimum turn-in time, these are all explained by the analysis. This is in stark contrast to the numerical analysis of the bicycle model in Adams that for all its accuracy and ability to model any of the effects we care to build in; it can never explain *why* a vehicle behaves as it does. The strength of the numerical model is that it explains *what* will happen. We can indeed include any effect we wish, and the model doesn't really get that much more complicated, a bit slower on the computer, certainly, but that's about it.

The best approach is of course to use both techniques. The analytical derivative analysis gets us very good approximations very quickly and easily. If greater accuracy is needed, then move over to numerically prepared models and use the analytical results to validate the numerical solution before extending them to include any manner of dynamic considerations as desired.

5.10 ACHIEVING NEUTRAL STEERING CAR WITH CRITICALLY DAMPED STEERING

In earlier work, we have seen that the neutral steering car with understeer gradient zero is the best from the point of view of completing corners. In the material above, we saw that a steering response function damping ratio of one makes for the best steering response. The obvious question is whether we can achieve both together. They both can be described in derivative terms, so if it is indeed achievable, it should be possible to find a description in derivative terms.

We start by looking at the expressions for zeta and the understeer gradient. Starting with the understeer gradient we have

$$US = \frac{m}{l}\left(\frac{a}{C_R} - \frac{b}{C_F}\right)$$

The equations take the value zero when $bC_R = aC_F$.

If we now look at the equations governing the steering system damping ratio, zeta, we have

$$\varsigma = \frac{c}{C_C}$$

and so,

$$\varsigma = \frac{-\left(N_R + \dfrac{IY_\beta}{mV}\right)}{2I\sqrt{\dfrac{\left(N_\beta + \dfrac{Y_\beta N_R}{mV} - \dfrac{N_\beta Y_R}{mV}\right)}{I}}}$$

This equation involves not only the values for a and b and C_F and C_R as appear in the equation for the understeer gradient but also the velocity, the vehicle mass and the polar moment of inertia. A method for ensuring that we achieve both and understeer gradient of zero and a zeta value of one is therefore to first produce a neutral steering car using a, b and C_F together with C_R and then to adjust only the mass and polar moment until zeta is one.

Looking at the equation for zeta and noting that, if we have a neutral steering car, as we desire, then the Y_R and N_β terms will be zero and so

$$\text{Thus, } \varsigma = \frac{-\left(N_R + \dfrac{IY_\beta}{mV}\right)}{2I\sqrt{\left(\dfrac{Y_\beta N_R}{ImV}\right)}}$$

We can now adapt this equation for zeta remembering that the derivative term N_R is velocity-dependent and so

$$\varsigma = \frac{-\left(\frac{N_R}{V} + \frac{IY_\beta}{mV}\right)}{2I\sqrt{\left(\frac{Y_\beta N_R}{ImV^2}\right)}} = \frac{-\left(N_R + \frac{IY_\beta}{m}\right)}{2I\sqrt{\left(\frac{Y_\beta N_R}{Im}\right)}}$$

Squaring the top and bottom and substituting $\psi = I/m$,

$$\Rightarrow \varsigma = \frac{\left(N_R + \psi Y_\beta\right)^2}{4\psi Y_\beta N_R}$$

Now, we desire that the damping ratio should be unity and so the top and bottom rows must be equal, hence,

$$\left(N_R + \psi Y_\beta\right)^2 = 4\left(\psi Y_\beta N_R\right)$$

$$\Rightarrow \psi^2 Y_\beta^2 + 2N_R\psi Y_\beta + N_R^2 - 4\psi Y_\beta N_R = 0$$

$$\Rightarrow Y_\beta^2\psi^2 + \left(2N_R Y_\beta - 4Y_\beta N_R\right)\psi + N_R^2 = 0$$

$$\Rightarrow Y_\beta^2\psi^2 - 2N_R Y_\beta\psi + N_R^2 = 0$$

This being a quadratic, we may find the solution for ψ to be

$$\psi = \frac{2N_R Y_\beta \pm \sqrt{\left(2N_R Y_\beta\right)^2 - 4Y_\beta^2 N_R^2}}{2Y_\beta^2}$$

Inspection of the root term shows that it equates internally to zero; there will therefore be one real root to the equation given by

$$\psi = \frac{2N_R Y_\beta}{2Y_\beta^2}$$

$$\Rightarrow \psi = \frac{N_R}{Y_\beta}$$

Thus, for a car that is neutral steering, we shall also have a critically damped steering response if the ratio of I/m equates to N_R/Y_β. In practice, achieving the desired ratio may be problematic. Many formula and road car designs will already drive the designer to achieve the minimum weight, and once a sensible and pragmatic layout is chosen, the polar moment of inertia becomes pretty much fixed. Nevertheless, there is always scope to do the best one can, and other competitors and manufactures will have the same constraints, so all one needs to do is exceed their offerings to confer an advantage.

The dependence of steering system damping ratio, ς, with speed for a range of understeer gradient values is shown in Fig. 5.26—Dependence of zeta on speed for a range of understeer gradient values. In all cases, the polar-moment-to-mass ratio is N_R/Y_β. It can be seen that, when the understeer gradient is zero, the vehicle has a critically damped steering response throughout the speed range.

Further insight can be gained by looking at the dependence of steering damping ratio ς on the polar-moment-to-mass ratio ψ as shown in Fig. 5.27.

It can be seen that the value of zeta is not particularly sensitive to changes in ψ and that, as long as a vehicle is close, zeta will be acceptable.

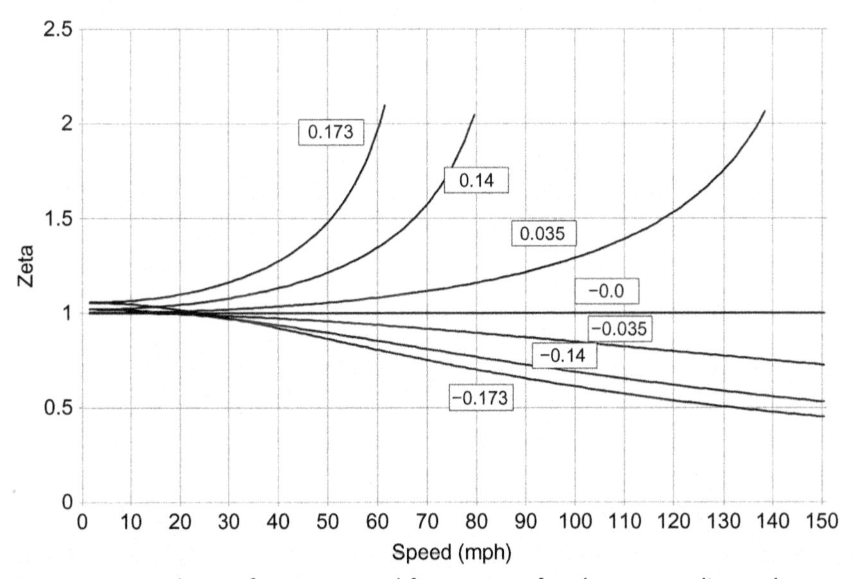

Fig. 5.26 Dependence of zeta on speed for a range of understeer gradient values.

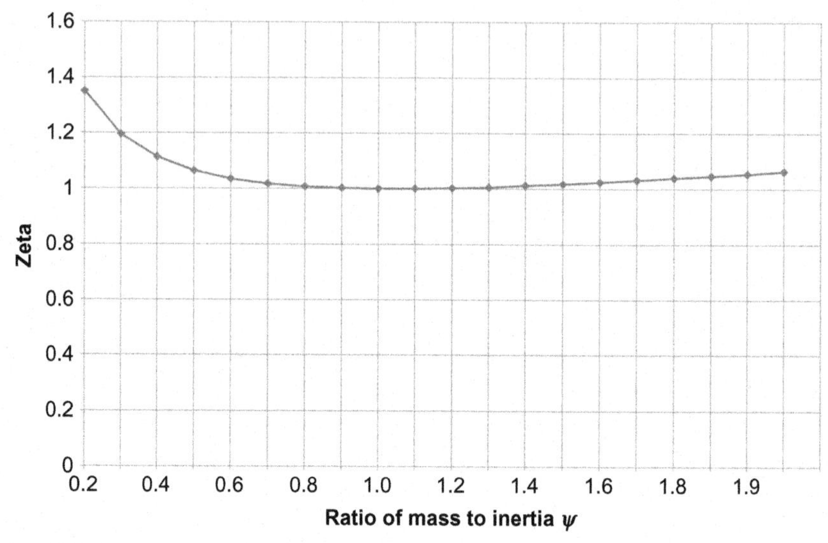

Fig. 5.27 Dependence of ς on ψ.

5.11 STEADY-STATE VEHICLE RESPONSE TO CONTROL AND STABILITY

In this section, we shall start to give consideration to the equations of motion of a vehicle developed above from the steady-state point of view:

$$\begin{aligned} mV\left(r+\dot{\beta}\right) &= Y_\beta\beta + Y_R r + Y_\delta\delta \\ I\dot{r} &= N_\beta\beta + N_R r + N_\delta\delta \end{aligned} \tag{5.22}$$

When we considered these before, we were concerned with the complete solution, one that allows us to develop a relationship between steering input and resulting yaw rate with time. In this section, we consider the separate topic of steady-state response. It may seem odd to do this having completed the above; after all, one can obtain the steady-state response simply by allowing the total response to run until it settles. However, it is helpful to be able to more simply and quickly determine overall characteristics, and this is possible by developing an understanding of the steady-state response.

We define the steady-state response as a vehicle in which the driver has made a steering demand, and all the transients have decayed away. Therefore, we can set $\dot{r}$, $\dot{\beta}$ and $\dot{v}$ to zero. Hence,

$$\begin{aligned} mVr &= Y_\beta \beta + Y_R r + Y_\delta \delta \\ 0 &= N_\beta \beta + N_R r + N_\delta \delta \end{aligned} \tag{5.23}$$

These equations relate the steering input δ to the yaw rate r through the derivative terms, the velocity v and the slide-slip β. For this simplified case, these equations are not differential and are simply a pair of equations balancing forces (in the first case) and moments (in the second). Since we wish to characterise the steady-state vehicle response, we are interested in ratios fthat relate outputs to inputs. Two commonly used control responses are the curvature response and the yaw rate response; each is considered below.

5.11.1 Curvature Response

This relates the radius of the corner to the steering demand. Ideally, this would be a linear relationship, and a doubling of the steering demand would result in a corner of half the radius. Thus, the ratio of interest here will be

$$\text{Curvature response} = \frac{{}^1/_R}{\partial} = \frac{1}{R\delta}$$

To make use of this ratio, we would first need to find an expression to which it equates. Once the value of this expression is known, we can then use the value obtained to determine R from any given δ or visa versa.

We start with

$$\begin{aligned} mVr &= Y_\beta \beta + Y_R r + Y_\delta \delta \\ 0 &= N_\beta \beta + N_R r + N_\delta \delta \end{aligned}$$

But since in steady-state turning $r = V/R$,

$$\begin{aligned} m\frac{V^2}{R} &= Y_\beta \beta + Y_R \frac{V}{R} + Y_\delta \delta \\ 0 &= N_\beta \beta + N_R \frac{V}{R} + N_\delta \delta \end{aligned}$$

$$\Rightarrow -Y_\delta \delta = Y_\beta \beta + \frac{1}{R}\left(VY_R - mV^2\right) \tag{5.24}$$

$$\& \quad -N_\delta \delta = N_\beta \beta + \left(\frac{1}{R}\right) VN_R \tag{5.25}$$

Now, to obtain the curvature response, we first eliminate β. By rearranging Eq. (5.25), we have

$$\Rightarrow \beta = -\frac{N_\delta}{N_\beta}\delta - \left(\frac{1}{R}\right) V\frac{N_R}{N_\beta}$$

Substituting

$$\Rightarrow -Y_\delta\delta = Y_\beta\left\{-\frac{N_\delta}{N_\beta}\delta - \left(\frac{1}{R}\right)V\frac{N_R}{N_\beta}\right\} + \frac{1}{R}\left(VY_R - mV^2\right)$$

$$\Rightarrow -Y_\delta\delta + Y_\beta\frac{N_\delta}{N_\beta}\delta = -Y_\beta\left(\frac{1}{R}\right)V\frac{N_R}{N_\beta} + \frac{1}{R}\left(VY_R - mV^2\right)$$

$$\Rightarrow \delta\left\{Y_\beta\frac{N_\delta}{N_\beta} - Y_\delta\right\} = \left(\frac{1}{R}\right)\left(\left(VY_R - mV^2\right) - Y_\beta V\frac{N_R}{N_\beta}\right)$$

$$\Rightarrow \delta\left\{Y_\beta N_\delta - Y_\delta N_\beta\right\} = \left(\frac{1}{R}\right)\left(N_\beta\left(VY_R - mV^2\right) - Y_\beta VN_R\right)$$

Hence,

$$\frac{\left(\frac{1}{R}\right)}{\delta} = \frac{\left\{Y_\beta N_\delta - Y_\delta N_\beta\right\}}{\left(N_\beta\left(VY_R - mV^2\right) - Y_\beta VN_R\right)}$$

$$= \frac{\left\{Y_\beta N_\delta - Y_\delta N_\beta\right\}}{V\left(N_\beta Y_R - mVN_\beta - Y_\beta N_R\right)}$$

$$\Rightarrow \frac{\left(\frac{1}{R}\right)}{\delta} = \frac{1}{\dfrac{V\left(N_\beta Y_R - Y_\beta N_R\right)}{\left\{Y_\beta N_\delta - Y_\delta N_\beta\right\}} - \dfrac{mV^2 N_\beta}{\left\{Y_\beta N_\delta - Y_\delta N_\beta\right\}}}$$

Substituting for the derivative expressions in the first term of the denominator yields

$$\frac{V\left(N_\beta Y_R - Y_\beta N_R\right)}{\left\{Y_\beta N_\delta - Y_\delta N_\beta\right\}} = a + b = l$$

Thus,

$$\frac{\left(\frac{1}{R}\right)}{\delta}=\frac{1}{l-\frac{mV^2N_\beta}{\left\{Y_\beta N_\delta - Y_\delta N_\beta\right\}}}$$

$$=\frac{\left({}^1/_l\right)}{1+\left(\frac{1}{l}\right)\left\{\frac{mN_\beta}{\left(Y_\delta N_\beta - Y_\beta N_\delta\right)}\right\}V^2}$$

from which

$$\frac{\left(\frac{1}{R}\right)}{\delta}=\frac{{}^1/_l}{1+KV^2} \tag{5.26}$$

where the term K, known as the stability factor, (not to be confused with the k of Eq. 5.16) is given by

$$K=\left\{\frac{mN_\beta}{\left(Y_\delta N_\beta - Y_\beta N_\delta\right)}\right\}\left\{\frac{1}{l}\right\} \tag{5.27}$$

As desired, the curvature response equates to a given number. That number, the right-hand side of the curvature response equation, is dependent on V^2 and the derivative terms. We can see that, for a car in a steady-state (constant speed) corner, operating below the limit of lateral acceleration at a given speed, if the steering input were suddenly doubled, the reciprocal of the radius would also have to double and thus the corner radius would have halved.

5.11.2 Yaw Rate Response

This relates the yaw velocity to steering demand:

$$\text{Yaw velocity response} = \frac{r}{\delta}$$

As with the curvature response, we need to develop an expression to which the yaw velocity response equates. In this case, the algebra is straightforward. Starting with Eq. (5.26) and multiplying by V,

$$\frac{\left(\frac{V}{R}\right)}{\delta}=\frac{{}^V/_l}{1+KV^2}$$

Since $r = V/R$

$$\Rightarrow \frac{r}{\delta} = \frac{V/l}{1 + KV^2} \tag{5.28}$$

where again the term K, known as the stability factor, is given by Eq. (5.27).

5.11.3 Responses of the Over-, Neutral- and Understeering Car

Having developed expressions for the curvature and yaw rate response, we need now to determine how the responses depend on the derivative terms. For example, how do the responses change with speed? In particular, how does the response vary with over- or understeer?

5.11.4 Stability Factor *K*

We start by realising that the K term in Eqs (5.26), (5.28) dominates the response. If K is positive, the denominator will always be positive, and the response in both the case of yaw and curvature will reduce as speed increases. If, however, K is negative, there is the possibility that the denominator can become very small, even zero, leading to an infinite response in both cases. Examination of the expression for K shows that it takes its sign from the numerator. This is because the denominator

$$Y_\delta N_\beta - Y_\beta N$$

that substituting for the vehicle parameters corresponding to each derivative term is

$$\begin{aligned} &= -C_F(aC_F - bC_R) - (C_F + C_R)(-aC_F) \\ &= -aC_F^2 + bC_FC_R + aC_F^2 + aC_FC_R \\ &= lC_FC_R \end{aligned}$$

In our sign convention, C_F and C_R are both negative and we see that the denominator is always positive, leaving only the numerator to dictate the sign of K. The numerator takes the sign of the derivative term, N_β, given by

$$N_\beta = aC_F - bC_R$$

This term has the same sign as the understeer gradient and is therefore itself a measure of understeer. There are then three possible cases for K; these are examined below.

5.11.5 Characteristic Speed and Positive *K*

The effect of *K* on the curvature response starts with an examination of Eq. (5.26). If *K* is positive, then as the speed increases, the denominator gets larger, and so, the right-hand side gets smaller. Thus, for a given radius corner, ∂ must get larger with velocity, and we have an understeering car. As the velocity increases, the dominator smoothly increases, and the understeering car will simply become more and more understeering as velocity increases. This is shown in Fig. 5.28.

The effect of *K* on the yaw rate response is obtained from Eq. (5.28). Since *V* appears on the top and the bottom, we see that for very small values of *V* the denominator is approximately one and the curvature response increases linearly with *V*. As *V* increases, the bottom row, with its squared term in *V*, will dominate, and so, the yaw rate response will decrease. In between, there will be a local maxima. Looking at Fig. 5.27, this can be seen. The speed at which the maxima occurs is called the characteristic speed and is given by

$$V_{char} = \sqrt{\frac{-N_\beta(VY_r) + Y_\beta(VN_r)}{N_\beta m}} = \sqrt{\frac{1}{K}}$$

The proof for this relationship is not easy but we can demonstrate that it is true in the following way:

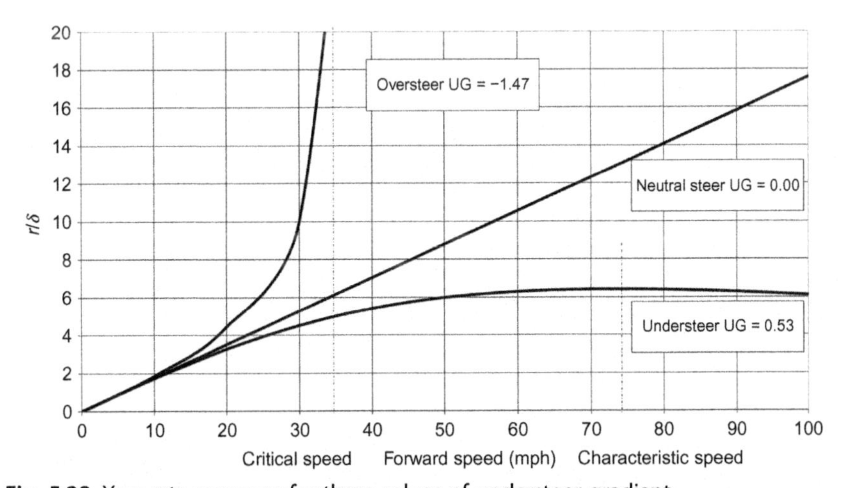

Fig. 5.28 Yaw rate response for three values of understeer gradient.

$$\sqrt{\frac{-N_\beta(VY_r)+Y_\beta(VN_r)}{N_\beta m}}=\sqrt{\frac{1}{K}}$$

$$\Rightarrow\frac{-N_\beta(VY_r)+Y_\beta(VN_r)}{N_\beta m}=\frac{1}{K}$$

$$\Rightarrow\frac{-N_\beta(VY_r)+Y_\beta(VN_r)}{N_\beta m}=\frac{1}{\left\{\frac{mN_\beta}{(Y_\delta N_\beta-Y_\beta N_\delta)}\right\}\left\{\frac{1}{l}\right\}}$$

$$\Rightarrow\frac{-N_\beta(VY_r)+Y_\beta(VN_r)}{N_\beta m}=\left\{\frac{(Y_\delta N_\beta-Y_\beta N_\delta)}{mN_\beta}\right\}l$$

$$\Rightarrow -N_\beta(VY_r)+Y_\beta(VN_r)=(Y_\delta N_\beta-Y_\beta N_\delta)l$$

Re-expressing the derivative terms in vehicle parameter terms, we can compare the left and right-hand side. The left-hand side equates to

$$\begin{aligned}
&=-(aC_F-bC_r)(aC_F-bC_r)+(C_F+C_R)\left(a^2C_F+b^2C_r\right)\\
&=-\left(a^2C_F^2-abC_FC_r-abC_FC_r+b^2C_r^2\right)\\
&\quad+a^2C_F^2+b^2C_rC_F+a^2C_FC_R+b^2C_r^2-a^2C_F^2\\
&\quad+abC_FC_r+abC_FC_r-b^2C_r^2+a^2C_F^2+b^2C_rC_F\\
&\quad+a^2C_FC_R+b^2C_r^2=-a^2C_F^2+a^2C_F^2-b^2C_r^2\\
&\quad+b^2C_r^2+2abC_FC_r+b^2C_rC_F+a^2C_FC_R\\
&=2abC_FC_r+a^2C_FC_R+b^2C_rC_F
\end{aligned}$$

whilst the right-hand side equates to

$$\begin{aligned}
&=(-C_F)(aC_F-bC_R)-(C_F+C_R)(-aC_F)\times l\\
&=\left(-aC_F^2+bC_RC_F+aC_F^2+aC_RC_F\right)(a+b)\\
&=(bC_RC_F+aC_RC_F)(a+b)\\
&=abC_RC_F+a^2C_RC_F+b^2C_RC_F+abC_RC_F\\
&=2abC_RC_F+a^2C_RC_F+b^2C_RC_F
\end{aligned}$$

Thus,

$$V_{CHAR} = \sqrt{\frac{-N_\beta(VY_r) + Y_\beta(VN_r)}{N_\beta m}} = \sqrt{\frac{1}{K}} \tag{5.29}$$

5.11.5.1 QED

The characteristic speed is therefore another way in which the understeering car can be characterised. In general, we seek to provide a driver with a linear steering response, and so, understeer of any kind is undesirable. In reality, a neutral car is an impossibility, and a better way to express the design requirement is that, if a car must be understeering, the characteristic speed should be much above the speed at which the car is to be driven. In racing, this is indeed practical, but in road cars, other pressures often lead to heavily understeering cars. Some road cars have characteristic speeds as low as 40 mph.

5.11.6 The Neutral Steering Car and Zero *K*

If K is zero, then the curvature response reduces to

$$\frac{\left(\frac{1}{R}\right)}{\delta} = {}^1/_l \Rightarrow \delta = \frac{L}{R}$$

and the yaw rate response to

$$\Rightarrow \frac{r}{\delta} = \frac{V}{l} \Rightarrow R = \frac{V}{r}$$

Both of which correspond to Ackermann steering geometry and show that corner radius is independent of the speed. Thus, $K=0$ corresponds to a neutral steering car. We obtained this relationship earlier from a geometric consideration of the steering car. We can now see that this very desirable linear relationship is a special case applying only to the neutral steering car.

5.11.7 The Critical Speed Negative *K*

The denominator of Eqs (5.26), (5.28) is given by

$$1 + KV^2$$

and so, if K is negative, then as the speed increases, the denominator gets smaller and smaller, eventually reaching zero. Thus, both the curvature response and yaw rate response become infinitely large when this happens.

The speed at which this happens is called the critical speed. This would correspond to an infinite curvature response, a tiny steering input giving, theoretically, a zero radius corner. This is clearly unstable behaviour, and the oversteering car, unlike the understeering car, has a discontinuity at this speed.

For the oversteering car, we can determine the critical speed by setting the denominator in Eqs (5.26), (5.28) to zero:

$$KV_{Crit}^2 = -1$$

$$\Rightarrow V_{Crit} = \sqrt{\frac{-1}{K}} \tag{5.30}$$

where again

$$K = \left\{ \frac{mN_\beta}{\left(Y_\delta N_\beta - Y_\beta N_\delta\right)} \right\}$$

In general, we seek to provide a driver with a linear steering response, and so, oversteer of any kind is undesirable. However, a K value of zero (a neutral steering car) is unachievable in practice, and a more helpful way to express our design aim is that, if the car is to be oversteering, the critical speed should be far above the top speed at which the car is to be driven.

The general response of a vehicle to steering input and its stability as described above can be summarised in the figures below. These show the response of an understeering, neutral steering and oversteering vehicle in both curvature response and yaw rate response (Fig. 5.29).

5.11.8 Neutral Steer Point and Static Margin

The neutral steer point is defined as that point along the length of the vehicle at which an applied side force produces no steady-state yaw velocity. It can be an important concept for the designers of the aerodynamic package. For example, it is desirable for both road and racing cars that a sudden gust of wind from one side or the other should not produce a steering input (Fig. 5.30).

In the figure, the neutral steer point can be seen to be a distance d behind the front wheel. If we assume that the car is in steady state with $\partial = 0$ and $r = 0$ so therefore $\dot{\beta} = 0$ and $\dot{r} = 0$, then the equations of motion developed above

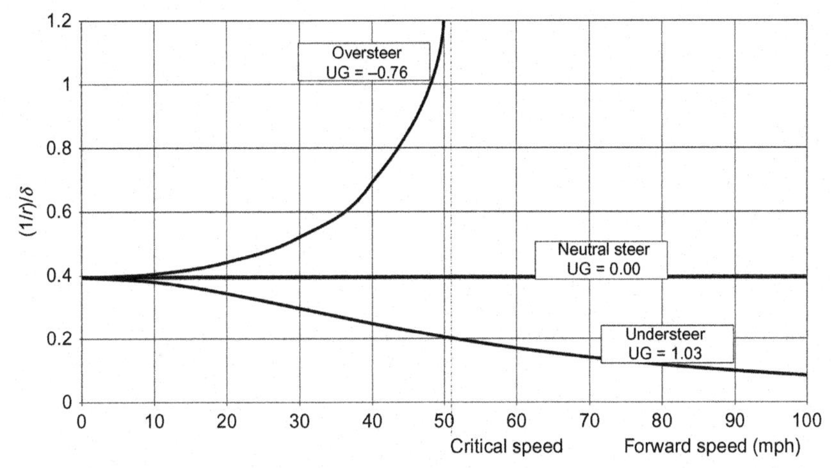

Fig. 5.29 Curvature response vs speed for three values of US gradient.

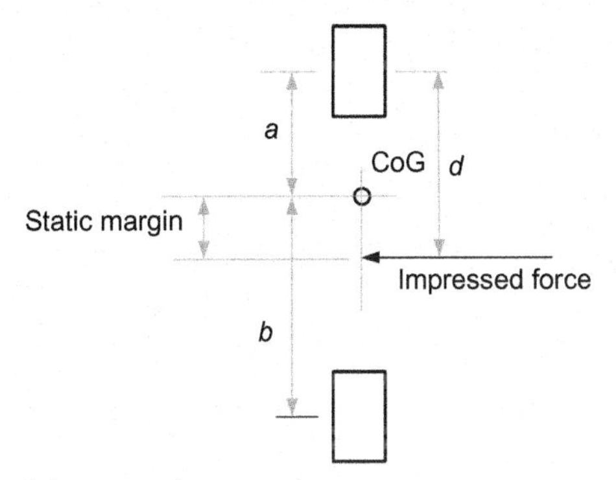

Fig. 5.30 Location of the neutral steer point.

$$mV\left(r+\dot{\beta}\right) = Y_\beta \beta + Y_R r + Y_\delta \delta$$

$$I\dot{r} = N_\beta \beta + N_R r + N_\delta \delta$$

Reduce to the following when the lateral force and torque generated by the lateral force are considered:

$$0 = Y_\beta \beta + F_0$$

and

$$0 = N_\beta \beta - (d-a)F_0$$

Hence,

$$\beta = (d-a)\frac{F_0}{N_\beta}$$

And so,

$$0 = F_0 + Y_\beta\left((d-a)\frac{F_0}{N_\beta}\right)$$

$$\Rightarrow 0 = F_0\left(1 + \frac{Y_\beta}{N_\beta}(d-a)\right)$$

$$\Rightarrow d = a - \frac{N_\beta}{Y_\beta}$$

that is often normalised to the wheelbase to produce a more general expression for the neutral steer point, *NSP* being the fraction of the wheelbase behind the front wheel:

$$NSP = \frac{d}{l} = \frac{a}{l} - \frac{N_\beta}{Y_\beta}\frac{1}{l}$$

The term 'static margin', *SM*, is used to further locate the neutral steer point. The *SM* is defined as

$$SM = \frac{d-a}{l}$$

that from above

$$\Rightarrow SM = NSP - \frac{a}{l}$$

$$\Rightarrow SM = \frac{-N_\beta}{Y_\beta}\frac{1}{l} \tag{5.31}$$

5.11.9 The Tangent Speed

When we considered Fig. 5.4, we met the concept of the tangent speed. This is the speed at which the centre of the corner has moved forwards sufficiently that it is level with the centre of gravity. At this speed, the vehicle is travelling in the same direction that it is pointing. Thus, the centre line of the vehicle makes a tangent to the radius of the corner, hence the name.

We shall consider the tangent speed here from two further points of view: firstly, to visualise the effect by considering the attitude a vehicle adopts as it goes around a corner with increasing speed, and secondly, to derive, using the derivative method, an equation for the value for the tangent speed.

Tangent speeds are typically around 30 mph for road cars and 80 mph for open-wheel racing cars such as F1. Consideration of the geometry of Fig. 5.4 (which was drawn far from scale for clarity) shows that the tangent speed would involve modest slip angles, certainly <8 degrees, and so, we should not be surprised that cars pass through the tangent speed before the tyres saturate.

We start in Fig. 5.31 at the coordinate origin, '*A*', where the short solid line shows the initial position of the vehicle. As time progresses, the car speeds up. Initially, there is no steering input, and the vehicle moves vertically upward in the figure, and the *y*-coordinate value increases. After travelling around 5 m, the vehicle reaches point '*B*', and a step-steer demand is made. The car enters a steady-state corner. In the simulation, the steering input remains constant once applied, and the speed very slowly increases, resulting in a negligibly small transient so that the vehicle can always be approximated by the steady-state condition.

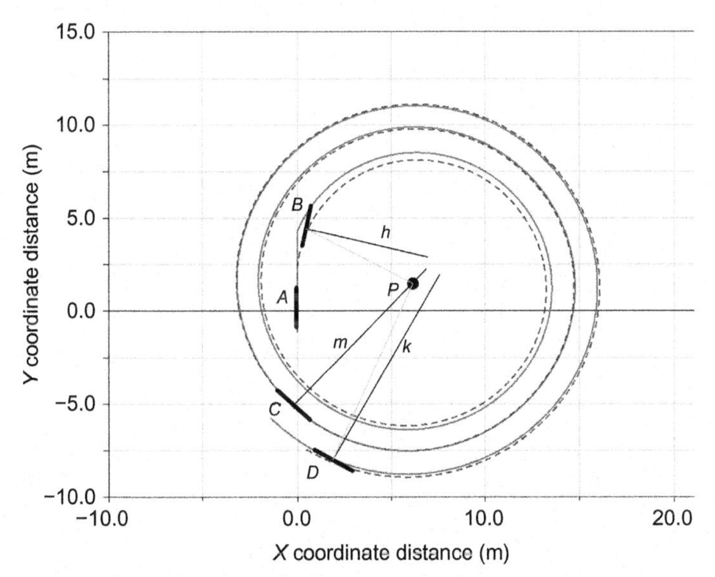

Fig. 5.31 Locus of wheel paths to find tangent speed.

The solid line shows the path taken by the front wheel and the dotted one that taken by the rear. It is clear to see that to begin with and at very low speed the rear wheel tracks a corner radius smaller than that of the front as we should expect. Point '*P*' shows the corner centre, and the grey line *B*–*P* shows the radius that pertains. The line *h* is drawn at right angles to the axis of the vehicle, and we can see that it passes ahead of the corner centre and the vehicle is *nose out* at this time. As the speed increases, we see that the paths taken by the front and rear wheels get closer together until at roughly position, '*C*', and the paths are the same, corresponding to the tangent speed. The line '*m*', drawn at right angles to the car, passes through the corner centre, and the car is pointing tangentially to the radius it is following. After this, the paths reverse, with the blue line for the rear taking the larger radius. By point '*D*', the car is *nose in* and will continue to increase the *nose-in* angle as the speed increases. Eventually, the forces required to maintain equilibrium are larger than can be generated by the tyres, and the car loses its ability to track the desired path. When this happens, there are two options: For the understeering car, where lateral force reduces at the front first, the front end moves outwards, the radius drastically increases, and the equilibrium above is lost; or for the oversteering car, where lateral force reduces at the rear first, the rear end moves outwards, the radius drastically decreases, and the car spins with the back end going out. This is the limit behaviour we have met before.

5.11.9.1 Derivation of the Tangent Speed

To derive an expression for the tangent speed, we need to obtain an equation for the vehicle side-slip angle in terms of the derivative coefficients. We may then set this equal to zero as is required at the tangent speed and solve to find the vehicle speed at which this occurs.

Starting with the equations of steady state,

$$\begin{aligned} mVr &= Y_\beta \beta + Y_R r + Y_\delta \delta \\ 0 &= N_\beta \beta + N_R r + N_\delta \delta \end{aligned}$$

and rearranging to eliminate r, we have

$$\begin{aligned} N_R r &= -N_\beta \beta - N_\delta \delta \\ \Rightarrow r &= \frac{-N_\beta \beta - N_\delta \delta}{N_R} \end{aligned}$$

Substituting

$$mV\left(\frac{-N_{\beta}\beta - N_{\delta}\delta}{N_R}\right) = Y_{\beta}\beta + Y_R\left(\frac{-N_{\beta}\beta - N_{\delta}\delta}{N_R}\right) + Y_{\delta}\delta$$

$$\frac{-N_{\beta}\beta mV}{N_R} - \frac{N_{\delta}\delta mV}{N_R} = Y_{\beta}\beta - \frac{Y_R N_{\beta}\beta}{N_R} - \frac{Y_R N_{\delta}\delta}{N_R} + Y_{\delta}\delta$$

$$\Rightarrow \frac{-N_{\beta}\beta mV}{N_R} - Y_{\beta}\beta + \frac{N_{\beta}\beta Y_R}{N_R} = + \frac{N_{\delta}\delta mV}{N_R} - \frac{Y_R N_{\delta}\delta}{N_R} + Y_{\delta}\delta$$

$$\Rightarrow \beta\left(\frac{-N_{\beta}mV}{N_R} - Y_{\beta} + \frac{N_{\beta}Y_R}{N_R}\right) = \partial\left(\frac{N_{\delta}mV}{N_R} - \frac{Y_R N_{\delta}}{N_R} + Y_{\delta}\right)$$

and so,

$$\Rightarrow \beta(-N_{\beta}mV - N_R Y_{\beta} + N_{\beta}Y_R) = \partial(N_{\delta}mV - Y_R N_{\delta} + N_R Y_{\delta})$$

$$\Rightarrow \frac{\beta}{\delta} = \frac{N_R Y_{\delta} + N_{\delta}(mV - Y_R)}{(-N_{\beta}mV - N_R Y_{\beta} + N_{\beta}Y_R)}$$

that relates the side-slip velocity β to the Ackermann steering angle ∂ and a velocity-dependent collection of derivative terms. At the tangent speed, the vehicle side-slip angle β is zero, and the top row above would therefore also have to equate to zero; hence,

$$N_R Y_{\delta} + N_{\delta}(mV - Y_R) = 0$$

So,

$$N_{\delta}(mV - Y_R) = -N_R Y_{\delta}$$

$$\Rightarrow V = \frac{1}{m}\left(\frac{-N_R Y_{\delta}}{N_{\delta}} + Y_R\right)$$

Remembering that N_R and Y_R are velocity-dependent and we must divide by the vehicle velocity when we come to use them, we may produce an equation for the critical speed V_T but must remember that, when using it in this case, the values for Y_R and N_R are the coefficients for these two derivative terms and are divided by the tangent velocity not a general V:

$$V_{TAN} = \frac{1}{m}\left(\frac{-N_R Y_{\delta}}{N_{\delta}V_{TAN}} + \frac{Y_R}{V_{TAN}}\right)$$

$$\Rightarrow V_{TAN} = \sqrt{\left(\frac{1}{m}\right)\left(\frac{-N_R Y_{\delta}}{N_{\delta}} + Y_R\right)} \qquad (5.32)$$

For a small hatchback, the tangent speed is generally <40 mph; for a formula one car, it is rather higher, a bit over 70 mph.

5.12 WORKED EXAMPLE OF APPLYING VEHICLE DERIVATIVES ANALYSIS

Having developed all the understanding above, we are now in a position to realise one of the great benefits of the derivative analysis technique, which is that it allows a very simple and quick method of assessing a vehicle performance and comparing one vehicle with another. The derivative numbers themselves may be compared one car with another, and with time, one gets very familiar with the numbers and what they mean. Even this rapid hand calculation can benefit from embedding in a spreadsheet to allow rapid adjustment of parameters, and an example is shown in Fig. 5.32—Derivative analysis comparison of two cars. In column B, we see the parameter in question, in column C its units and in column D its value. Column E shows the source of the given number. This may simply be a given piece of data or may show the equation in the chapter used to determine the number. The table shows data for two cars, and these are discussed below. The two data columns 'G' and 'I' are the made by first filling in the parameters desired into column 'D', and the spreadsheet then calculates the output data. Column 'D' is then copied into 'G', values only, to be a record of that particular car. The data were then updated to the second car and then copied into column 'I' to form a record of that car. This makes it easy to compare column 'G', for the hatchback, with column 'I' for the open-wheel racing car.

Car A is an approximation of a small hatchback, whilst Car B is an idealised single-seater racing car. Car A is much heavier than B, and its tyres have much lower cornering stiffnesses. In the 'derived parameters' section, we see that whilst A is a significantly understeering car (to be expected with a front engine hatchback) car B is exactly neutral. This becomes even more noticeable when we look at the value for K in row 18. In the case of A, it is sufficiently large that the resulting critical/characteristic speed is 62.8 mph, a reasonably good value (the spreadsheet uses the positive value of K, and one must remember that, for an understeering car, such as A, this means a 'characteristic' speed whereas for the oversteering car it is a 'critical speed'). For car B, by contrast, the characteristic/critical speed is over 5000 mph! There is clearly no chance that this car will operate at such a speed and it is so far above the working speed range that we can be sure that none of

	A	B	C	D	E	F	G	H	I
4									
5		Input Parameters			Source		Car A - hatchback		Car B - Idealised
6									
7		Mass	kg	450	data		900		450
8		Polar Moment of inertia	kg.m^2	700	data		1100		700
9		Step Steer input	deg	5	chosen		5		5
10		CofG - front axle (a)	m	1.1	data		0.9		1.1
11		Wheelbase (l)	m	2.5	data		2.6		2.5
12		Cornering stiffness f	N/deg	-8800	data		-1100		-8800
13		Cornering stiffness r	N/deg	-6920	data		-900		-6920
14									
15		derived parameters							
16		CofG - rear axle (b)	m	1.4	=l-a		1.7		1.4
17		Understeer Gradient	deg/m/s^2	0.00	Eqn 5-5		1.85		0.00
18		K	(m/sec.)^2	1.65E-07	Eqn 5-27 & 28		1.27E-03		1.65E-07
19		Char / Crit speed	mph	5503.2	Eqn 5-29 & 30		62.8		5503.2
20		Static Margin	m	0.0	Eqn 5-31		0.1		0.0
21		Tangent Speed	mph	118.4	Eqn 5-32		37.5		118.4
22		Actual I/m		1.56	I/m		1.22		1.56
23		Desired I/m Nr/Yb		1.54	Nr/Yb		1.75		1.54
24									
25		Derivatives							
26		Yβ		-900689.65	Eqns 5-9		-114591.56		-900689.65
27		Yr/V		458.37	Eqns 5-9		30939.72		458.37
28		Yδ		504202.86	Eqns 5-9		63025.36		504202.86
29		Nβ		458.37	Eqns 5-9		30939.72		458.37
30		Nr/V		-1387199.58	Eqns 5-9		-200076.86		-1387199.58
31		Nδ		554623.15	Eqns 5-9		56722.82		554623.15
32									
33		Output Data	Speed/mph						
34		Speed/mph	20	60	Given		60		60
35		Speed m/s	8.94	26.82	conversion		26.82		26.82
36		Undamped Nat freq/Hz	35.45	74.26	Eqn 5-20		7.68		74.26
37		Damping ratio	1.00	1.00	Eqn 5-19		0.75		1.00
38		Damped Nat Freq/Hz	#NUM!	0.11	Eqn 5-21		0.81		0.11
39		St St. Radius of corner/m	28.65	28.65	Eqn 5-26		56.97		28.65
40		St St. Yaw rate/ degs/sec.	17.88	53.64	Eqn 5-28		26.98		53.64

Fig. 5.32 Derivative analysis comparison of two cars.

the problems will be experienced and the car will appear completely neutral over its entire speed range. The tangent speed is much higher for car B. The actual ratio of polar moment to mass is much closer to the desired for car B, and this allows car B to be both neutral steering and have a steering response that is critically damped. Car A does not achieve this (see row 37—*damping ratio*).

The derivative terms themselves are noticeably different. Y_R and N_β are very small and make the car very close to neutral (the number of the derivative terms can easily be quite large so rather than look at absolute values its best to compare). Car B has values here that are around one-hundredth those of car A. The undamped natural frequency for car A is just <8 Hz, but its damped natural frequency is 0.81 Hz meaning that we should expect it to have a yaw rate response with a cycle length of just 1.23 s. It is a significantly damped response, damping ratio 0.75, and the consideration of a graph of such a response shows that you only get around one cycle visible before the response is settled.

The response of car B is much faster. The undamped natural frequency is 74.27 Hz. However, it is the fact that the response is critically damped that is important. One can conclude that car B will have a much faster response and will be settled in a fraction of the time that car A takes simply from this, but to aid visualisation, Fig. 5.33 shows the two responses, and indeed, car B has completed the corner entry in around 100 ms, whereas car B takes nearly a second. Whilst car B is a little contrived to display excellent performance, this level is broadly achievable, and an F1 car will respond similarly. This analysis makes clear what an amazing drive such a car would be!

Finally, we see the difference in the corner that results from the 5 degrees of steering input that has been supplied to both cars. In the case of car B, the corner radius is roughly half that of car A and the yaw rate roughly twice.

Such an analysis brings a rational, clear way to assess one vehicle against another and with practice is achieved quickly and easily. There is no better way to develop this ability than to make your own spreadsheets to perform these analyses, and the website for the book has further guidance and tutorials to help.

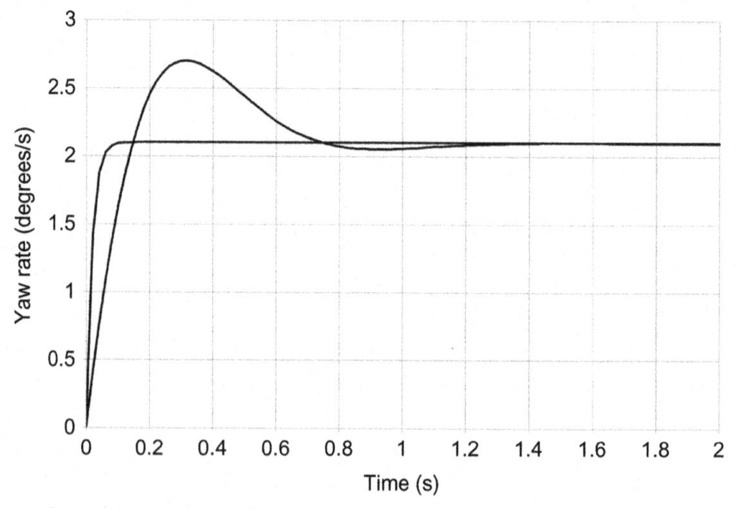

Fig. 5.33 Derivative analysis of steering response against time for two cars.

5.13 QUESTIONS

1. A vehicle is travelling around a corner of radius 34 m at a speed of 37 mph. What is the yaw rate?
2. For the vehicle in (1), if the wheelbase is 2.9 m, what is the Ackermann steered angle?
3. A vehicle is cornering at 45 mph around a 38 m radius corner. The vehicle has a mass of 1100 kg and cornering stiffnesses C_F=2500 N/degree and C_R=1900 N/degree. Determine the slip angle front and rear and produce a CAD drawing for the situation similar to Fig. 5.3 but drawn to scale.
4. Why is it that in Fig. 5.5 a vehicle can never reach positions around D_5 onwards?
5. Explain in your own words how the forward/aft location of the centre of gravity for a vehicle with equal cornering stiffness tyres front and rear affects the limit behaviour of the vehicle.
6. How is the situation in (5) modified if the centre of gravity is moved forwards or backwards?
7. What is the name given to the cornering car when $\alpha_F = \alpha_R$ and why is this desirable?
8. Why are the names *oversteer* and *understeer* logical?
9. A car is cornering with an Ackermann steering angle of 1.0 degrees. The wheelbase is 1.9 m, and the corner radius is 100 m. The speed is 30 m/s. What is the understeer gradient?
10. A vehicle has a wheelbase of 2.6 m, an '*a*' value of 1.1 m, mass 850 kg and forwards velocity of 16 m/s and is travelling around a corner of radius 22 m. The cornering stiffnesses are 1010 N/degree at the front and 830 N/degree at the rear. What is the steering angle required?
11. Using a spreadsheet analysis such as in Fig. 5.22, determine the time taken for the following car to settle within 5% of its final yaw rate. Mass 1200 kg, polar moment of inertia, 1280 kg m^2, '*a*' value 0.9 m, wheelbase 2.5 m and C_F=1100 N/degree and C_R=1000 N/degree.

5.14 DIRECTED READING

Genta [12] gives a very detailed and mathematical treatment of the derivative analysis.
Cossalter [14] deals entirely with motorbikes and offers a bicycle analysis of cornering that is actually for two-wheeled vehicles.

Gillespie [11] offers a treatment of cornering that is readily accessible and makes a good place to start.
Milliken W.F. and Milliken D.L. [9] offer excellent and very practical chapters in the text that the reader should find easy to move onto after this chapter if more is wanted.
Segal [5] authored a paper where the derivative approach was first applied to vehicle dynamics.

5.15 LEARNING PROJECTS

- A vehicle has a wheelbase of 2.1 m and is negotiating a corner of 22 m. Use a CAD package to produce a diagram similar to Fig. 5.2 for the situation but drawn to scale.
- For a group of cars you are interested in, research the dynamics data required and determine the critical speeds (they will have to be oversteering cars).
- Produce a spreadsheet analysis of a vehicle of your choice to analyse an understeering car as shown in Fig. 5.22.
- Extend the spreadsheet to allow for an understeering, neutral steering or oversteering car to be entered. You will need to first make the spreadsheet recognise each case and then select the correct equation appropriately.
- Produce an Adams bicycle model as shown in Fig. 5.21. Use it with the same data as used in the spreadsheet above and compare the graphs of yaw rate so obtained. Quantify the error over a number of cases. Why are the responses so similar and yet not identical?
- Use the Adams bicycle model and adjust it to active a neutral steering car with critical damping.
- For a vehicle of your choice and for which you can obtain dynamic data, produce a graph of Zeta vs ψ as shown in Fig. 5.26. You can use data similar to examples in the text if desired.
- Make an Adams bicycle model and produce your own analysis comparing the steady-state yaw and curvature response as determined by the derivative analysis and the Adams simulation. Compare the results and examine the reasons for the level of agreement.
- Use the Adams model developed to demonstrate to your own satisfaction the characteristic speed for an understeering car and the critical speed for the oversteering car.

- For a range of vehicles you are interested in, determine the six-derivative coefficients and compare them. Make your own quick estimates of the performance in each case and then proceed to analyse them using the derivative spreadsheets and Adams bicycle models above.

5.16 INTERNET-BASED RESEARCH AND SEARCH SUGGESTIONS

- The YouTube website has dozens of videos explaining the oversteer and understeer effect with a vehicle in motion. Research a collection of these and establish agreement between the reading above and observation of the videos.
- Skid and yaw marks on a road surface can be used to determine the speed of the vehicle leaving the marks. Research this possibility and its relevance to accident investigation.
- Use YouTube to find video explanations of how a differential works; there are some very clear presentations available.
- Perform a Google search on 'limited-slip differential', 'spool differential', 'active differential', 'active yaw control' and the manufacturers 'Torsen' and 'Quaife' and review the state-of-the-art offerings.

CHAPTER 6

Braking

6.1 INTRODUCTION

I once saw a very good racing driver getting into an F1 test car for his first drive. I wondered what he would do first; the car was going to be entirely new to him; there would be so much to learn.

'Right', he said lowering himself in, 'Let's test the brakes!'

So much safety depends on the correct operation of the brakes that it does make sense to test them first. In road cars, there is legislation governing the design of brake system, and although the system is simple enough in terms of overall functioning, there is a lot to them, and brakes play a large part in vehicle performance.

On typical circuits, racing drivers spend only around 10% of the time with the brakes on. When they are on, they tend to be fully on and decelerations are high. It follows that a roughly a 10% improvement in braking performance (an enormous amount!) would equate to only a 1% overall improvement in vehicle performance. On the other hand, having a braking deceleration that is only marginally greater than the opposition will enable our driver to brake later and spend less time at low speed. Even small advantages in braking performance can lead to much high finishing positions when margins are small. Thus, we conclude that making improvements in lap-time performance by improving braking performance will be hard but that any small gain will be welcome.

In road cars, the contribution to the vehicle product performance is clearly still a vital part of the package. No one would buy from a manufacturer that had a reputation for poor brakes. On the other hand, no one buys a particular model of car because the brakes are exceptionally good. It is simply expected that they will be optimal. It is easy enough to produce brakes that can lock up any wheel, and in fact, no more can be done than this for a passive braking system. Usually in racing, passive systems are mandatory, but in road cars, this is not so at all. The advanced braking system (ABS) that prevents a vehicle from skidding no matter how hard the brakes are applied, is normal. To improve handling, there are braking systems that brake one side more than the other to beneficially control the yaw response to steering

http://dx.doi.org/10.1016/B978-0-12-812693-6.00006-7

input. Thus, racing and road car drivers simply expect the braking system to be as good as it is possible to be.

In this chapter, we shall examine the dynamics of braking and come to understand how it affects dynamic performance.

6.1.1 Braking Energy and Temperatures

In this section, we shall make some overall estimates of the quantities of energy involved in braking. Braking systems almost always involve a frictional contact between a surface driven by the wheels and another held stationary with respect to a part of the vehicle. Thus, under braking, the kinetic energy of vehicle in motion is dissipated as heat rise in braking materials that once hot must shed their heat into the airstream.

We start by estimating the energy to be dissipated in a road car slowing down from 40 mph (18.6 m/s). For a 1 tonne car, this is

$$E_{TOT} = \frac{1}{2}mv^2 = \frac{1}{2} \times 1000 \times 18.6^2 = 0.17\text{MJ}$$

If it takes the car 5 s to slowdown, the rate of energy delivery is

$$\frac{0.17 \times 10^6}{5} = 34000 = 34\text{kW}$$

If we now assume that the friction material of the brakes directly warmed up by braking weighs 2 kg per wheel and has a specific heat capacity of 450 J/kg/K, then we may estimate the temperature rise of this material to be as follows:

$$H = MC\Delta\theta$$

$$0.17 \times 10^6 = 4 \times 2 \times 450 \times \Delta\theta$$

$$\Rightarrow \Delta\theta = 47^0$$

Thus, for a road car, the quantities of energy involved are not so great, and the thermal mass of the discs themselves are sufficient to absorb much of the energy. We can compare this with a racing car of mass 500 kg decelerating from 160 mph (72 m/s) to rest, for which the energy lost is

$$E_{TOT} = \frac{1}{2} \times 500 \times 72^2 = 1.3\text{MJ}$$

Such a car can do this in around 2 s giving a rate of energy delivery

$$\frac{1.3 \times 10^6}{2} = 650\text{kW}$$

If we again assume the friction material of the brakes to be 2 kg per wheel and specific heat capacity 450 J/kg/K, then we may estimate the temperature rise to be as follows:

$$H = C_{\text{carbon}} \Delta \theta$$

$$1.3 \times 10^6 = 4 \times 2 \times 450 \times \Delta \theta$$

$$\Rightarrow \Delta \theta = 361^0$$

In fact, this is a very poor estimate of the temperature rise because it all happens too fast for the heat to flow into the body of the carbon disc that is normally used. What actually happens is that the surface temperature rises very quickly to around 850°C. Once this temperature is reached, the surface is in temporary equilibrium with the heat generated flowing into the surface of the disc and, from there, into the airstream that is ducted over them.

Thus, for racing brakes, we need materials that can shed energy very quickly by heat loss. From a design point of view, this means racing brakes must be very well ventilated and run as hot as we can arrange. It is the ability of carbon fibre to run at high temperature that makes it the material of choice. The choice of carbon fibre is normally associated with weight saving, but this does not apply here; indeed, carbon disc brakes are actually very heavy.

In terms of braking capacity, it must be remembered that it is the contact patch that generates the longitudinal force that decelerates the car, and so, the requirement of the braking system is that it can lock up all four wheels. More than this is pointless, though the well-designed system will easily be able to do this under the highest Mu values whilst also providing the ability for a driver to demand much smaller levels of braking in a controlled way if desired, so that the brakes operate smoothly over a range of driver pedal pressure. It is no good if the wheels lock up at the slightest pressure. Drivers need to be able to control the braking effort applied to the wheels accurately and smoothly; for example, during a turn, a good driver will be able to keep the brakes on the point when the inside wheel is just about to skid (the most that should be supplied). In a road vehicle where the total weight can vary very significantly, for example, a lorry, it is necessary to provide a brake system with variable gain so that, regardless of vertical load on the tyres, a smooth application to full brake pedal pressure results in a smooth increase of braking effort without locking up too early regardless of whether the vehicle is empty or fully laden.

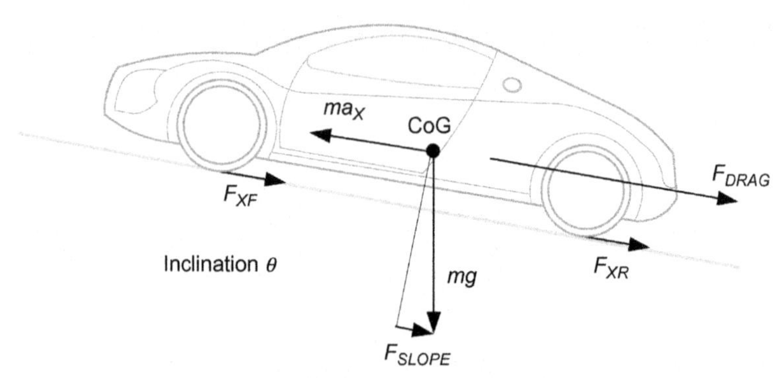

Fig. 6.1 Free body diagram for a decelerating vehicle.

6.1.2 Dynamics of Braking

Fig. 6.1 shows the braking forces acting on a vehicle going up an incline.

For a vehicle of weight mass m,

$$ma_X = F_{XF} + F_{XR} + F_{\text{DRAG}} + mg\sin\theta$$

For a vehicle with negligible aerodynamic drag (slow or very streamlined) braking on a flat road, the terms F_{XF} and F_{XR} will dominate. With a total brake force F_{TOT} being the sum of the components, we may write the following:

$$Ma_X = F_{TOT}$$

$$\Rightarrow \frac{F_{TOT}}{M} = \frac{dV}{dt}$$

Thus (assuming that F_{TOT} is constant with time), we can integrate between an initial velocity of V_I to a final velocity of V_F over a braking period of t:

$$\int_{VI}^{VF} dV = \int_0^t \frac{F_{TOT}}{M} dt$$

$$\Rightarrow \Delta V = \frac{F_{TOT}}{M} t$$

which relates change in velocity to braking force, vehicle mass and braking period, from which the time to stop T_S can be estimated (noting that at this time $V_F = 0$) and is given by

$$\Delta V = \frac{F_{TOT}}{M} T_S$$

$$\Rightarrow T_S = \frac{V_I}{F_{TOT}/M}$$

$$\Rightarrow T_S = \frac{V_I}{D_X}$$

where D_X is the deceleration value. Also since

$$\frac{F_{TOT}}{M} = \frac{dV}{dt}$$

$$\Rightarrow V = \frac{dx}{dt} \Rightarrow dt = \frac{dx}{V}$$

$$\Rightarrow \frac{F_{TOT}}{M} = V\frac{dV}{dx}$$

$$\Rightarrow \int_{V_I}^{V_F} Vdv = \frac{F_{TOT}}{M}\int_0^X dx$$

$$\Rightarrow \frac{1}{2}\left(V_I^2 - V_F^2\right) = \frac{F_{TOT}}{M}X$$

where X is the distance travelled during the deceleration. When the deceleration is to complete rest, V_F is zero and the stopping distance SD is given by

$$SD = \frac{V_I^2}{\left(2\frac{F_{TOT}}{M}\right)}$$

$$\Rightarrow SD = \frac{V_I^2}{2D_X}$$

Worked Example

It is desired that a car of mass 1200 kg should be able to stop within a time of 2.5 s from a speed of 50 mph (23 m/s). Calculate the total braking force required and the stopping distance. Comment on whether an ordinary road car is likely to be able to achieve this:

$$2.5 = \frac{23}{\left(\frac{F_{TOT}}{1200}\right)} \Rightarrow F_{TOT} = \frac{23 \times 1200}{2.5} = 11\,\text{kN}$$

$$SD = \frac{23^2}{\left(2 \times \frac{11000}{1200}\right)} = 28.8\,\text{m}$$

Deceleration is $a_X = \frac{23}{2.5} = 9.2 = 0.94\,g$ probably achievable from a road tyre but marginal and certainly not in the wet.

6.1.2.1 Effect of Road Inclination

We can estimate the deceleration that results from road inclination alone by considering Fig. 6.1 and ignoring the drag and braking forces, thus

$$F_{SLOPWE} = W \sin\theta \Rightarrow F_{SLOPE} \approx mg\theta$$

$$\Rightarrow mg\theta = ma_X$$

$$a_X = g\theta$$

Thus, a 1:10 hill (0.1 rad inclination) will provide ~0.1 g of deceleration. For road cars, this would be very noticeable since even the peak decelerations would be generally rather <1 g. For racing, where the peak deceleration for cars with downforce are around 4 g, it would be a much smaller quantity but very likely to be significant if a small advantage in time can be gained.

6.1.3 Ideal Brake Force Ratio

It might seem that analysing the dynamics of a car under braking would be opposite to the acceleration case that we have considered before and would come down only to getting rubber with the highest μ available and then pressing the brake pedal until the tyres are just about to slip, therefore operating at that slip value that provides the largest possible longitudinal force. In fact, it isn't that simple and the reason for this is that unlike forward acceleration where normally only one axle is active, when the car is braking, all four wheels are involved. We have already seen that under braking, when weight is transferred to the front axle. If then, we simply arrange for the brake system to supply the same retarding torque to all four wheels then the driver will continue to apply pressure to the brake pedal until the rears just begin to skid and will then not increase the pressure further. This is clearly less than optimal since the front wheels with their extra load are capable of producing a much greater retarding force more than this. What is needed is a system whereby the front brakes apply a greater pressure to the pads than do the rear ones and in addition, the apportioning of pressure results in all four wheels being just on the point of locking up at the same time. Then, we shall have maximum deceleration. We begin our consideration of this condition with Fig. 6.2.

In the figure, all the relevant forces acting during braking are set out. We wish to find out the ideal ratio of F_{XF} to F_{XR}. This is the ratio that ensures the

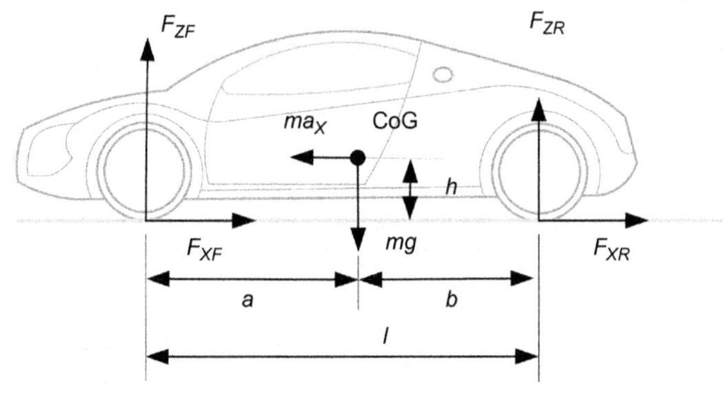

Fig. 6.2 Wheel loads braking.

two forces take the largest value that they can, given that the vertical load is more on the front than the rear. We start by noting that

$$F_{XF} = \mu F_{ZF}$$

and

$$F_{XR} = \mu F_{ZR}$$

Since we are determining the peak deceleration that pertains when both axles are developing the largest longitudinal force that they can just before skidding, the total force is the sum of that from both axles:

$$\begin{aligned} m\hat{a}_X &= \hat{F}_{XF} + \hat{F}_{XR} = \mu(F_{YF} + F_{YR}) = \mu mg \\ \Rightarrow \hat{a}_X &= \mu g \end{aligned}$$

Next, noting that the ideal brake force ratio is defined as the ratio F_{XF} to F_{XR}, we shall therefore start by finding expressions for each of these terms and then determine the ratio. To determine F_{XF}, we start by taking moments about the rear contact patch:

$$\begin{aligned} ma_X h + mgb &= F_{ZF} l \\ \Rightarrow F_{ZF} &= \frac{1}{l}(m\hat{a}_X h + mgb) \\ \Rightarrow F_{ZF} &= \frac{mg}{l}\left(\frac{\hat{a}_X}{g}h + b\right) \\ \Rightarrow F_{ZF} &= \frac{mg}{l}(\mu h + b) \\ \Rightarrow F_{XF} &= \frac{\mu mg}{l}(\mu h + b) \end{aligned}$$

Now to find F_{XR},

$$F_{ZR}+F_{ZF}=mg$$

$$\Rightarrow F_{ZR}=mg-F_{ZF}$$

$$\Rightarrow F_{ZR}=mg-\frac{mg}{l}(\mu h+b)$$

$$\Rightarrow F_{ZR}=mg-\frac{mg}{l}\mu h-\frac{mg}{l}b$$

$$\Rightarrow F_{ZR}=\frac{\mu mg}{l}\left(\frac{l}{\mu}-h-\frac{b}{\mu}\right)$$

$$\Rightarrow F_{XR}=\mu\frac{\mu mg}{l}\left(\frac{l}{\mu}-h-\frac{b}{\mu}\right)$$

$$\Rightarrow F_{XR}=\frac{\mu mg}{l}(l-\mu h-b)$$

From this, we can now find the ratio of F_{XF} to F_{XR}:

$$IBFR=\frac{F_{XF}}{F_{XR}}=\frac{\frac{\mu mg}{l}(\mu h+b)}{\frac{\mu mg}{l}(l-\mu h-b)}$$

$$\Rightarrow IBFR=\frac{(\mu h+b)}{(l-\mu h-b)}$$

Or more simply, since $l-b=a$,

$$\Rightarrow IBFR=\frac{(b+\mu h)}{(a-\mu h)}$$

Thus, we can see that the front brake force should be larger than the rear brake force by an amount that depends on the geometry of the vehicle and the coefficient of friction for the tyres.

6.1.3.1 Graphical Presentation

Considering the equation for the ideal brake force ratio derived above

$$\Rightarrow IBFR=\frac{(\mu h+b)}{(l-\mu h-b)}$$

and remembering that for a racing car, b will be around half of l and that h will be much smaller than b, we can see that as μ increases the bottom line

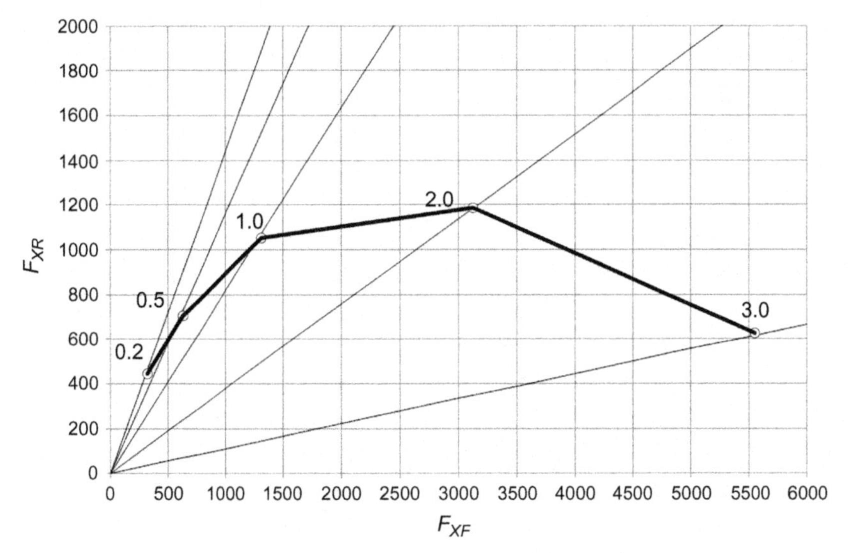

Fig. 6.3 Ideal rear brake force versus front brake force for a range of μ values.

will get smaller. In addition, as μ increases, the top line will get larger, and so, the brake force ratio will increase as the grip gets larger. This is to be expected on a surface with excellent grip; since the deceleration will be high, a lot of weight will be transferred to the front wheels, and they will need far more torque applied to them to get them to the point of slipping than will the rears.

To illustrate this further, Fig. 6.3 shows an example. The car in question has the following:

Wheelbase	l	1.6 m
C of G rear	b	0.6 m
C of G height	h	0.28 m
Weight	w	200 kg

The figure shows lines for the brake force ratio that would be needed for a wide range of μ values, together with points that show the maximum brake force that can be generated at that μ value.

To understand the graph, we start by looking at the five straight lines for each μ value. For example, if the car were being driven on a surface with a μ value of 1.0 and the front brake force was 1000 N, then the brake biasing system should ensure that 800 N of braking effort is generated at the rear. This is all very well, but it does not tell us how much the actual amount that should be generated at each axle. However, it is a simple enough matter to calculate this for each μ case using the equations for F_{XR} and F_{XF} developed above. These were used for the racing car values listed above and five points plotted. Thus, for example, if the current road surface has a μ value of 2.0, then the best that we can ever do will be to have around 2800 N of braking effort from the front and 1200 N from the rear. We can see from these lines that as μ increases, not only does the total amount of braking force increase but also the fraction generated from the front also increases as we would expect. Finally, in practice, μ values for soft sticky slick racing tyres on a hot day with good quality tarmac may be as high as 1.5. By contrast, treaded road tyre on poor roads may have a μ value as low as 0.3.

By considering the diagram we can also see the effect of a change in conditions. If, for example, a racing car was set up for a current μ value of 1.0 on a dry sunny day, then a considerable fraction of braking effort would be on the front. If the conditions then worsened with rain and μ fell to 0.4, then too much braking effort would be diverted to the front for the new μ value, and it would be the fronts that lock up first. Note that one can have a need for F_{XR} to be larger than F_{XF} if the weight is towards the rear.

Considering the diagram, we can see that if the Mu value were to reach a high enough value, in this case perhaps around 4.0, then it would be possible for the ideal braking force at the rear to become zero. This corresponds to the situation where the braking is so hard that the rear of the vehicle lifts off the ground. For the vehicle represented in Fig. 6.3, a Mu value this high cannot be achieved in practice and long wheelbase vehicles with a low centre of gravity do not trip over their front wheel under heavy braking. They skid. However, any motorcyclist will tell you that for vehicles with short wheelbases and a rather higher centre of gravity, it is all too easy to go over the handle bars under heavy braking.

6.1.4 Stability Under Braking

We have seen that in an ideal situation all wheels would lock up together. In reality, this is rather difficult to achieve and one axle generally locks up before the other. The stability of the car under braking is dramatically affected by this (Fig. 6.4).

In the figure, we start in part (A). The car is travelling in a straight line, and the brakes are applied. Brake bias is implemented, and there is more braking effort from the front wheel as shown by the large arrow at the front. In part (B), the car has encountered some disturbance in yaw; this might be a gust of wind or one wheel riding over a small bump causing a small steering

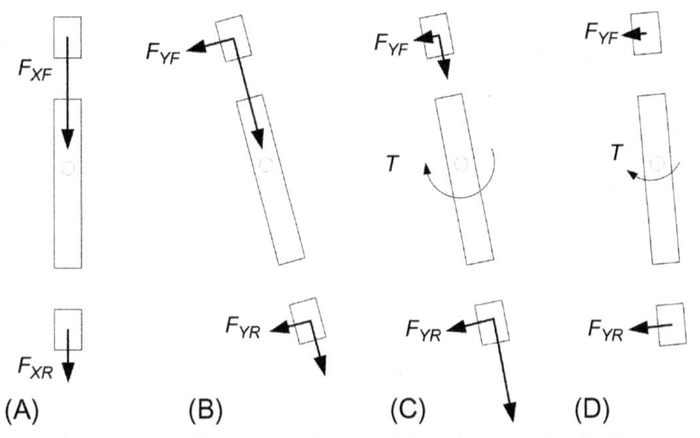

Fig. 6.4 Less rear braking effort makes for a stable vehicle under braking.

input. Such very small inputs are always present, for example, driving along a straight road the driver can sense constant but very small input disturbances steering the wheel by tiny amounts all the time. If the driver applies the brakes and brakes hard enough that the front wheels exceed their maximum lateral force-generating capacity, the lateral force generated is much smaller than at the rear. This is shown in position (C). Even without any steering input being present, this causes a torque to be generated about the centre of gravity as shown. Thus, the car accelerates in yaw, turning into the disturbance and so straightening up. As it does so, the slip angle at the rear decreases and with it the torque restoring the car to the straight ahead position and the car is therefore stable under the disturbance, reacting in such a way as to correct the perturbation. The car is shown returning to is straight ahead position in (D).

In Fig. 6.5 below, the same situation is presented, but this time, the rear is receiving more braking effort.

In the figure, we start in part (A). The rear braking effort is larger than the front, and the driver is continuing to apply the brakes. In part (B), the rear wheel has locked up and is therefore producing a much smaller lateral force shown by the smaller arrow. The car has again encountered some disturbance in yaw, and the front and rear tyres are at a small slip angle. Since the front is well within its limit, it produces significant lateral force. The rear, however, produces almost none because it is skidding. Thus, there is a yaw moment on the car, and it is accelerating in yaw. Left to its own devices, it will move to a larger slip angle as seen in part (C). Here, the slip angle at the front has grown even larger, and so, there is even more lateral force. The rear

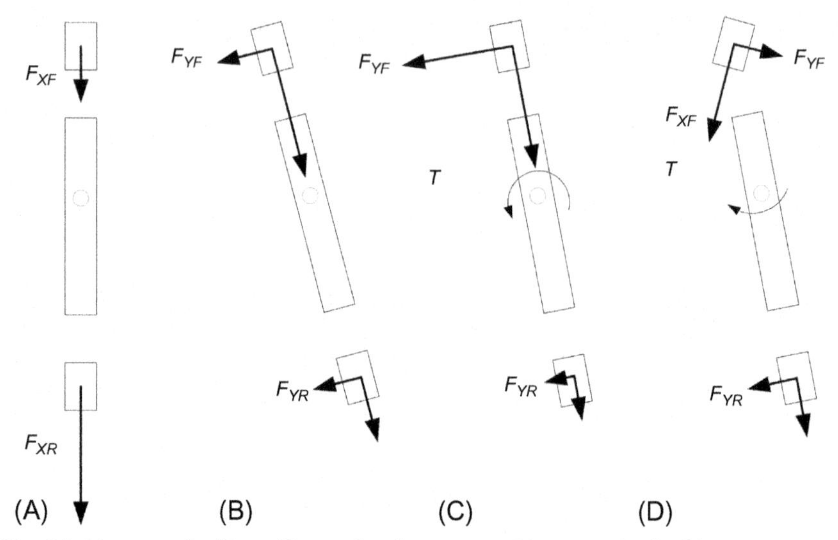

Fig. 6.5 More rear braking effort makes for an unstable car under braking.

is still skidding and producing virtually no lateral force. Clearly, this situation is self-exciting, and so, the tiniest initial disturbance in yaw will rapidly grow into a huge yaw excursion, and the car will spin out if the rears remain locked. A very undesirable and dangerous situation this is too. The process continues until the car has swung right around with the back ahead of the rear. Most cars 'catch' themselves at this point and adopt a new controlled forward path since they have now become cars in which the brakes are biased to the front! Part (D) of the diagram shows what the driver would have to do to recover here. The front wheel would have to be turned a quite large amount much past the dead ahead position. What's needed is for the force F_{YF} to become sufficiently large to accelerate the car in yaw out of trouble (clockwise in the diagram). However, the useful moment generated by force F_{XF} is being offset to some extent by the moment generated by the braking force F_{XF}. Any good driver will tell you that to recover from position (C), the wheel must indeed be turned as shown in (D), but the brakes must be released, which causes helpful F_{YR} to yaw the car ahead again and also to remove the unhelpful F_{XF}.

Once control is regained, braking can resume. Most road drivers, however, simply see the brake pedal as a 'get out of trouble' pedal and simply press very hard when control is lost when in fact the opposite is required before braking can start again. We therefore normally keep the brake bias a little towards the front so that if the driver does get a little over zealous with the brakes, the car is at least stable and he or she is very much more likely to recover control. It is also for this reason that attentive fans of F1 and other racing will often see a little puff of smoke from the occasional front tyre but never at the back.

6.1.5 Brake Bias Bar

From the above, we can see that it would be a great advantage to be able to adjust the brake force ratio easily and quickly on a vehicle. If, for example, the weather is hot and grip very good, a much higher forward bias will be needed than when it is cold and damp. A common way to adjust the brake for ratio is the 'brake bias bar', and a schematic of the way it works in shown in Fig. 6.6 below.

In the figure, the brake pedal, 1, is forced forwards by the driver's foot. Inside the brake pedal, a swivel mounting, 2, permits the brake bias bar, 3, to pivot, and this ensures good load sharing between the two pistons. The brake bias bar, 3, is threaded, and the thread engages on the swivel mounting, 2. It can be rotated by the driver via a flexible link, 4. As it rotates, it winds itself to the left or right as the driver chooses. In doing so, the

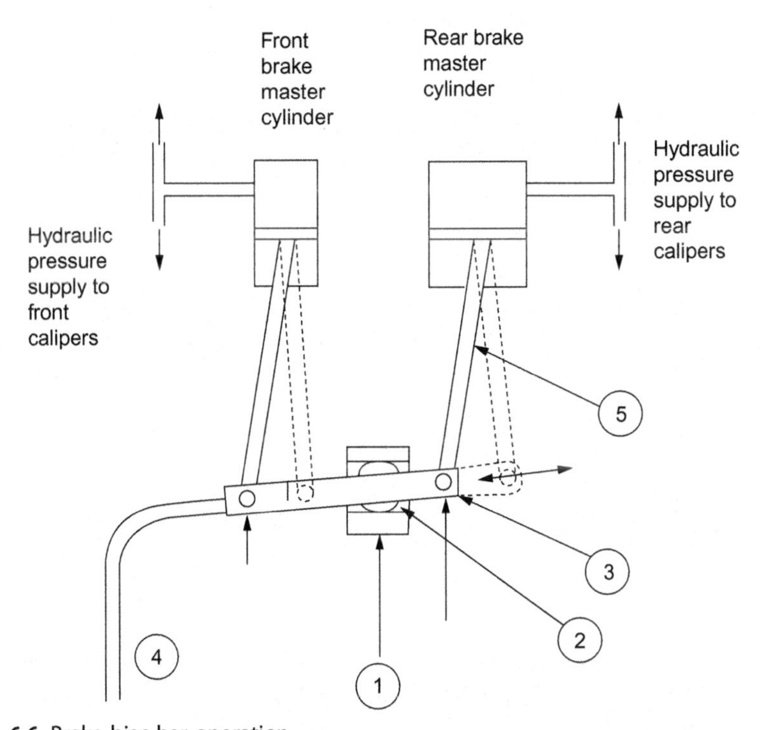

Fig. 6.6 Brake bias bar operation.

geometry of the system is changed, and the load passed into each master cylinder is also changed. This adjusts the pressure in the master cylinder that in turn applies greater or less pressure to the callipers. Since under most racing condition, grip is very good it is normal to use a smaller diameter master cylinder for the front brake circuit than for the rear. This means that with the brake bias bar in the middle position and the geometry of the system good from a load bearing point of view, the fronts already have a significant bias, and the bar adjusts the bias from this starting point. The flexible link is brought up to the cockpit where a rotary control allows adjustment. Some high-value formula cars have self-adjusting systems that can adjust themselves for each corner depending on the instantaneous dynamic conditions.

In road car design, such driver adjustments are not permitted, and the brake force ratio is set to a value that is sensible under low Mu. The resulting deoptimisation when Mu is high must simply be accepted.

6.1.6 Advanced Braking System

An alternative method of ensuring that all wheels on a racing car lock up together with skidding is to use ABS. In this method, wheel speed sensors detect how fast each wheel is rotating. If any one wheel suddenly slows down a vast amount compared with the rest and the car was recently in motion, then it is a safe conclusion that the wheel involved has started to skid. When the ABS system detects this, it operates a pressure control value that reduces the hydraulic pressure at the calliper braking that wheel until it regains the speed comparable with the others. When this happens, the valve shuts again, and the pressure starts to rise. The system cuts in and out repeatedly until the brake demand is sufficiently small that skidding stops occurring. This gives rise to the juddering that the vehicle experiences when the system cuts in. The system is particularly effective in road cars where drivers experiencing difficulty are not skilled enough to modulate the brakes and tend merely to pressure very hard when in difficulty. A skidding wheel produces much less braking force than one on the point of skidding, and the ABS brings the tyre back to this situation for the driver. In almost all racing, ABS is now banned even though it had it origins in the sport.

6.2 QUESTIONS

1. If you were designing a regenerative braking system what sort of things would you need to know before design can commence?
2. Why do better brakes make for reduced lap times?

3. A university lecturer in vehicle dynamics took his large 4 × 4 off-road vehicle to the garage for its 2-year service only to be told that the rear discs were worn and needed replacing but the fronts were fine. Why was he unhappy and asked to see the discs for himself?
4. Produce one sentence that succinctly describes why there is an ideal brake force ratio.
5. The brakes don't actually stop the car, so what does?
6. A racing car can decelerate at an average value of 16 m/s between 100 mph and 0. Determine the stopping time and distance.
7. Determine the difference in stopping distance for a vehicle on a flat road and a 10% gradient. The vehicle has a mass of 1300 kg, an Mu value front and rear of 0.46, a centre of gravity height of 0.42 m, a wheelbase of 1.9 m and an '*a*' value of 0.65 m and is operating with an ideal brake force ratio. Compare this with the situation when the brake force ratio is equal front to rear.
8. For a vehicle with the following data, determine the Mu value required for the vehicle to lift off the rear wheel under heavy braking.

Wheelbase	l	1.6 m
C of G rear	b	0.6 m
C of G height	h	0.28 m
Weight	w	200 kg

6.3 DIRECTED READING

Gillespie [11] has a whole chapter on braking and deals with the ideal brake force ratio together with many other braking effects.

Genta [12] offers a theoretical treatment and takes several ideas presented here onto further analysis.

Cossalter [14] again includes material on braking that should be easily accessible after completing this chapter together with its suggestions for reading and further work.

6.4 LEARNING PROJECTS

- For a collection of various vehicles ranging from high-performance racing cars and road cars to very modestly performing road cars, estimate the 0–60 mph braking times and distances and comment on the range of values found when compared with the range of other performance indicators such as 0–60 mph acceleration times.

- For a range of vehicles of interest, produce a plot showing ideal brake force ratio against vehicle mass and labelling each point with the vehicle type.
- Produce a model in Adams to demonstrate the stability under braking discussed in this chapter. Use the model to demonstrate the behaviour shown in Figs. 6.4 and 6.5.
- Aerodynamic downforce can significantly increase the vertical load on a tyre. Produce an analysis of how downforce affects stopping distance and braking times. You can start by approximating the downforce to an average value. This can then be refined to deal with the downforce as a function of speed. You can assume a quadratic dependence to get going. This can then be refined by conducting computational fluid dynamic (CDF) analysis of a real car to produce a downforce versus speed lookup table and use this to produce a much improved estimate of braking performance.

6.5 INTERNET-BASED RESEARCH AND SEARCH SUGGESTIONS

- Research the operation of a disc brake. What steps are needed to determine a relationship between brake pedal pressure and braking force developed at the wheel? (You may assume a value for Mu of the disc pad of 0.2 and should balance torques acting on the wheel and consider the ratio of piston sizes between the master cylinder and the calliper. Ignore expansion of the brake lines under pressure.)
- Power-assisted brakes are often used on road cars that mean that modest brake pedal pressures are able to generate large braking effort. Research the methods used to produce power assistance, explaining how they work. Why are some road cars fitted with electrical systems and others with pneumatic ones driven from the inlet manifold of the engine?
- Research how the ABS works.
- Produce a spreadsheet analysis of vehicle braking performance. The approach taken here should be similar in nature to the approach taken to straight-line acceleration in Chapter 4. You should start with a simple approach that determines basic performance. This should then be extended to include more sophisticated methods and considerations. You should be able to develop a similar numerical approach involving the determination of the time taken to slow down through each increment of speed loss. As with the straight-line example, you could include the effect of more subtle effects, such as rotating inertia, wind speed and downforce as a function of speed.

CHAPTER 7

Suspension Kinematics

7.1 INTRODUCTION

As a vehicle travels over terrain, undulations in the terrain must be accommodated. Even roads of very good quality cause the contact patch to rise and fall as the vehicle passes over them. We divide the treatment of this suspension movement into two separate areas; these are *suspension dynamics* and *suspension kinematics*. Suspension dynamics, which deals with the forces and motions that result from the road inputs, is the subject of Chapter 8. Here, we learn how, for example, the choice of spring and damper functions affect suspension performance and develop methods to optimise such values. The second topic, suspension kinematics is the subject of this chapter.

Suspension kinematics is the consideration of the motions of the suspension without explanation of their origin. For example, as the wheel rises and falls, it is normal for the camber to change. The way in which it changes is dictated by the geometry of the suspension. The position of the wheel within its available travel is determined by the suspension dynamics, but it is the kinematics that defines the map within which the suspension moves. For these reasons, it makes sense to consider the two topics separately.

A generalised suspension is shown in Fig. 7.1. Normally, the motion of the wheel up and down is controlled by rigid links giving precise control to the alignment of the wheel as it rises and falls. Clearly, the exact way in which it is controlled depends on the actual mechanism placed inside the 'cloud', and the layout the designer has given it.

In this chapter, we shall start with a long look at all the different kinematic behaviours that we seek to control, for example, camber change under bump and droop or roll centre location and kingpin inclination or castor angle, to name but a few.

Once we have an understanding of suspension kinematics, we shall be in a position to examine the different design approaches on offer and assess how well each configuration is able to provide good control of the outputs. Some are much better than others, but then again, it depends on what is wanted from the vehicle. All this is addressed in Section 7.3.

Performance Vehicle Dynamics
http://dx.doi.org/10.1016/B978-0-12-812693-6.00007-9

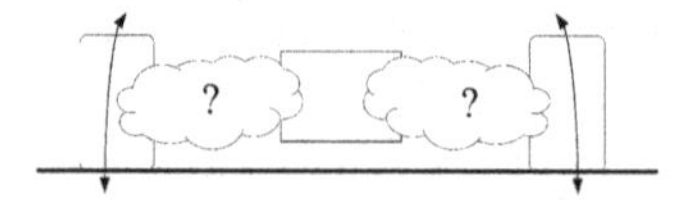

Fig. 7.1 A suspension controls the kinematic movement of the wheels.

7.2 SUSPENSIONS KINEMATICS

We shall start the consideration of suspension kinematics with a treatment of camber control.

7.2.1 Camber Control

Camber is the name given to the inclination to the vertical made by the tyre and is shown in Fig. 7.2. The presence of camber causes camber thrust as we met in Chapter 2. We define camber as being positive when the top of a wheel is further out than the bottom.

7.2.1.1 Fundamentals of Camber Kinematics

When a car corners, the body will roll, and weight will be transferred to the outside wheel. We have seen that camber thrust acts on a wheel experiencing camber and results in a force in the direction of lean. Fig. 7.2 shows a cornering car in the front view. The outside wheel is shown and clearly, to make the best use of camber thrust, we desire that this wheel goes into negative camber under roll. A suspension will normally be symmetrical in the front view, and this means that the inside wheel will experience positive camber that seems undesirable. However, since weight transfer means that much less vertical load applies to the inside wheel, and it generates very little lateral force. Thus, the gains resulting from arranging for negative camber on the outside wheel much outweigh the losses associated with the consequent positive camber on the inside wheel.

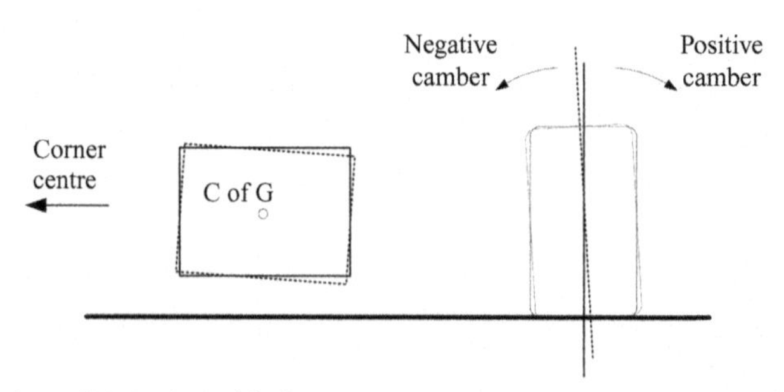

Fig. 7.2 Under roll, it is desirable for a suspension to cause negative camber.

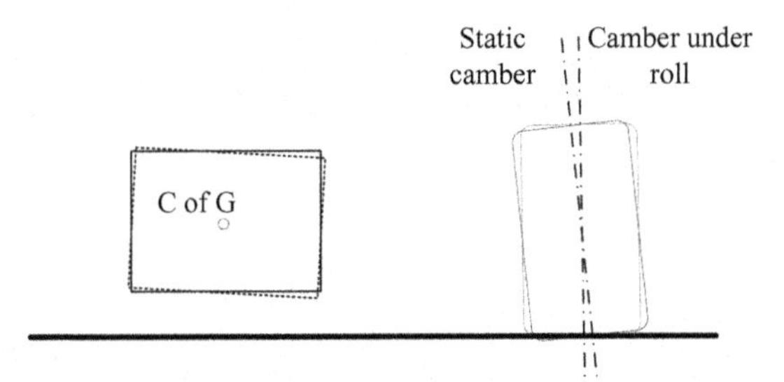

Fig. 7.3 Static camber allows for camber gain under roll.

In practise, it can be problematic to arrange for a suspension to behave in this idealised way, and a compromise is to arrange for *static camber* as shown in Fig. 7.3, in which the wheel is in negative camber normally, but under roll, the outside wheel camber increases until it is in the desired position. The fact that the wheel is in a poor orientation from a camber point of view when travelling straight ahead is not actually a problem since lateral force is not required at this time. However, a significantly cambered wheel compromises tractive effort, and so whilst it is common to make use of static camber at the front, it is of much less use at the rear.

As the wheel rises and falls, it is desirable to have the camber change such that the outer wheel goes into negative camber as the vehicle corners in order to make the best use of camber thrust. This control of camber, sometimes called *camber compensation*, is achieved by detailed design of the geometry being used. Some geometries are amenable to permitting camber compensation, whilst others are not. When analysing the kinematic performance of a given suspension, it is common to characterise the design with graphs such as that shown in Fig. 7.4 that can then be related to estimations of the amount of body roll and tyre characteristic graphs to settle on the ideal camber compensation approach.

In Fig. 7.4, we see that when the wheel is at *design point* meaning the undisplaced position (zero on the x-axis) where it is to be found under constant straight-ahead conditions, the camber is small and negative. Under roll, when the body has rolled sufficiently far for the wheel to have risen to around 6 mm of bump, we can see that the camber has risen to nearly zero, and the wheel will be upright.

7.2.1.2 Camber Considerations for Suspension Selection

When we come to consider the advantage that any given suspension configuration has to offer, we need to consider the following points.

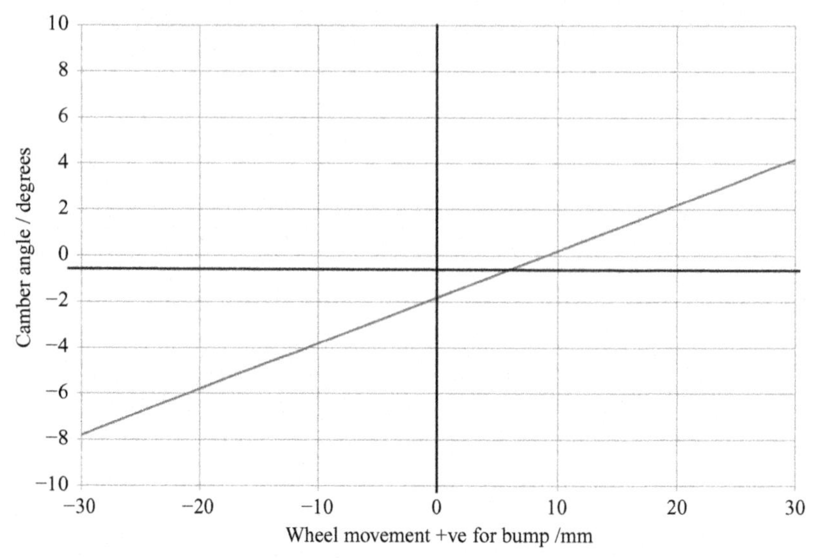

Fig. 7.4 Camber change as a function of wheel movement.

Firstly, the extent to which the given configuration facilitates camber control. Ideally, we should be able alter the design to gain any desired amount of camber change under roll. It should offer the option to design in static camber. The level of dimensional control necessary to obtain the benefits of camber management is demanding. Consideration of cambered tyre curves shows that as the camber angle changes by just a few degrees, the whole camber response has passed. In general, one needs to be able to control any variable to at least an order of magnitude better than one significant unit of the characteristic in hand. So here, where one degree is significant, we need to control camber to around a tenth of a degree. For this reason, suspension configurations that intrinsically offer stiffness because of their layout are to be preferred.

7.2.2 The Roll Centre

Anyone who has sat inside a car whilst it corners is aware that at the same time it will roll. Understanding roll is important because it will affect the performance of the tyre, for example, if the body rolls, then the tyre experiences camber change. This affects the amount of camber thrust. On top of this, there is the need to understand roll fundamentally. The very name *roll centre* makes it sound like it is the place about which the body rolls when cornering. Curiously, however, this is not the case, though, as we shall see, its name still has a logic to it.

To understand roll, we shall do exactly as one should when applying Newtonian dynamics. We want to understand how the body moves in response to cornering forces, and so, we shall resolve all the relevant forces acting on the body into an equivalent set acting on the centre of gravity. Any vertical component will cause the body to rise and fall, and any horizontal component will accelerate the car laterally and keep it in circular motion about the corner centre. Any torque will serve to rotate the car in the front view, that is, to roll it.

7.2.2.1 Kinematic Roll Centre

One of the earliest treatments of the roll centre was by Lanchester [1], which refers to a *sideways location* of a suspension. Dixon [8] gives an excellent treatment of the emergence of the roll centre concept, pointing out that the notion does not have its origin in any single paper but seemed to emerge generally and was, by the 1930s, in common usage.

Lanchester [1] noted that a force applied to a car as shown in Fig. 7.5 caused the car to roll as expected and that if the position of the force was high on the chassis, rotation would be in the anticlockwise direction, but if lowered sufficiently, it would be in the clockwise direction (in the view shown).

In between these positions, there is a *null point* where no rotation occurs. If one thinks this through by starting at the lateral force F at the tyre, it follows that this *sideways location* must be level with the point at which the lateral force F is applied to the chassis. This is because if it is above or below this point, then roll will occur, and here, it doesn't.

Under cornering, the centripetal force acts on the centre of gravity, and so if this *sideways location* could somehow be placed at the same height as the

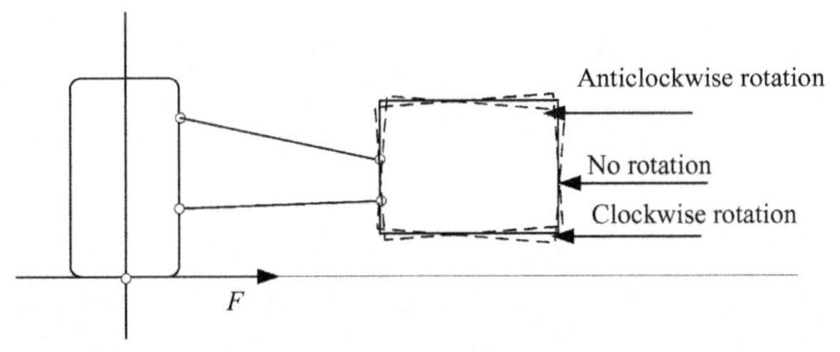

Fig. 7.5 Lanchester's *sideways location* of a suspension.

centre of gravity, there would be no moment and so no body roll. But where is this point and how does one find it?

To answer this, we start by making use of the Arnold–Kennedy theorem. The theorem is named after its two independent discoverers, Arnold (1872) and Kennedy (1886). It states the following:

> *The three instant centres shared by three rigid bodies in relative motion to one another (whether or not connected) all lie on the same straight line.*

The theorem can easily be proved by contradiction. Consider Fig. 7.6. In the figure, objects 1 and 2 are free to rotate with respect to a *ground* object, 3. Object 1 rotates with respect to 3 about position A, and object B rotates with respect to 3 about position B. If we now consider where the point about which object 1 rotates with respect to object 2, we could guess it to be at position C. If it was to be there, then the velocity of point C, considered as belonging to object 1, would be given by the vector V_1. If it was considered to be belonging to object 2, then its velocity would have to be V_2. Clearly, the same point can't have two velocities at the same time, and so, the only place it can be is where V_1 and V_2 point in the same direction. This can only be on the line A–B. Thus, all three centres of rotation lie on the same straight line as stated by the Arnold–Kennedy theorem.

To apply the theorem to a vehicle, we consider Fig. 7.7. We are dealing here with a car that is stationary, not travelling towards the viewer in the figure. Clearly, in a real analysis, the car will indeed be travelling, but for the time being, we consider this simplified case.

Points 7 and 8 are the points about which the tyres are considered to rotate in the front view with respect to the ground. We are assuming, for the time being, that the vehicle acts as if there is a hinge between each tyre and the ground.

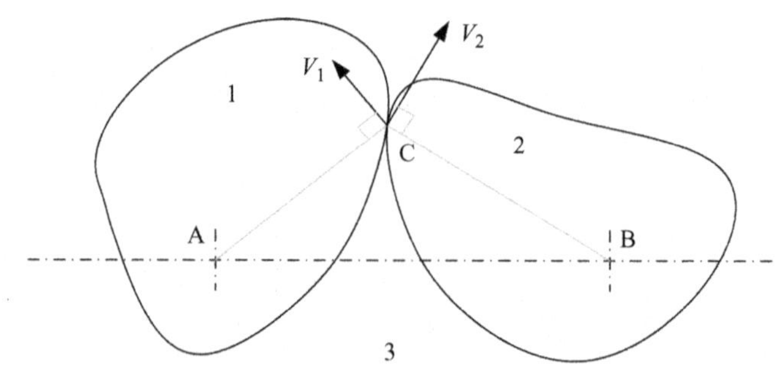

Fig. 7.6 Proof of the Arnold–Kennedy theorem.

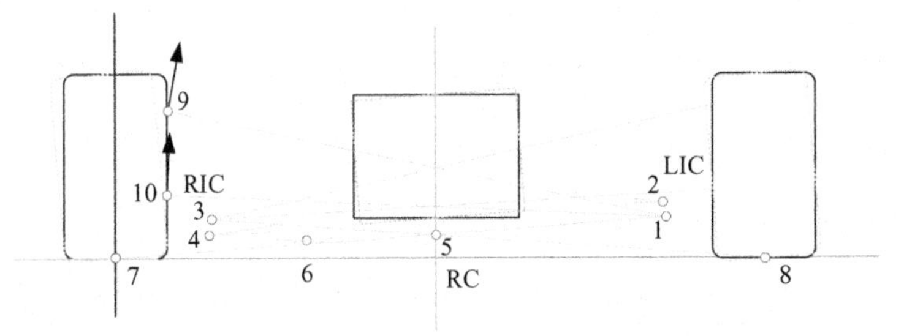

Fig. 7.7 Determination of the kinematic roll centre for a stationary car.

The velocity at points 9 and 10 are shown, and these are at right angles to the wishbone links by definition. Thus, if we now draw lines that extend the wishbones, for the wheel on the left in the figure, these lines meet at the left-hand instant centre (LIC), and this is the instant centre of that wheel with respect to the body. A similar process produces the right-hand instant centre (RIC). Thus, we have the instant centres of the wheel-ground and wheel-body combinations. Using the Arnold–Kennedy theorem, we can now draw a line from point 1 to 7 and know that the instant centre of the body ground must be on this line somewhere. Drawing in the line 3–8, we also know that the body ground instant centre must be on this line too, and the only place it can therefore be is at 5.

If we now allow the body to roll to the dotted position, the LIC will move, perhaps to position 2, and the RIC to position 4 giving a new kinematic roll centre at 6. This movement is sometimes called *roll centre migration*.

It initially sounds as if it is a serious matter. If the roll centre (meaning here the position about which the body rotates in the front view under cornering then) can move to position 6 and perhaps beyond, then the car body will continue to rotate about this point and could rise and fall significantly if, for example, the roll centre was to move much outside the track. In fact, this is not what happens at all for the vehicle at any forward speed. The analysis here is only applicable to a car at rest where the joint between the tyres and road may be reasonably approximated to be a *hinge joint*. Even at rest, this is a questionable assumption since under lateral load the tyre will plunge.

The website associated with this text has pictures taken at Oxford Brookes University which demonstrate exactly this. A formula student car is rocked manually back and forth whilst a time lapse photograph was taken. For modest forces, the car is clearly seen to roll about the roll centre as determined by the Arnold–Kennedy theorem, and experiment certainly seems to confirm that the body roll about the roll centre.

7.2.2.2 Kinematics of a Car With Forward Speed

A car is a free body, and wheels are free to scrub over the road surface, etc. Thus, if we want to understand how the car body rolls, we must replace all the forces acting on it by equivalent forces acting at the centre of gravity and determine the resulting motion. This is what using Newton's dynamic framework means.

In Fig. 7.8, we see that if a car rolls, then there will be a consequent movement of the contact patches. For real suspensions, such movements are very small, of the order of millimetres, but the important point here is that the contact patches are free to scrub in this situation and not in reality fixed as if hinged, as we assumed above.

Imagine a rolling tyre as shown in Fig. 7.9. In part '1', the tyre is in the straight-ahead position, and no lateral movement has been imposed. If we then move to part '2' where it has, we see that the tread is distorted in response. The segment B–C in part 1 corresponds to the segment D–E in part 2, where it has be displaced laterally. This causes a lateral transient force

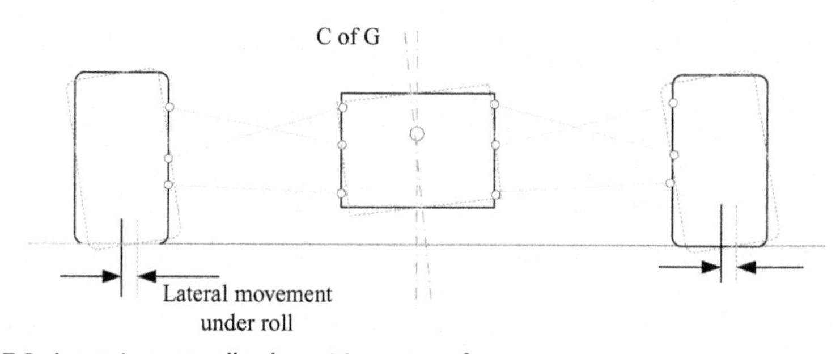

Fig. 7.8 A moving car rolls about it's centre of mass.

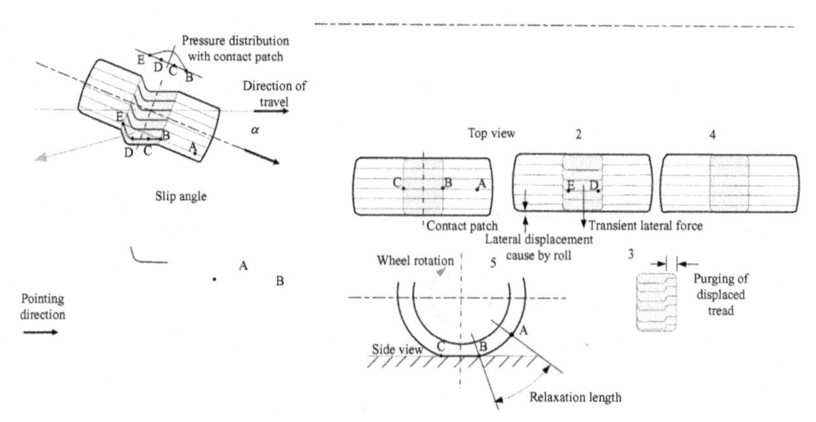

Fig. 7.9 The rolling tyre offers no resistance to track change.

transient because this situation does not pertain for long. As new rubber enters the contact patch, it does so from the laterally displaced position. Part 3 shows this process in progress. By part 4, the contact patch has been completely purged of deformed tread, and the transient lateral force has gone. All this is accomplished in what is often called the *relaxation length* as shown in part 5, a length approximately equal to the contact patch length but can be longer if the imposed displacement is larger or goes on for longer. At 60 mph, it will take approximately around 0.002 s for this to happen, and so, we conclude that lateral force generation at the contact patch in response to roll is negligible. Thus, for a vehicle in motion, the tyres will freely scrub over the road surface, they are not hinged to the surface and the body is free to rotate about its centre of gravity in the front view under cornering forces.

If any further proof is needed, on the website associated with the book, there are pictures taken at Oxford Brookes University that demonstrate this. A car was placed on the four-post rig with PTFE-loaded polythene between the tyres and the pads of the rig. In addition a high-frequency small amplitude sine wave was supplied, and the car was made to 'float' on the rig's pads. In response to a roll input, the car was clearly seen to roll about its centre of gravity. This was in marked contrast to the experiment above where the stationary car is seen to rotate about the Arnold–Kennedy determined roll centre.

7.2.2.3 The Force-Based Roll Centre

Armed with this understanding of the rolling car, we may now proceed to examine the significance of the roll centre from a force point of view.

7.2.2.4 Consideration of Symmetrical Lateral Tyre Force

We shall start by working out where the lateral force developed at the tyre is applied to the vehicle body. Clearly, this must depend on the geometry of the suspension being used. However, it is possible to replace any suspension with an equivalent single link transmitting a force and a torque to the body. We shall analyse this general case first and later see how each configuration relates to it.

In Fig. 7.10, we see that the general suspension is represented by the equivalent link 1–2. This link will transmit a force acting along its length, but it will also transmit the torque that results from the lateral force F, and we can see that $M_1 = Fa$ the way in which this torque is transmitted to the vehicle body needs careful treatment. The torque is transmitted by the linkages of the suspension. In the figure, we approximate all those elements above the link 1–2 to be transmitted through the imaginary link 3–4

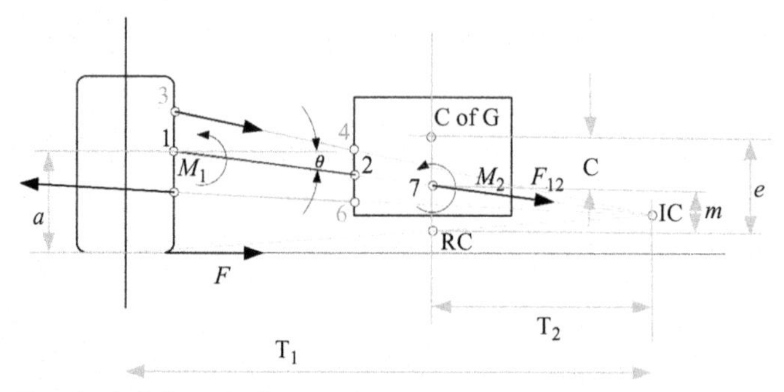

Fig. 7.10 The *single link* equivalent suspension.

and all those below to be transmitted through the imaginary link 5–6. These links have an alignment and must by definition point towards the instant centre, about which the wheel rotates in bump and droop with respect to the body. This is the rationale behind lines 3 IC and 5 IC. Since the vertical separation reduces as one moves from the wheel to the body, so too will the torque. At the instant centre, for example, the torque applied by the forces in the links 3–4 and 5–6 would be zero since there is no separation at this point. Thus, at the centre line of the car, the torque M_2 is given by

$$M_2 = M_1 \frac{T_2}{T_1}$$

Thus, it is the case that the force acting on the body, originating from the contact patch for F, is communicated through link 1–2, and this consists of two parts; firstly, the force in the link acting in line with it and the torque M_2. These act together to produce a torque acting on the centre of gravity of the body, M_{TOT}, given by

$$M_{TOT} = cF_{12}\cos\theta + M_2$$

But taking moments

$$Fa = M_1$$

$$F_{12}\cos\theta = F$$

So

$$M_{TOT} = c \times F + M_2$$

But from above,

$$M_2 = M_1 \frac{T_2}{T_1}$$

However,

$$M_1 = Fa$$
$$\Rightarrow M_2 = Fa\frac{T_2}{T_1}$$

And by geometry,

$$\frac{a}{m} = \frac{T_1}{T_2}$$
$$\Rightarrow a = \frac{T_1}{T_2}m$$
$$\Rightarrow M_2 = F\frac{T_1}{T_2}\frac{T_2}{T_1}m$$
$$= Fm$$

And so the total torque applied at the body is

$$\begin{aligned} M_{TOT} &= cF + M_2 \\ &= cF + Fm \\ &= F(c + m) \\ &= Fe \end{aligned}$$

This is a surprising result; it means that the force acting to rotate the vehicle body about its centre of gravity (which is communicated through the suspension) is given by the product $F \times e$. Looking again at the figure, we see the distance e is the vertical distance from the centre of gravity to the roll centre. Thus, the body will move exactly as if the horizontal force F is applied at this 'roll centre'. This agrees perfectly with Lanchester's [1] observation of a *sideways location* above.

Since this horizontal force, originating at the tyres, is applied to the vehicle body at the roll centre, we can see that changing the vertical position of the roll centre will have an effect on the suspension performance. The horizontal force at the roll centre produces a moment about the centre of gravity, and if the roll centre is lowered, the moment will be larger, and the body will roll more. Conversely if it is raised, the moment will be smaller, and there will be less roll. If placed on the centre of gravity, there will be no roll at all. This sounds desirable, but in fact, it isn't. To understand why we must first realise that in the treatment above, we have shown that a horizontal force at the tyre may be taken to be applied to the body, horizontally and the roll centre and that this will cause the body to roll. However, we have omitted the effect of the moment M_2 and the vertical component of the suspension force F_{12}. In Fig. 7.10, we see that it acts in an anticlockwise sense and so serves to raise the car body vertically upwards, whist the vertical component of F_{12} acts downwards. In addition, we have not considered the effect of the other side of the car or the fact that the tyres do not only have a horizontal force but also a vertical one. On top of this, for the cornering car, there will be very different magnitudes for both the vertical and horizontal loads left to right.

Thus, the analysis so far is not complete; it does however explain Lanchester's *sideways location* and further shows that the height of the roll centre affects suspension performance. Placing the roll centre on the centre of gravity will eliminate roll, and the further it is from it, the large roll will be.

A last point of note is that the location of the roll centre is exactly the same as determined by the Arnold–Kennedy theorem even though this is not the centre of rotation in roll. This is because in the Arnold–Kennedy theorem, the lines were drawn to show directions at right angles to motions, and in the force-based roll centre analysis, lines are drawn to show the line of action of forces. These are the same lines and produce the same construction even tough the results have different meaning. This fact is a likely explanation for confusion that is sometimes found surrounding the meaning of the roll centre where it can, wrongly, be thought to be the centre of rotation under roll for the vehicle at speed. If it were, then this would be a good reason to design the suspension to avoid significant lateral migration since it would confer jacking. In fact, lateral migration of the roll centre is not important and has no meaning for the vehicle at any speed.

7.2.2.5 Consideration of Asymmetrical Combined Tyre Loads

We start the consideration of asymmetric and combined loading by first resolving the vertical and horizontal tyre loads into components that act towards the instant centre and perpendicular to a line form the contact patch to the instant centre. This is shown in Fig. 7.11 for one side of the car. Once we have this foundation, we will consider the whole axle.

The vertical and horizontal forces F_{YL} and F_{ZL} are resolved into components F_1 and F_2. These are perpendicular and parallel to a line from the contact patch to the LIC. The force F_2 generates no tension or compression in the suspension links but causes movement in the wheel compressing the suspension spring. The link 3–4 is the suspension link and can only

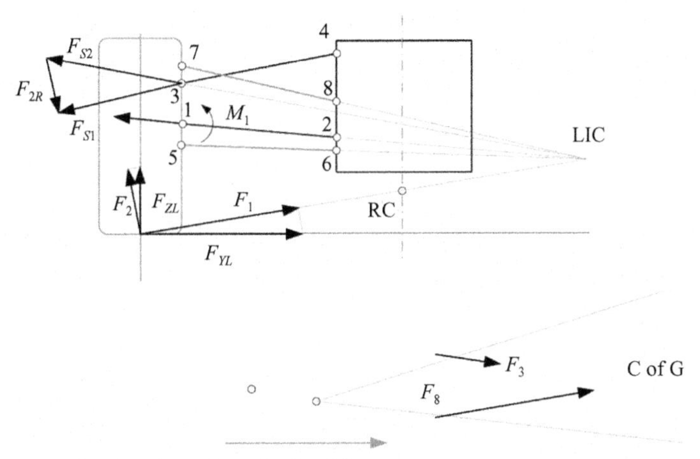

Fig. 7.11 Resolution of combined wheel forces.

produce an axial force. Thus, the component F_2 must be reacted by the force F_{2R}. In generating this force, the suspension link also produces the force F_{S2}, and this force produces tension in the suspension links 7–8 and 5–6. These can be equated to a force and moment acting in the equivalent suspension link 1–2. Thus, the wheel force F_2 is completely reacted in a loop consisting of the pushrod-upright-links-body. F_2 therefore plays no role in the vertical, horizontal or rolling motion of the body.

We now apply this analysis to a car in the front view, where we deal with asymmetry and consider both the vertical and horizontal tyre forces.

In the figure, we start with the forces F_{YL} and F_{ZL}. We resolve these into two components, F_1 and F_2. The component F_2 is tangential to a line from the contact patch to the LIC. This means that it produces no force in the suspension links that is transmitted to the body. It does produce a compression force in the spring linkage, but this is reacted within a loop consisting of the wheel, the spring linkage and the suspension linkages and is not transmitted to the body. The moment produced by the spring linkage is exactly balanced by the moment produced by the suspension linkages, and no net force is applied to the body.

The component F_1 is applied to the body at the roll centre by the force F_3. This must be the case because the horizontal component of F_1 seen above must be the force F_4 and the force F_3 is the only force that acts in the direction required and has the horizontal component F_4.

The right-hand side of the diagram represents the inside of the vehicle, where the vertical tyre load and consequently the horizontal force are smaller because of weight transfer. Here, we have forces, F_{ZR} and F_{YR}. These may be treated in exactly the same way to produce the force F_8. The lower part of the figure shows the forces F_3 and F_8 resolved into horizontal and vertical components F_{JACK} and F_{ROLL}. The force F_{ROLL}, as we have seen, makes a moment with the centre of gravity and acts to roll the vehicle body. The force F_{JACK} acts vertically on the body and serves to raise it vertically. The position of the roll centre will clearly affect the magnitude of these components and so the performance of the suspension; this is the next topic for consideration.

If we consider for a moment how the application of the analysis in Fig. 7.12 would work out for the symmetrical case we considered in Section 7.2.2.4, we can see that forces F_3 and F_8 would be equal in magnitude, and there would be no jacking force, as was concluded then. Thus, we see that the jacking force is a direct result of two things; the inclination of forces F_3 and F_8 together with the difference in their magnitude.

We can now advance the analysis presented above to the situation where roll has begun to evolve and the car body makes a significant angle to the road.

This is shown in Fig. 7.13. There are now two separate points *g* and *h* that are the points at which the forces F_1 and F_6 are communicated to the body as forces F_3 and F_8. It is at this point that again the difference between the kinematic roll centre and the forced-based roll centre becomes clear. In the kinematic case, it is point *j* that matters; this is the point about which body rotation occurs in the front view for the stationary car. In the force-based case, where we are dealing with a car at speed, it is points *g* and *h* that matter, and body roll occurs about the centre of gravity. Point *j* has no significance

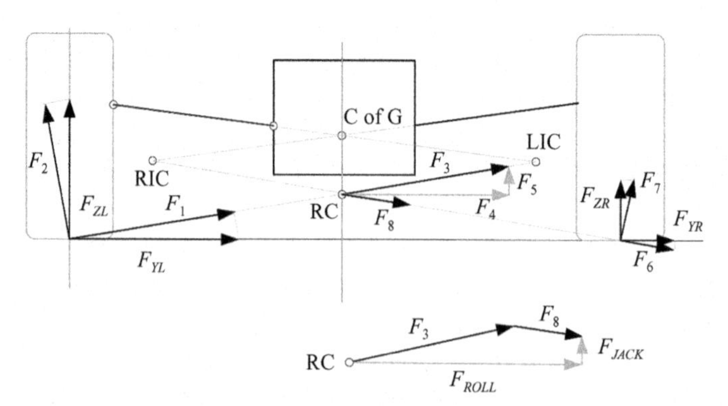

Fig. 7.12 Asymmetric roll with combined loading.

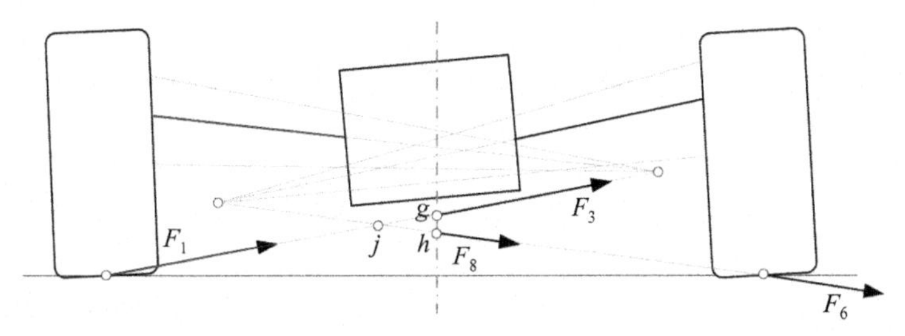

Fig. 7.13 Asymmetry during roll.

for the rolling car. The amount of roll and jacking is determined by the horizontal and vertical components of F_3 and F_8 as before.

7.2.2.6 Control of Roll and Jacking Using Roll Centre Position

We now consider the effect of moving the position of the roll centre in Fig. 7.12 and can see that there are five interesting possibilities for its location.

Roll Centre Above the Centre of Gravity

The large angle made to the horizontal in this case will make for a large jacking force in the upwards direction, and the car body will rise significantly when cornering. This is undesirable from the point of view of weight transfer, and, if relevant, aerodynamic downforce since increasing the ride height will compromise downforce. The rolling component will be a very significant fraction of the size of the lateral forces if one considers correctly the proportions of the diagram. Since the rolling force is above the centre of gravity, the car body will roll 'into' the turn, meaning the top of the car will move towards the corner centre. The negative effects of jacking outweigh any increase in perceived comfort, and normally, vehicles are not designed in this way.

Roll Centre on the Centre of Gravity

The rolling force makes no moment about the centre of gravity, and the car does not roll in a corner. The jacking force is still significant, and the body rises, which is undesirable.

Roll Centre Between the Centre of Gravity and the Ground

As the roll centre is placed lower and lower to the ground, the jacking force reduces, and the undesirable upwards movement of the body decreases.

At the same time, the amount of roll increase. This is easily compensated for by stiffening the antiroll bar, whereas jacking has no solution other than lowering the roll centre. For these reasons, a low roll centre is desirable.

Roll Centre on the Ground

It is entirely possible to place the roll centre on the ground, and some designers do this in order to eliminate jacking. Others prefer to keep the roll centre slightly above the ground on the basis that once the roll centre divides into the two force application points *g* and *h* in Fig. 7.13, it is better to keep their vertical locations from moving much so that the car equilibrium does not change. It is desirable to keep the roll centre from moving below the ground, see below, and having it a little above at the design point is an easy way of achieving this.

Roll Centre Below the Ground

With the roll centre below ground, the rolling moment becomes even larger, and the vehicle will roll more. The jacking force now has a downward magnitude however, and so, the vehicle body will move downwards under cornering. Given that the body will already have been designed to ride as low as possible for reasons of weight transfer minimisation, aerodynamic performance and clearance of the expected terrain, negative jacking is very undesirable. Suspensions are not normally therefore designed this way.

7.2.2.7 Roll Centre Considerations for Suspension Selection

From the material above, it is clear that a desirable attribute of any given suspension geometry is that it facilities control over the roll centre location. We are very likely to want the roll centre to be low to the ground to avoid jacking and also that the force application points for the left and right side don't move significantly up or down during roll. If it is desired to produce and antiroll suspension, despite the attendant jacking, then we need the ability to control the roll centre location to near the centre of mass. When we come to consider suspension geometries, the ability of any given geometry to control the roll centre location in this way will need attention.

7.2.3 Pitch and Squat Analysis

It is in general true that any analysis that can be applied in the front view can also be applied in the side view. In Fig. 7.14, we see exactly the same analysis as above, but this time applied to the side view. In this case, we need worry

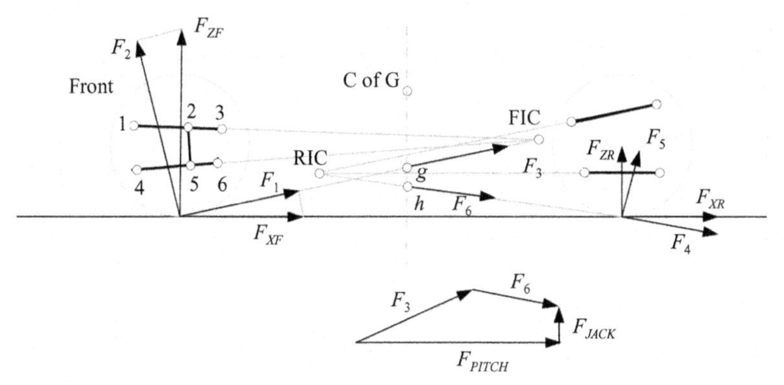

Fig. 7.14 Antisquat and antidive geometry.

about the equivalent of track change in the side view since the rolling wheels offer no resistance to wheelbase change. The force-based approach is just as applicable, and we start with the front wishbones mounted at the body at '1–3' and '4–6'. The outboard mounting to the upright is '2'. Thus, as the wheel rises and falls, the velocity of point '2' must be perpendicular to the line 1–3, and so, the front instant centre, FIC, must lie on the line extending 1–3. By a similar argument, it must lie on the extension of the line 4–6, and this determines it. Point '*g*' is found using the same arguments as above for the front view roll centre, and the force F_1 is applied to the body at this point as F_3. Similarly, force F_4 is applied to the body as F_6. These are then combined in the insert diagram where the resultant horizontal force makes a moment about the centre of gravity.

This serves to pitch the car, causing it to rotate in the side view to be front end down, as is familiar to any occupant of a decelerating car. The vertical component of F_3 and F_6 jacks the car. Just as before, the vertical height of the points *g* and *h* dictates how much jacking and pitching result. The five regions of roll centre location apply exactly as before, and pitching can be eliminated by placing the points *g* and *h* on the centre of gravity. However, this will be at the expense of much increased jacking. Placing these points on the ground, or at least near to it, ensures no jacking, and the resulting pitch can be controlled by suspension spring stiffness. In the case of roll in the side view, the antiroll bar can be used to provide additional roll stiffness without adding bump stiffness. In the side view, this is not possible since making an antipitch bar requires the communication of very small movements over long distances with the vehicle, and mounting stiffnesses cannot be made practically high enough. This is less of a problem since body roll causes camber change that needs limiting to small values, but pitch causes only a tiny amount of upright rotation in the side view that is not problematic anyway. Pitch causes bump and droop front and rear, and it is important that the camber change under this bump, and droop is controlled using the suspension geometry.

There are three commonly used applications for the ideas set out in Fig. 7.14; these are considered below.

7.2.3.1 Antisquat Geometry

This is the name given to the tendency of the rear end of an accelerating vehicle to lower itself. Firstly, we notice that the horizontal force is coming only from the driven axle. If this is the front axle, then force application point g is the relevant one. In this case, the tractive force would be in the opposite sense to F_{XF} and would be communicated as the force F_3 of negative magnitude, which would rotate the front end of the car upwards and the rear down. If the driven axle is the rear, then force application point 'h' is used, and the pitch under acceleration is in the same sense.

7.2.3.2 Antidive Geometry

Sometimes, vehicle dynamicists describe a suspension as 'antidive', meaning that the front axle is designed to prevent the front of the vehicle from descending under braking. In this case, it is point g that controls the effect, and its position is determined by the height of the front instant centre. Again, placing point g on the centre of gravity will eliminate dive. However, elementary consideration of the situation shows that this would mean the anchorages on the vehicle body would have to make a significantly upward slope as one travels backwards. This in turn means that they would carry significant bump force through them that would not be communicated to the body by the suspension spring. This is very undesirable since it leads to unnecessarily strong linkages and a much rougher ride.

7.2.3.3 Antipitch Geometry

Antipitch is the name given to a suspension geometry, front and rear that prevents pitch under braking. This involves placing the force application points g and h on the centre of gravity. As before, this produces greatly increased jacking, and most designers would find this too disadvantageous.

Lastly, it should be noted that whilst it would be irrational to have different suspension geometries on the left and right side of a vehicle, it can be entirely logical to have different suspension geometries front to rear. When this happens, the pitch centre can still be determined using the approach shown in Fig. 7.14, but the construction for each end must clearly be the appropriate one for the geometry being analysed.

Simultaneous Elimination of Pitch and Jacking

In the front view, it is very unlikely that a designer would accept an asymmetric suspension design. This would mean that left-handed corners would be taken in a different way to right-handed ones. An exception would be oval racing although the rules don't normally permit much asymmetry. In the side view, we have no such constraints, and the designer is free to make the rear different to the front if desired. This option makes possible the simultaneous elimination of pitch and jacking (Fig. 7.15).

We can eliminate jacking by ensuring that the sum of F_3 and F_6 in the vertical direction is zero. This can be achieved, as before, by making them horizontal. However, if a different value of θ and ϕ are permitted, then we can also achieve no jacking as long as

$$F_3 \sin\theta + F_6 \sin\phi = 0 \tag{7.1}$$

Noting that for the decelerating car with brake bias F_1 of the total applied to the front and total braking force $F_{TOT} = F_{XF} + F_{XR}$ we have

$$F_3 \cos\theta = \zeta F_{TOT} \quad (F_3 = F_1)$$

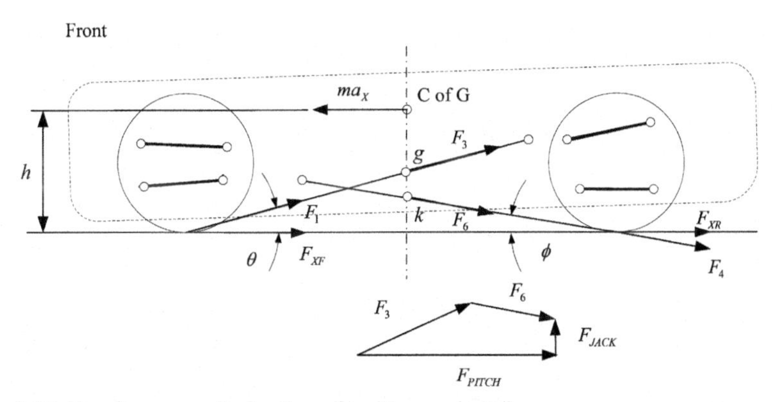

Fig. 7.15 Simultaneous elimination of jacking and pitch.

and

$$F_6 = \cos\phi = (1-\zeta)F_{TOT} \quad (F_6 = F_4)$$

Substituting these into Eq. (7.1),

$$\frac{\zeta F_{TOT}}{\cos\theta}\sin\theta = \frac{(1-\zeta)F_{TOT}}{\cos\phi}\sin\phi$$

$$\Rightarrow \frac{\zeta}{1-\xi} = \frac{\tan\phi}{\tan\theta}$$

Thus, if this relationship is made to hold, there will be no jacking. To eliminate roll, we need to ensure that the weight transferred (gained) at the front is exactly opposed by the vertical component of F_1 and the vertical component of F_4 exactly opposed by the weight transferred (lost) at the rear. Taking moments about a point on the ground under the centre of gravity,

$$ma_X h = F_1 \sin\theta\frac{l}{2} + F_4 \sin\phi\frac{l}{2}$$

From above, substituting for F_1 and F_4,

$$ma_X h = \frac{\zeta F_{TOT}}{\cos\theta}\sin\theta\frac{l}{2} + \frac{(1-\zeta)F_{TOT}}{\cos\phi}\sin\phi\frac{l}{2}$$

$$\Rightarrow \frac{2h}{l} = \zeta\tan\theta + (1-\zeta)\tan\phi$$

Thus, if this relationship holds, there will be no pitch, so substituting from above for no jacking,

$$\frac{2h}{l} = \frac{\zeta(1-\zeta)\tan\phi}{\zeta} + (1-\zeta)\tan\phi$$

$$\Rightarrow \tan\phi = \frac{h}{l(1-\zeta)}$$

Similarly, deriving an expression for θ,

$$\tan\theta = \frac{h}{l\zeta}$$

Thus, we achieve no pitching and no jacking if both the following hold:

$$\tan\phi = \frac{h}{l(1-\zeta)} \quad \tan\theta = \frac{h}{l\zeta}$$

Such a vehicle would have suspension linkages with considerable inclinations, and this would lead to significant forces transmitted through, them making for larger linkages to resist them and a rougher ride, as some undulations in the road are transmitted to the body via the unsprung path. This may well make such a design undesirable depending on the application.

7.2.4 Modelling roll and pitch

One can use the force diagram above to easily estimate the amount of pitch, roll or jacking that would result from a given suspension. One needs a knowledge of the roll gradient (meaning how many degrees of roll result from a given lateral acceleration), the pitch gradient (same but for pitch) and the vertical suspension rate front and rear. With these known, the values of the forces F_{JACK}, F_{ROLL} and F_{PITCH} are used to produce roll, pitch and jacking displacements.

Another very common method is to use multibody code such as Adams or other suspension simulation software and make informed adjustments using the knowledge above to achieve desired and acceptable design performance.

7.2.5 kingpin Inclination

In the suspension of an axle with steering, there will need to be an axis about which the wheel rotates as it steers. One might have this axis arranged vertically, and indeed, a lot of people think it is. In general, however, it is designed to be at a small angle to the vertical, much as shown in Fig. 7.16. The name derives from the heavy steel pin used on agricultural vehicles but is still used today even when no actual 'pin' is used.

In the figure, we see the straight-ahead position shown on the left. As steering is applied, we move to the situation shown on the right. The inclination of the kingpin axis means that as steering is applied, the bottom of the tyre must necessarily move downwards a little, and indeed, the figure shows the contact patch on the right has moved a little below ground. The more kingpin inclination there is, the larger the effect will be. It makes no difference whether the steering is to the left or right; the wheel moves downwards in both cases. Clearly, the wheels cannot in reality move downwards—instead, there is a vertical thrust force generated causing the body of the car to rise. This effect biases the wheels towards the straight-ahead position, and if the steering is released, it tends to realign to the straight-ahead position. When a car is at rest, the friction of the tyre on the road generally

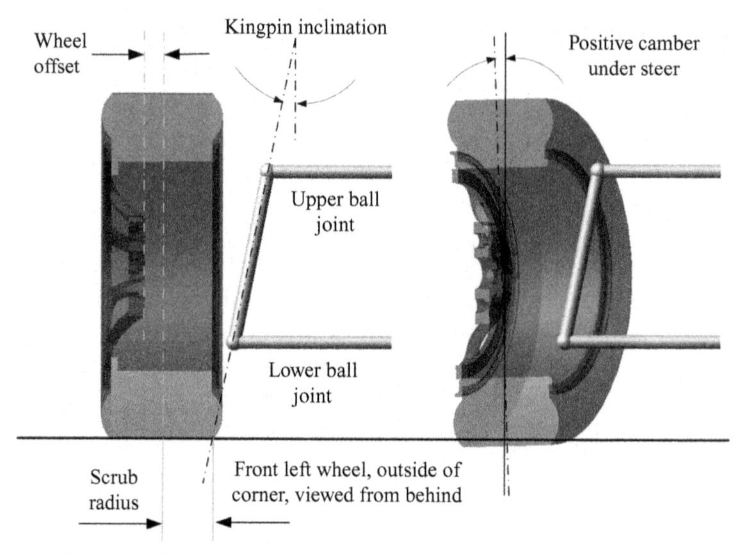

Fig. 7.16 Kingpin inclination.

prevents the tyres returning to the straight-ahead position after a steering input. However, when the car is rolling, resistance to steering movement is much less, and the self-return action can easily be felt. Most people are familiar with this self-centering action, and drivers often let the wheel slip through their hands when wanting the wheel to return to the straight-ahead position, even though it is considered by some to be poor practise for safe driving. With this force feedback coming through the steering wheel, drivers are aware of the steering wheel position in a way other than of simply being able to see it. Indeed, drivers don't generally look at a steering wheel; they know and can feel what steering demand they are currently making.

From this, we conclude that there is no 'correct' setting for kingpin inclination. However, the weight of a human and of a vehicle, the friction of the rubber and other affects all conspire together to affect the design. Once a car has less than around 4 degrees of kingpin inclination, the self-centring action is too weak, whilst much above 10 degrees means that it is too strong for the driver to control accurately and becomes tiresome with time. Kingpin inclination is not the only source of self-alignment in the steering system; however, as we shall discover later, it is an important source since it operates all the time. In cars with high aerodynamic downforce, the amount of self-centering action varies with speed. This is because at high speed, a greater vertical load is placed on the wheels, and so the self-aligning action caused by the kingpin inclination is proportionately greater. Kingpin inclination is therefore introduced to a suspension to provide a low-speed self-centering action, which aids good driving.

However, in addition to this benefit, kingpin inclination also brings a disadvantage. It has the very serious drawback is that it introduces positive camber to the outside wheel when a car is cornering. Scrutiny of the right-hand part of Fig. 7.16 shows this. The positive camber is on the outside wheel, the one with much increased vertical load due to weight transfer. In high performance, this is serious, but even for a vehicle of modest performance, there is no need to allow this to happen since it is easily cured by the introduction of castor.

7.2.6 Castor

To introduce castor, we incline the kingpin axis when viewed from the side. This is shown in Fig. 7.17, where the castor plane is inclined at a small angle to the vertical. The castor axis cuts the ground plane in advance of the contract patch. The effect takes its name from to design of small undriven wheels on devices such as a supermarket trolley where the wheels, actually called castors, always trail behind the axis about which they steer.

The beneficial effects of castor can be seen in the figure. On the right, the wheel is shown with no steering input pointing straight ahead. On the left, it

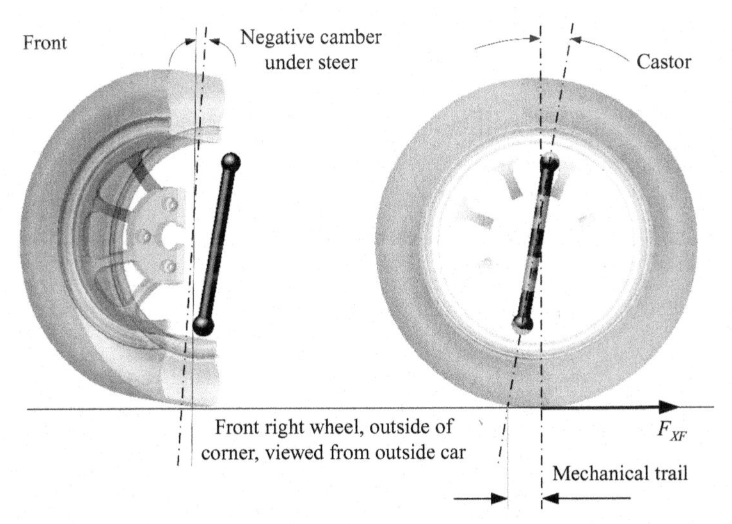

Fig. 7.17 Castor.

has been steered, and we clearly see that the castor angle has introduced negative camber. There is no kingpin inclination in the figure.

If we now compare the effects in Figs. 7.16 and 7.17, we can see that if kingpin is introduced first, undesirable positive camber results from steering but if castor is then added in, negative camber is introduced. By including both camber and castor, the designer can have the benefits of kingpin inclination but offset the disadvantageous camber gain. For this reason, it is common to introduce approximately equal amounts of both.

Looking again at Fig. 7.17, we can see that under the steering input the wheel has risen slightly. This is a source of instability since in the absence of a steering linkage, the wheel would quickly steer itself to find the point at which it has risen the most with respect to the body, seeming to offset the stability introduced by kingpin inclination. However, thinking of the wheel on the other side from the one shown, that will have moved downwards, and the effect across the whole axle will be very small. Thus, the stability introduced by kingpin inclination is not much removed by the instability of castor, but the positive camber is.

7.2.7 Mechanical Trail

In Fig. 7.17, on the right, we see the parameter, 'mechanical trail', defined. The presence of castor leads to mechanical trail and increases as more castor is added. The braking force F_{XF} is shown, and if we imagine that the vehicle is being steered whilst braking, then this force will produce a moment about the steering axis, serving to bring the wheel to the straight-ahead position. The wheel will be stable. Indeed, if the steering linkage fails and steering is lost, the front wheels will keep the car pointing ahead, and braking will be controlled. For this reason, it is normal practise to include some castor.

Considering again Fig. 7.17 but imagining now how it would work for tractive effort on a front-wheel-drive vehicle with pure castor, here, the direction of F_{XF} would be reversed, and in this case, tractive effort will as a result negate the castor action. It doesn't produce instability since the forward pointing force always points towards the castor axis.

7.2.8 Wheel Offset and Scrub Radius

Referring to Fig. 7.16, we can see two more parameters defined. Firstly, the 'wheel offset'. This is the distance between the face of the rim that mounts onto the hub and the centre line of the tyre. It is a property of the rim itself and is important in that it contributes with kingpin inclination to give the 'scrub radius'. If one is seeking to achieve a given scrub radius, it must be accounted for.

The wheel offset is the distance between the point on the ground where the kingpin axis cuts the ground and the centre line of the tyre. If the scrub radius is large, then as the wheel is steered, the tyre takes a wide circular path around the kingpin axis. If instead the offset is zero, often called centre point steering, then the wheel rotates about the centre of the contact patch as it steers. Such a wheel takes up much less working space inside, which can be a consideration within bodywork.

The scrub radius is important when considering steering. Imagine a vehicle heading forwards. When the brakes are applied, the braking force will make a moment about the kingpin axis. This will be exactly balanced by the same moment in the other wheel, and the force within the steering links will communicate the force from one side to the other. Thus, whilst there is a large tension force within the steering links, there is no tendency to steer. However, if the suspension is not symmetric from one side to the other, perhaps because one side is in bump and the other in rebound, then the scrub radius on one side may be larger than the other, and the combined braking effort will produce a steering input. In a front-wheel-drive vehicle, accelerating hard whilst cornering, the same effect may again produce a steering input. Such a situation is called *torque steer* or *brake steer* and is clearly undesirable.

7.2.9 Steering Geometry

7.2.9.1 Ackermann Geometry

In Chapter 5, we met the idea of Ackermann geometry. It describes the very desirable vehicle steering arrangement in which the corner radius, wheelbase and steering input are related by $l = \partial R$. Fig. 5.2 shows the geometry for a bicycle model. However, a real suspension normally has four wheels, and in this chapter on suspension kinematics, we need to see how this geometry can be achieved. Fig. 7.18 shows all the possibilities for a steering geometry. In the figure, we see that the Ackermann geometry is in the centre and is, in fact, only one member of a family of possibilities. In the case of the Ackermann geometry, the lines drawn at right angles to the front wheels must meet at the corner centre level with the rear axle, called *full Ackermann*. However, they could be made to meet anywhere on the line *a–b*. If they do so, on the corner side of the vehicle, it is said to be Ackermann geometry but not *full*. It is sometimes expressed as percentage or fraction often based on

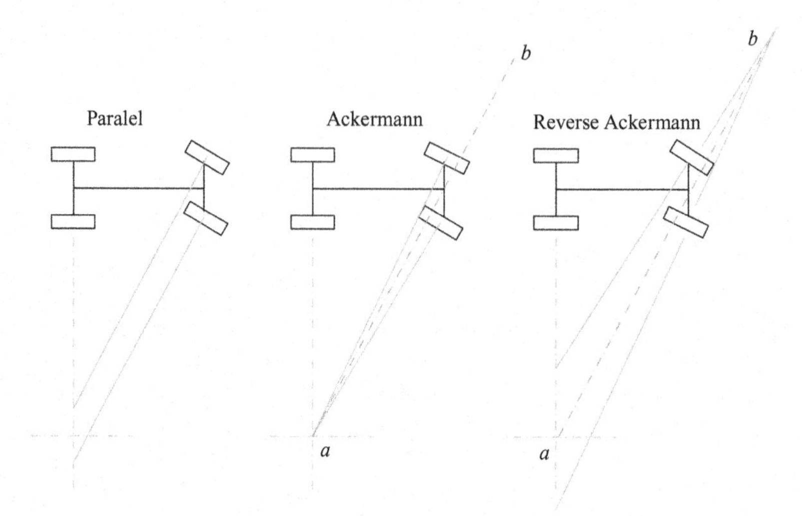

Fig. 7.18 Steering geometry and Ackermann angles.

how far forward along the centre line of the car the meeting point has moved. If the lines meet on the other side of the vehicle, the geometry is described as *reverse Ackermann* again sometimes expressed as percentage; 100% means the lines shown meet one wheelbase in front of the front axle. In between this is the special case of parallel geometry, shown on the left.

The steered axle detail often used to give Ackermann geometry is shown in Fig. 7.19. Here, the line *a–b* and its counterpart meet at the centre of the rear axle. As long as this it true, it is often claimed Ackermann geometry will result. In fact, this is not the case, and the geometry only provides an approximation to Ackermann geometry. It is a very good approximation for small angles, but much beyond around 15 degrees, it starts to deviate. It is a simple enough matter in CAD or a kinematics package to lay the steering out at

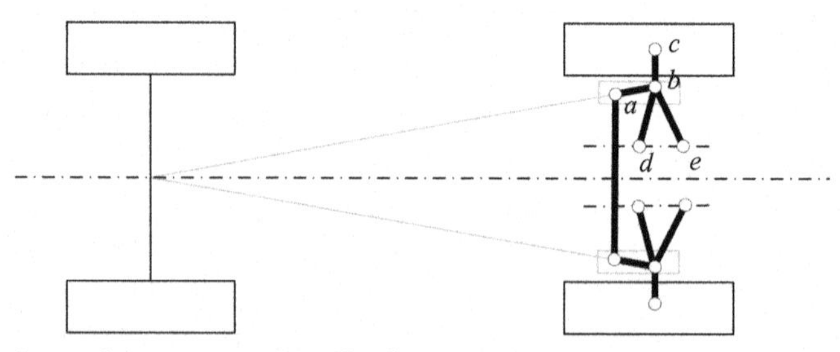

Fig. 7.19 Approximate generation of Ackermann geometry.

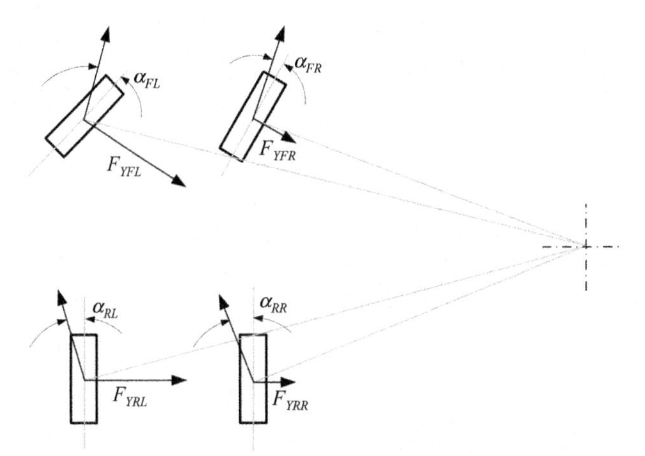

Fig. 7.20 Reverse Ackermann for racing.

various positions and determine where the tangent lines cross and so how close the approximation is.

In road cars where cornering speeds are generally low enough that weight transfer is not overwhelming, the Ackermann geometry is a good solution. However, in racing, the conditions are different. Both slip angles and weight transfer are, or at least should be, much larger. Most of the load will be on the outside wheels, and they are therefore able to generate more lateral force. This can only happen if they are at a larger slip angle, as shown in Fig. 7.20. For the inside wheels, the reverse is true. Thus, the outer tyres should be steered more than the inner, exactly as the reverse Ackermann steering convention provides.

There would be benefit in having the rear outer wheel 'steered' more than the rear inner, too, but without a steering system for the rear axle, there is only roll steer from the suspension movement available for creating this effect and it is difficult to generate large differences in slip angles between the rear wheels. Thus in racing, more is gained using 'anti-Ackermann' steering at the front axle than is lost through the attendant drag at low-speed corners.

7.2.9.2 Bump Steer

When a steered wheel rises or falls from its design point, it is possible for it to steer if the suspension geometry is not correctly designed. In Fig. 7.21, we see how this can happen. In the top view, the steering arm *c–d* is seen to be

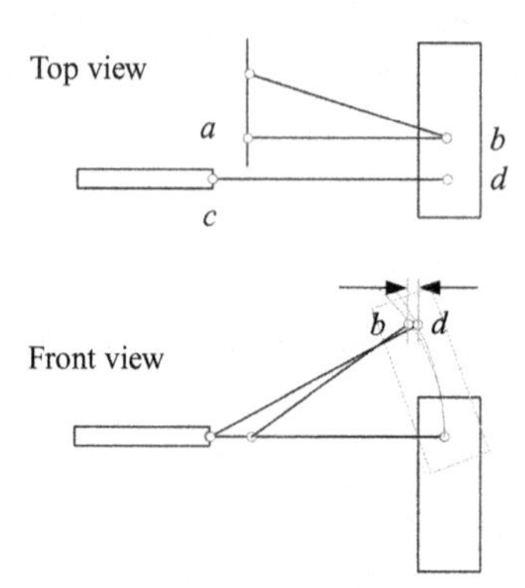

Fig. 7.21 Suspension geometry resulting in bump steer.

longer than the wishbone link *a–b*. The consequence of this in bump is seen in the front view.

Here, point *b* follows an arc of smaller radius than point *d*. Thus, point *d* becomes further out than point *b* under bump, and as a result, the front wheel will steer. The amount of movement shown in the front view is greatly exaggerated to make the effect clear. In large open-wheel racing cars, the effect must be made negligible. One method of determining the amount present is to support the car on stands, remove the suspension springs, attach a mirror to the disc and then shine a laser onto the mirror. If it is then raised and lowered, movement of the reflected spot betrays any bump steer present. One approach to adjusting it is to clamp the steering rack to its mounting on a shim. Once the measurement is taken, the shim can be replaced by one of exactly the thickness required to eliminate bump steer. This makes the adjustment easy.

A similar effect can be experienced at the rear where the steering link is replaced by a toe link, keeping the upright in the correct position. If the toe link makes a different arc under bump, then bump steer will result at the rear and is equally undesirable.

7.2.10 Roll Steer

Roll steer is the term that describes any steering of a wheel that results from roll of the chassis. The nature of the effect comes from exactly the same geometrical effects as bump steer. In that case, the steering input results from potentially just one wheel rising or falling in response to a bump in the road surface. For roll steer, all wheels are involved; the outer two rises with respect to the body as it rolls, and the inner two falls.

7.2.11 Wheel Rates Spring Rates and Installation Ratios

A vehicle suspension will include, at some point, a spring to provide compliance to the wheel as it rides over the road surface. In some configurations, the spring may be a very direct connection running from the body directly onto the control arms. In others, it may be via a more complex path. In Fig. 7.22, such a situation is shown. The wheel can be thought of as having its own spring rate, as distinct from the rate of the suspension spring. If, for example, there is a great deal of mechanical advantage in the linkages, it may well be that the wheel rate is very much less than the spring rate. We define the wheel rate as being the vertical force applied at the wheel divided by the resulting displacement. From Fig. 7.22,

$$k_{WHEEL} = \frac{F_W}{\Delta_{WHEEL}}$$

When one gets into a car and it settles on its springs to a slightly deflected position, it is the wheel rate that governs the movement.

The linkage transmitting wheel displacement into spring compression will result in a spring deflection Δ_S, and there will be a relationship between Δ_S and Δ_W called the installation ratio.

We define this installation ratio (sometimes called motion ratio), IR, as

$$IR = \frac{\Delta_{SUS}}{\Delta_{WHEEL}}$$

Often, when designing a suspension, it is useful to be able to determine the spring rate required to provide a given wheel rate. A relationship between the spring rate and wheel rate can be derived as follows:

Firstly, balancing moments about point '*c*'

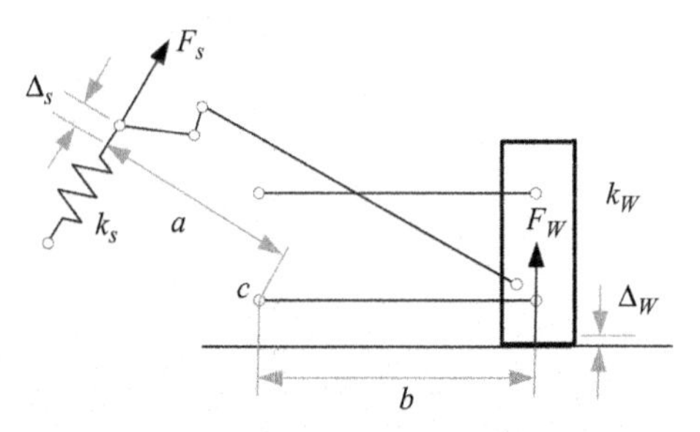

Fig. 7.22 Wheel rates, spring rates and installation ratios.

$$F_S a = F_W b$$
$$\Rightarrow k_S \Delta_S a = k_W \Delta_W b$$

And so

$$\frac{k_S}{k_W} = \frac{\Delta_W}{\Delta_S} \frac{b}{a} = \frac{1}{IR} \frac{b}{a}$$

But

$$\Delta_W = \Delta_S \frac{b}{a}$$
$$\Rightarrow \frac{b}{a} = \frac{1}{IR}$$

So

$$\frac{k_S}{k_W} = \frac{1}{IR^2}$$

It should be noted in this analysis that it has been assumed that the installation ratio remains constant during the wheel travel. Often, however, it isn't constant. Consider Fig. 7.23. The spring is lying nearly horizontal at design point shown by position *a–c*. A unit of wheel rise will produce a much smaller unit of compression in the spring, and an extremely stiff spring will be needed to support the vehicle. However, by the time the wheel has risen to position *b–c*, it is horizontal. At this point, a unit of wheel rise will produce no compression at all in the spring, and no matter how stiff it is made, the wheel rate will have dropped to zero. Clearly, it will have changed during the process. One can produce a graph of the function by analysing on CAD the suspension at various positions.

In Chapter 8—suspension dynamics—we shall see that a rising rate suspension is desirable, and it is common to make use of the necessary suspension geometry to produce a rate that is rising naturally as a result of the suspension design. One feature that makes this particularly easy is the bell crank, as we see below.

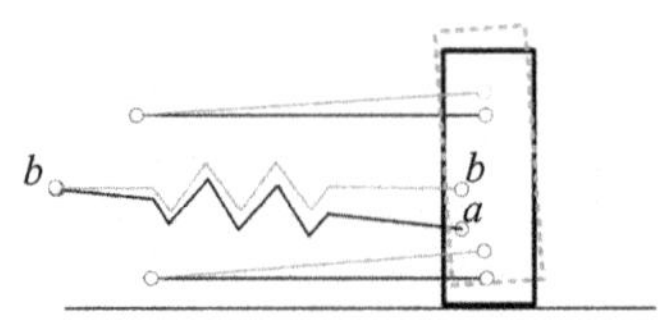

Fig. 7.23 A falling rate suspension.

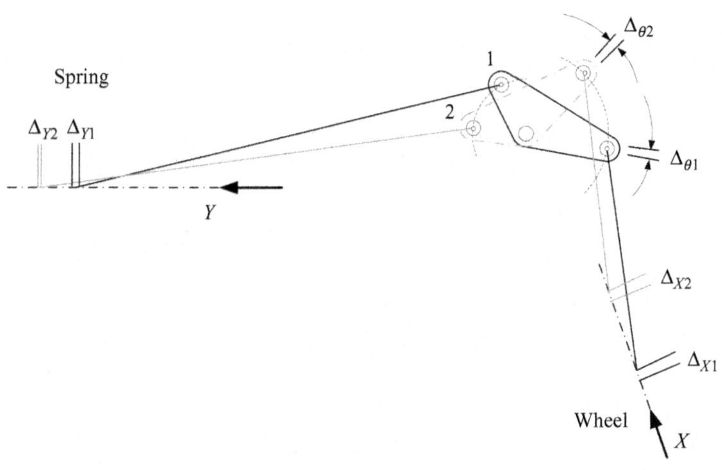

Fig. 7.24 Bell crank and variable installation ratio.

7.2.11.1 Bell Cranks

Fig. 7.24 shows part of a suspension. The wheel moves an amount Δ_{X1} as a result the bell crank rotates an amount $\Delta_{\theta 1}$. This in turn means the spring is compressed by amount Δ_{Y1}. If the wheel now rises sufficiently that the bell crank is in position 2 and we again input a small displacement Δ_{X2} of very similar in magnitude to Δ_{X1} we see that the consequent compression in the spring is now only Δ_{Y2}, which is much less than Δ_{Y1}. Thus, the suspension has a falling rate. If the position of the spring and wheel are reversed, then it becomes a rising rate, and it is possible to achieve pretty much any desired function.

A graph of the installation ratio through the suspensions working displacement range can easily be determined by use of CAD or a kinematics package.

7.2.12 Track to Wheelbase Ratio and Centre of Gravity Location

The positioning of the front and rear axles is one of the most fundamental design decisions to be made. For a vehicle with equal cornering stiffness tyre front and rear, it is best to place the centroid of the suspension on the centre of gravity as shown in Fig. 7.25. If the design of vehicle requires the centre of gravity to be away from the centroid of the suspension, then tyre sizes can be made different front to rear to keep the car neutral steering.

The ratio of track to wheelbase influences high-speed stability. Long cars, with higher yaw inertia, are more stable at high speed and respond less in yaw to the same yaw inputs from the road through the suspension. Short

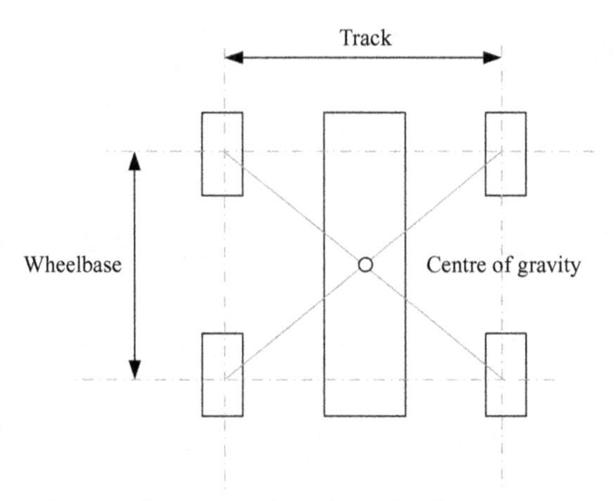

Fig. 7.25 Centre of gravity location and track to wheelbase ratio.

cars are less stable and respond to steering inputs more quickly. In many cases, the designer will have little or even no control over the track and wheelbase. In racing, it is common for the regulations to specify them. In road car design, there is generally more flexibility, but road dimensions, together with the need to house an engine and a given number of occupants, generally mean the dimensions are largely dictated.

7.3 SUSPENSION CONFIGURATIONS

In this section, we shall consider the main candidates from which a suspension configuration may be chosen for the vehicle in hand. The treatment of each configuration is very similar to make comparison easier. For some suspension geometry parameters, all the suspension configurations offer good control, for example, kingpin inclination; these topics need no new treatment. In the case of pitch and squat, the appropriate version of Fig. 7.15 can be used to determine the characteristic for any of the configurations below, and indeed, for any combination, since whilst it is unlikely that a designer would want a different suspension geometry left to right, it is entirely possible that a different one, front to rear, would be desirable.

7.3.1 Beam Axle

General design: The beam axle consists of a rigid structure that connects the wheel on one side to the wheel on the other. Fig. 7.26 shows this schematically. If the axle is driven, then the structure contains the differential and

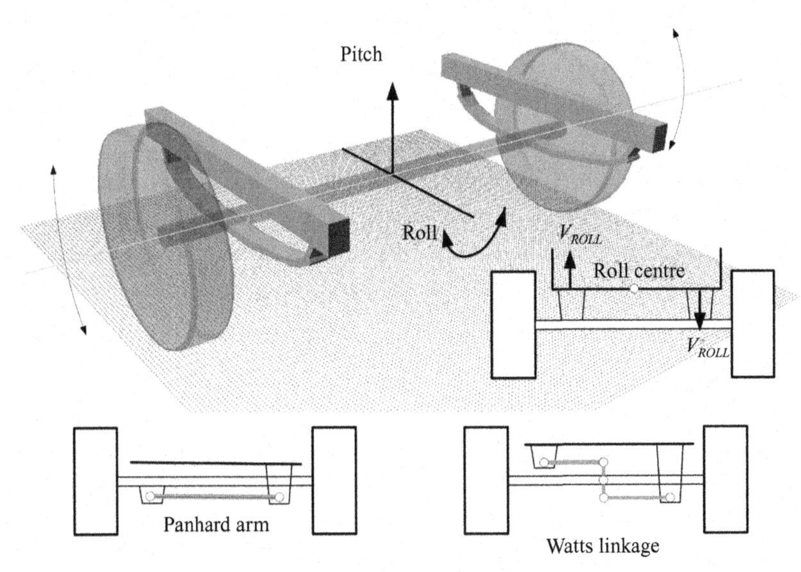

Fig. 7.26 Beam axle.

drive shafts and is normally used at the rear. If it is used at the front, then the wheels must be made to steer, and this is achieved by having uprights that mount to the beam and can rotate about the kingpin axis. The design is inherently rugged, making it ideal for use on light trucks, pickups and lorries. The axle can be mounted directly to the chassis by leaf springs as shown in the figure. If the lateral restraint offered by these is insufficient then additional restraint can be added by either Panhard arm or watts linkage, also shown in the figure. The configuration is simple, cheap and easy to maintain, making it ideal for a very large number of vehicles.

Unsprung weight: The configuration makes for a heavy axle. This is not a problem in vehicles where acceleration is less important and where the added weight is a relatively small part of the vehicle.

Camber control: There is no control of camber. As a wheel rises to negotiate a bump, it will go into negative camber by an amount dictated by the track of the vehicle. Worse still, a bump on one side is communicated to the other, and a bump on one wheel will cause positive camber on the other. For these reasons, the beam axle is not suitable for performance vehicles.

Roll centre control: The roll centre for the beam axle is easily determined, and in the figure, the velocity of the points on the body at the top of the springs is shown. These are equal in magnitude, and the only point equidistant to the vectors and at right angles to them is the roll centre, as shown. Thus, the roll centre is high and fixed with all the attendant disadvantages.

Installation ratio: The beam axle offers no control over the installation ratio. If vertically mounted leaf springs are used, the installation ratio is unity. If coil springs were used and inclined at an angle a falling rate could be achieved, but this would be disadvantageous. A rising rate could only be achieved by including linkages above the axle to amplify the motion in a nonlinear way.

7.3.2 Swing Axle

General design: The beam axle consists of a rigid structure that connects the wheel on one side to the wheel on the other. Fig. 7.27 shows this schematically. The wheels are made to articulate about the inboard universal (or constant velocity) joints and move up and down in an arc as shown. The geometry is a step upwards from the beam axle because it offers independent movement of one wheel from the other. A bump on one wheel does not therefore disturb the other. However, the configuration has serious disadvantages and offers too little compared with the geometries below to make it a popular choice.

Unsprung weight: The configuration makes for reduced unsprung mass compared with the beam axle. The performance here depends on how well the linkages mounting the upright to the suspension are designed; certainly, there is the capacity to do and a double wishbone geometry.

Camber control: The performance in camber comes from two sources. Consider first the outside wheel for a cornering car as it rises. The wheel will adopt beneficial negative camber. However, the high roll centre means that the car will jack significantly. Consideration of Fig. 7.27 shows that as the body rises, lifting the inboard end of the drive shafts, both wheels will adopt positive camber. The positive camber from this source outweighs

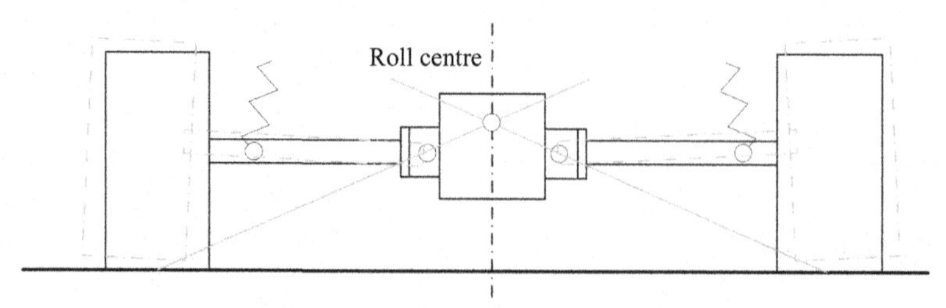

Fig. 7.27 Swing-arm geometry.

the negative camber from bump and the overall effect is a disadvantageous one. Positive camber on corner entry will lead to a loss of grip at the driven wheels. Presuming this to be the rear, problematic oversteer will result, and the vehicle is likely to spin out, and handling will be poor.

Roll centre control: The roll centre is high and cannot be controlled. A very disadvantageous feature.

Installation ratio: The swing axle offers no specific control over the installation ratio. As with the beam axle, if vertically mounted leaf springs are used, the installation ratio is unity, and the same comments made for the beam axle apply here.

7.3.3 Trailing Arms

General design: The trailing arm suspension is shown in Fig. 7.28. In version one, on the left, the upright *c* pivots about the centre lines at *a* and *b* and so rises and falls. If *a* and *b* are much above the axle centre line, then the wheel will move backwards slightly under bump, which serves to soften the bump and make bump negotiation good. In version two, on the right, the wheels rise and fall vertically as the arms pivot about the centre lines shown. The centre lines are horizontal. Such an arrangement can make for very compact packaging, making it a sensible choice, for example, for a light van where the rear suspension must be packaged under the payload space. Version one has poor resistance to lateral loads at the tyre from cornering, making version two, where resistance is much better, a more common approach.

Unsprung weight: This is much lower than an equivalent beam axle. Version one will always require larger sections to resist lateral loads than a wishbone suspension but can still be made to have an acceptable unsprung mass. Version two will always feature arms with some significant mass, often made

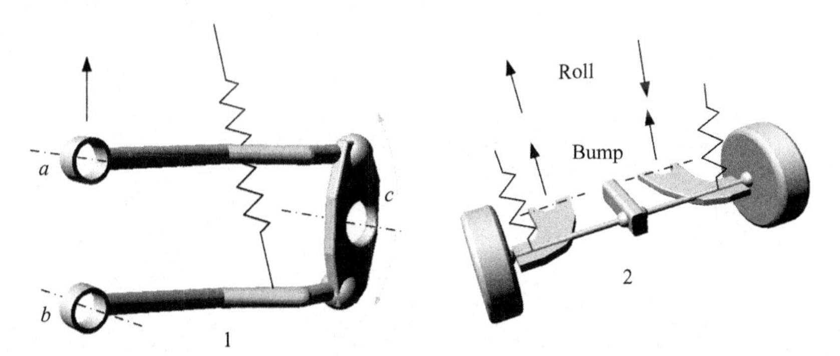

Fig. 7.28 Trailing arm suspensions.

out of pressed steel. If the axle suspended in this way is the driven axle, then half shafts (as shown) will be required. Since their length will change during bump, joints that accept plunge will be needed.

Camber control: As the wheel rises and falls, it will remain vertical, which is desirable, but under roll, the wheel vertical axis will remain parallel to a vertical axis through the vehicle resulting in undesirable positive camber at the outside wheel.

Roll centre control: With the chassis mounts rising and falling vertically, the roll centre is in the middle of the vehicle at the same height as the centroid of the body link mountings. This will clearly be some considerable height off the road but not as high as the swing axle above. It can't be controlled, therefore making the geometry less desirable.

Installation ratio: Again, the same comments made for the beam axle apply here.

7.3.4 Semitrailing Arms

General design: A development of the trailing arm is the semitrailing arm suspension. In this configuration, the axis about which the links pivot is inclined in all three planes. This is shown in Fig. 7.29. The comments made above for the trailing link suspension apply here. However, the inclination of the axis makes for much improved options for control of performance through design.

Unsprung weight: Same as above for training link.

Camber control: Imagine the top view shown in Fig. 7.29 is at the rear right of a vehicle cornering to the left, making this the outside wheel. With weight transfer, this wheel will rise towards the viewer. The inclination

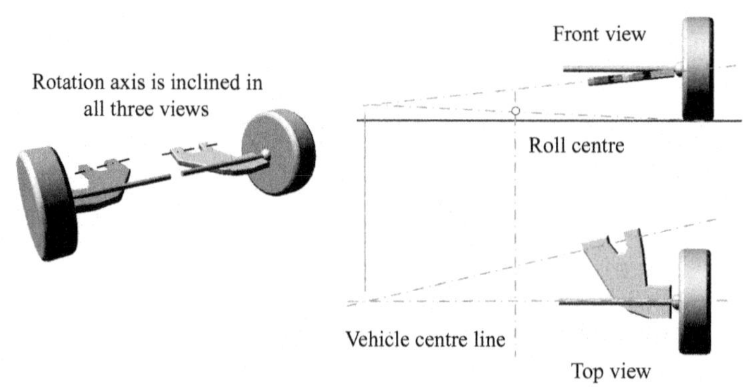

Fig. 7.29 Semitrailing arm suspension.

of the axis of rotation means that the wheel will develop negative camber, with the top of the wheel becoming closer to the centre line of the vehicle. This is clearly beneficial. In addition to this, the inclination of the axis can be used to make the wheel steer as it rises. This can then be used to provide a different steering response for high-speed corners, where the vehicle rolls more, and low-speed corners, where it rolls less.

Roll centre control: The roll centre location is shown in the figure, and it is clear that by adjusting the inclination of the arms and the angle made to the centre line in the top view, the designer can achieve any desired roll centre location, offering a good benefit.

Installation ratio: The same comments as above apply here.

7.3.5 MacPherson Strut

General design: The MacPherson strut is named after Earle S. MacPherson who designed the strut whilst working for the Ford Motor Company in 1949. The general layout of this geometry is shown in Fig. 7.30. A control arm runs from the upright to the chassis, *b–c*. The strut itself runs from *a–b* and slides telescopically, with a spring mounted outside to support the vehicle. The internal friction of the strut could be a serious issue but can easily be almost completely eliminated by ensuring that the axis of the spring passes through the centre point of the tyre. If the anchorage *b* is on this axis, then strut friction is almost eliminated, so too is bending in the control arm that is good practise. A great advantage of the geometry is the space it provides within the chassis, to mount an engine, for example, and this makes it a

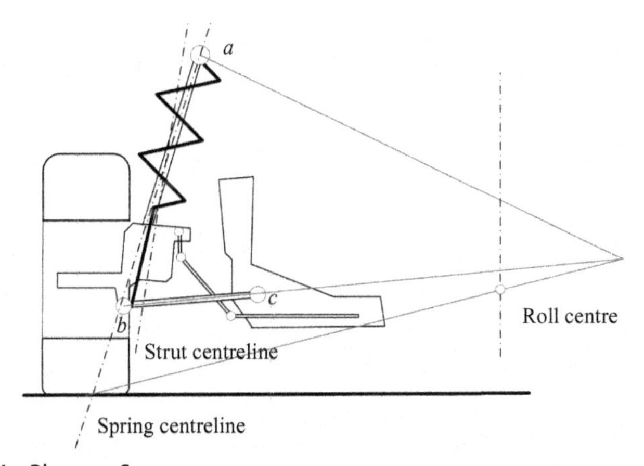

Fig. 7.30 MacPherson Strut.

very sensible choice for road cars with a front engine where it is now almost universally used.

Unsprung weight: Is easily kept to acceptable levels. However, the strut itself and upright are inevitably fairly bulky items and the geometry will never offer levels of unsprung mass reduction achievable with the double wishbone.

Camber control: Is not ideal. As the body rolls, so does the strut. The top mount *a* on the outside wheel in a corner will move outwards, causing undesirable positive camber.

Roll centre control: The construction for roll centre of the unrolled vehicle is shown, and we can see that the orientation of the strut and inclination of the lower control arm allow the designer good control of the roll centre and it can generally be placed as desired.

Installation ratio: This is not easily controlled. The orientation of the strut is largely fixed. It needs to be reasonably vertical but must be inclined to clear the wheel, as shown. The installation ratio is marginally less than one and is a weakly falling rate. Secondary springs and progressive bump stops can be used to assist here.

7.3.6 Double Wishbone ('A-Arm') Suspension

General design: The general layout of the double wishbone geometry is shown in Fig. 7.31. The layout offers excellent control of all important suspension parameters. Its linkages are all in pure tension or compression and can be made very light, in addition to which they can easily be designed as aerofoil sections, all of which makes the geometry the first choice for racing and in high-performance vehicles where the suspension is in the air stream. The design necessarily features a wheel mounted remotely from the chassis. This is desirable in racing, helping to minimise the polar moment and drag, but not so for a road car where a significant amount of empty space can soon result. The great flexibility that the geometry affords the designer, together with the opportunity to produce a very lightweight structure makes the double wishbone a first choice for racing where the formula allows.

Unsprung weight: It is easy to keep unsprung weight to an absolute minimum.

Camber control: Inclination of the wishbone in the front view gives good camber control. Camber change under bump and droop is easily determined from a CAD drawing. We can also use the double wishbone geometry to eliminate camber change altogether.

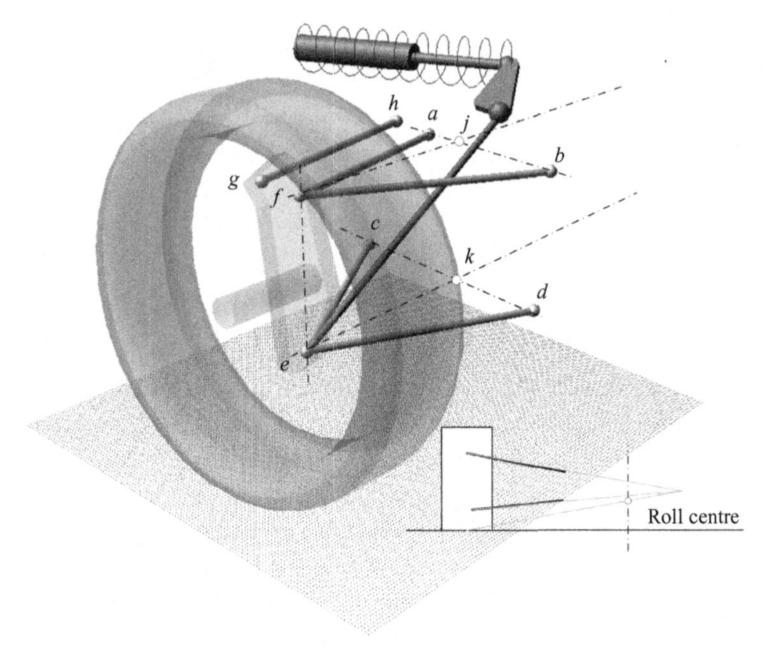

Fig. 7.31 Double wishbone suspension.

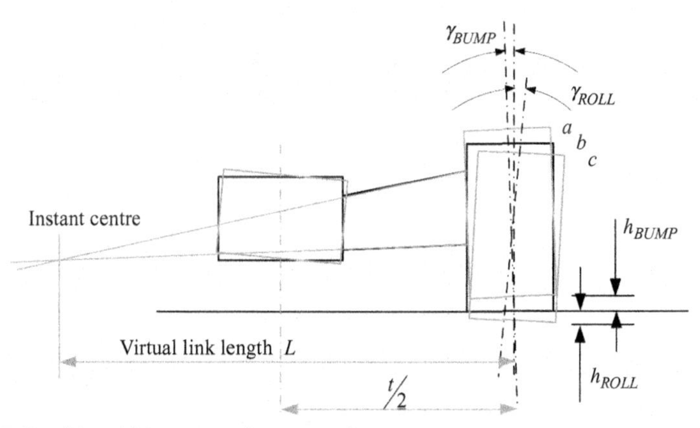

Fig. 7.32 Double wishbone camber control.

In Fig. 7.32, we see γ_{BUMP} and γ_{ROLL}. If these are made equal, then no camber change will result. If this holds, then $h_{BUMP} = h_{ROLL}$ and both quantities can be called h. Analysing the figure and remembering that in bump, the wheel rotates about the instant centre, whilst in roll, it rotates about the centre of the vehicle:

$$\gamma_{BUMP} = \tan^{-1}\frac{h}{l} \approx \frac{h}{l}$$

$$\gamma_{ROLL} = \tan^{-1}\frac{h}{(^t/_2)} \approx \frac{2h}{t}$$

Thus, for no camber change,

$$\frac{h}{l} = \frac{2h}{t}$$

$$\Rightarrow l = \frac{t}{2}$$

Thus, if the instant centre is placed on the midline of the vehicle, no camber change occurs under roll. This sounds like a very desirable effect but is generally avoided because the inboard anchorages for the wishbones become very close together, and mechanical design is compromised with very large loads and bending. If the instant centre were placed between the wheel and the body, negative camber would result and be highly desirable; however, the mechanical design necessary makes this very awkward indeed. The geometry also permits the designer to place the instant centre on the other side, outboard from the wheel giving even greater flexibility to the designer. The common approach used in racing is to have a significant amount of static camber and to arrange for the wheel to have the ideal camber setting under cornering. This reduces straight line drag and optimises lateral acceleration at the same time. The important point is that the geometry affords great flexibility to the designer.

Roll centre control: The geometry offers very good control of the roll centre. The inset shows how the roll centre is determined. The lines used are *f*–*j* and *e*–*k* from the main figure. Points *j* and *k* are determined by taking a line at right angle to the axis of the vehicle through points *f* and *e*. Where these lines cross, the lines *c*–*d* and *h*–*b* give points *j* and *k*.

Pitch and squat: The inclination of lines *a*–*b* and *c*–*d* provides excellent control of pitch and squat.

Installation ratio: The geometry is normally used with either a pushrod (as shown in Fig. 7.31) or pullrod that is then connected to a bell crank and from there to the spring damper. This linkage path allows excellent control of the installation ratio, and anything likely to be of use can be achieved.

All this makes the double wishbone geometry an excellent choice for racing but much less so for road cars. Some road car rear suspensions are manufactured this way. One problem with its adoption on road cars relates to manufacturing tolerances. To achieve the benefits of the configuration, very good positional control of the hard points is needed. In racing, tubs are generally CNC drilled, and jigs are used for accuracy. In mass production, the individual chassis must be assembled and fitted under production line conditions with generally less positional accuracy.

7.4 QUESTIONS

1. In which sense is a wheel cambered if the top leans inwards?
2. In Fig. 7.4, what value of static camber is present?
3. What is the difference between the *instant centre* and the *roll centre* as used in the description of suspension geometry above?
4. What is the difference between the roll centre of a stationary car on a garage floor and one at speed on the open road?
5. A vehicle is travelling at 70 mph on a motorway. The tyres are 0.87 m in diameter. How long will it take for the contact patch to be purged of tread and why is this important in terms of the roll centre?
6. What will happen in terms of roll and jacking, if the roll centre of a vehicle is moved from being very close to the centre of gravity to being just below the ground? Comment on the desirability of this move.
7. Why is it acceptable to design a suspension that eliminates both jacking and pitch but not so for jacking and roll?
8. Why would the value of kingpin inclination used in a vehicle often be very similar to the value of castor?
9. Which parameter affects *mechanical trail, castor* and *kingpin inclination*?
10. What is the main disadvantage of the swing-arm suspension geometry?
11. Compare and contrast the design opportunities of using a MacPherson strut with a double wishbone suspension.

7.5 DIRECTED READING

Dixon J. [15] This paper offers a very good example of the detailed geometric treatment given to the roll centre.

Staniforth [16] is a very readable text and gives a good treatment of different suspension designs with an emphasis on racing.

Dixon J. [8] has full treatment of suspension components and characteristics.

Blundell M. and Harty D. [17] offer an analysis-based examination and include obtaining solutions by matrix techniques and Adams.

7.6 LEARNING PROJECTS

- For a vehicle you are interested in, obtain (by either direct measurement, estimation from a photograph or technical manual data) the coordinates of the suspension hard points in the front view, and use these in a CAD or multibody code to obtain a graph of camber change in bump and droop.
- For the car use above, extend the same analysis for the determination of the roll centre.
- Construct in Adams a front view analysis of a collection of suspension geometries and use it to determine the effect on jacking on roll centre position. You will need to model the asymmetric lateral forces for this to work.
- Repeat the analysis above for the roll centre but for jacking and pitch in the side view.
- Produce a complete suspension analysis tool in Adams or a spreadsheet or similar for a front suspension that outputs vertical motion and camber change of the wheel under steer.
- Produce a CAD of multibody simulation of the steering geometry of a vehicle of your choice. Use the analysis to determine the level of accuracy with which the steering geometry shown in Fig. 7.19 approximates to Ackermann geometry. Start by producing a graph showing accuracy of agreement against steered angle.
- Produce a multibody code simulation of a suspension of your choice to determine the sensitivity of bump steer to rack height. Use the understanding gained to estimate the production accuracy with which the rack height should be controlled.
- Produce a quarter car model of a MacPherson strut and double wishbone suspension and use the models to compare and contrast the performance of the suspensions from a range of points of view.

7.7 INTERNET-BASED RESEARCH AND SEARCH SUGGESTIONS

- Search using *suspension configurations* and categorise all the different approaches you can either in terms of, or as variants of, ones we have meet or as entirely new ones.
- Search again for *camber compensation suspensions* and review all you can find, assessing whether the examples you find offer advantages and when they might be of use.

- Repeat the search for the following terms and assess the material you find:
  - Roll centre suspensions
  - Perfect suspension geometry
  - Bell cranks
  - Camber car
  - Leaning car design
- Compare the understanding we have gained for road car suspension with approaches taken to other vehicles such as tractors, lorries, motorbikes, trains and aircraft where there are similarities but also large differences.

CHAPTER 8

Dynamic Modelling of Vehicle Suspension

In Chapter 7, we saw how an understanding suspension kinematics is necessary for contact patch optimisation. For example, the amount of camber change under roll has a big effect on the amount of grip. However, it's not just the kinematics of the suspension that matters; the dynamics does too. Dynamics is the study of things in motion and provides not just a description of the motion, but an explanation of it as well.

In this chapter, we shall develop a series of dynamic models that start with the simplest possible approach and then build in complexity and accuracy. We shall use these models to answer all the dynamics questions we may wish to pose, for example, what would be the best spring or damper function to use in a suspension, or how may an anti-roll bar be specified.

The modelling involved starts simple and becomes more complex, and the aim of the chapter is to raise understanding to the point where the reader may then undertake analyses for themselves. The preparation of all the models necessary to completely parametrise all suspensions from all points of view is too large a task for this work. The approach we shall use is the 'lumped parameter method', and this is introduced below.

8.1 THE LUMPED PARAMETER APPROACH

A common and very fruitful approach in dynamics modelling is to divide the system in question up into discrete *lumped* parameters that offer a simplified model when compared with the continuous system seen in the real life. For example, in the figure below, the encastred beam on the right is subjected to a sinusoidally varying force F. This system could be modelled by considering the beam as a continuous elastic solid with internal damping and preparing complex analytical equations for the resulting motion. A numerical approach could be taken using dynamic finite element methods. However, by considering the system to be equivalent to the lumped parameter system on the right, a relatively swift analysis can be prepared using classical methods (Fig. 8.1).

Performance Vehicle Dynamics
http://dx.doi.org/10.1016/B978-0-12-812693-6.00008-0

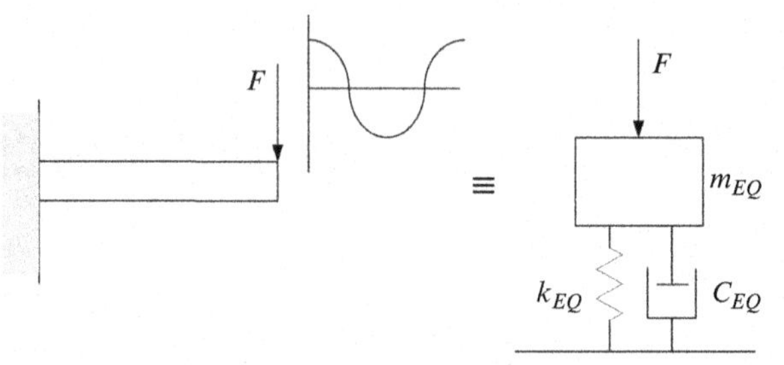

Fig. 8.1 The lumped parameter approach to modelling dynamic systems.

In the example above, we would need to find a way to model the elastic displacement under the load in the beam as a single spring constant, k_{EQ}, in the lumped parameter model on the right. In a similar way, we would need to find an equivalent damping value C_{EQ} to represent the damping distributed evenly within the beam and similarly for the mass m_{EQ}. Once these equivalent parameters are found, then a solution to the model can be prepared, based on equations of motion for the lumped parameter model, and from these, optimisation can begin. The questions we must now address are therefore what the lumped parameter model for a whole car should look like.

Fig. 8.2 shows a seven-degree-of-freedom-lumped parameter model of a vehicle and its suspension. This level of detail provides a very useful model, and much can be learned and usefully applied though its consideration.

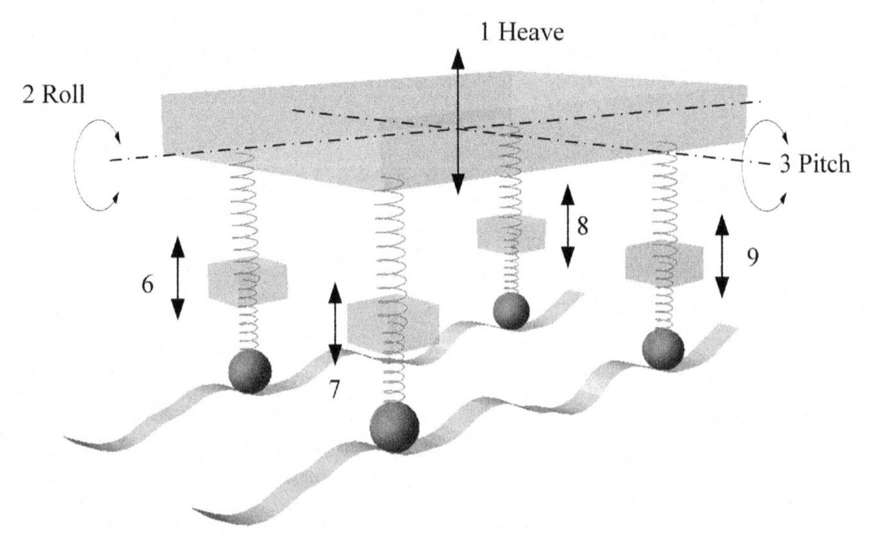

Fig. 8.2 Seven-degree-of-freedom approximation for a suspension.

In the model, the road provides a vertical displacement input that the contact patch is required to follow. The displacement of the contact patch with respect to the unsprung mass causes compression in the tyre, which develops a force. The unsprung mass is free to displace vertically. The unsprung mass is linked to the car body by a spring to model the suspension stiffness and normally a damper to model the damper unit. The body is free to move vertically in *heave* and to rotate about the centre of gravity in the side view, causing *pitch,* and in the front view causing *roll.*

In the figure, we can see that the road inputs might conceivably conspire to produce vertical motion only. Indeed, there will be a particular natural frequency in heave at which the road heave input will produce a maximum body heave output. Alternatively, the inputs might happen to oscillate by rising at the rear whilst falling at the front or by rising at the left whilst falling at the left. These possibilities give rise to a heave, pitch or roll motion, respectively. Again, there will be a natural frequency for each of these motions influenced by the stiffness damping and mass moment of inertia about each axis. One final possibility exists called *warp,* when the diagonal inputs move in antiphase, the left rear and right front rising whilst the right rear and left front fall. There is no resulting large-scale movement of the body in warp because it is a stiff structure designed to resist such deformation. A warp input for a symmetrical car with equal dynamic parameters all around would result in no body motion because the moment rotating the body in one direction is cancelled by a moment rotating it in the other, from the other end. However, such symmetry is rare, and although the displacements resulting from warp are generally smaller than other modes, because even if the moments don't cancel they oppose each other, it is still significant. On top of all this, there are the motions of the unsprung mass to consider, each of which will have their natural frequencies.

For a real car travelling over a real road surface, the input waveforms would clearly consist of many different frequency inputs, and there would be an input frequency spectrum. The inputs would very likely be a random mixture of displacements.

Thus, we see that for the seven-degree-of-freedom system, the output response is going to be a complex affair. The road will provide inputs over a wide frequency range, and the body will respond in four modes and have seven degrees of freedom. The overall response of the body would then be a superposition of these four modes. The uprights will have their own resonant frequencies that will affect the body response, for example, if the front uprights are being driven at their natural frequency, then their displacements will be large, and this will be communicated to the body that itself will show an increased amplitude of vibration at this frequency. So the seven-degree-of-freedom system is complicated, and it would be a mistake to start with it. Instead, we shall approximate the car to the simplest of all such systems, the single-degree-of-freedom system, and start there.

8.2 MODELLING A SUSPENSION AS A SINGLE DEGREE OF FREEDOM SYSTEM

To begin, we shall approximate the suspension to a single-degree-of-freedom system, SDoF, featuring either a harmonic input of constant amplitude or a step input, as shown below. In this approximation, the whole car is assumed to move vertically together, and there is no roll or pitch. The upright is not modelled, and the input from the road is assumed to be entirely in heave, with all four wheels moving as one. This is clearly a large approximation, but it gives a starting point and yields useful results despite its simplicity. Heave, is in fact, the largest of the road inputs. The system presented in Fig. 8.3 is familiar to most students of mechanical engineering and is the case of SDoF—base excitation.

The figure shows both the constant frequency harmonic input and the step input. In the harmonic case, the base moves in the x coordinate and is a sinusoidal function of amplitude X. The 'body' of mass m_{EQ} moves in the Y coordinate and also has a sinusoidal function, this time of amplitude Y. This output response of the body lags behind the base input by an angle, phy φ. In the case of the step input, the body responds to the step input of the base with an exponentially decaying sinusoid that settles to the same displacement as the step input.

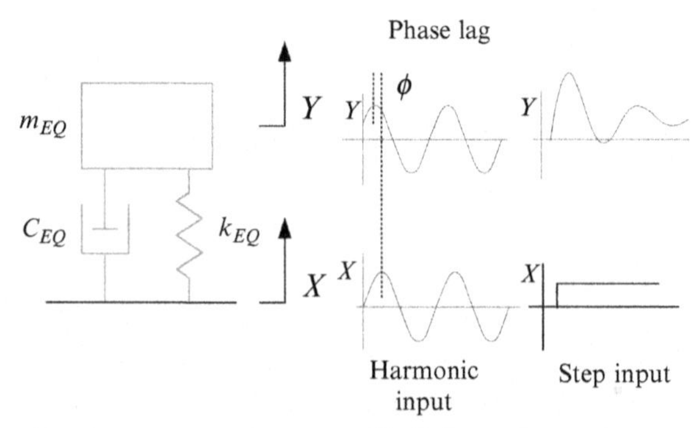

Fig. 8.3 Approximating a suspension to an SDoF dynamic system.

8.2.1 SDoF Base Excitation Equations of Motion and Modelling

To develop the equation of motion for the system, we start with the damper. This is a device that produces a force proportional to the speed at which it is being stroked and always acts to oppose the motion. Thus, the force in the damper is

$$F_D = C(\dot{y} - \dot{x})$$

where $(\dot{x} - \dot{y})$ is the difference in velocity between the body and the base and C is the damping coefficient. The spring produces a force that is proportional to the extension it experiences, and so, the force in the spring is

$$F_{SP} = k(y - x)$$

Remembering that the sum of all forces must be zero and that the inertial force on the mass is $m\ddot{x}$, the differential equation of motion governing the whole system shown in Fig. 8.3 is in the following:

$$m\ddot{x} + c(\dot{y} - \dot{x}) + k(x - y) = 0 \quad (8.1)$$

The solution of second-order linear differential equations is a topic in its own right, and here, we shall not be concerned with the mathematical proofs used. For these, if needed, a good text on the topic should be visited. Boyce and DiPrima [13] are ideal. For any revision on the fundamentals of dynamics, Rao [26] is ideal.

8.2.2 Classical Response of SDoF to Step and Harmonic Inputs

8.2.2.1 SDoF Transient Response to step input

When the system is provided with a step input and the base is not in motion, the equation of motion reduces to

$$m\ddot{x} + c\dot{x} + kx = 0$$

and the form of response is crucially dependent on a parameter called the *damping ratio*, which is given the symbol zeta, ζ. The damping ratio is defined as the ratio of the damping present in the system to a quantity called the critical damping. The value of critical damping, C_C is given by

$$\Rightarrow C_C = 2\sqrt{kM} \quad (8.2)$$

where k is the value of the spring constant and M is the mass of the body. So,

$$\zeta = \frac{C}{C_C} \quad (8.3)$$

If X is less than one, the system in described as *underdamped*. If m_{EQ} is one, then the system is *critically damped*. If φ is larger than one, it is *overdamped*. We shall now look at the response of the system in these three regimes of response.

Underdamped Transient Response

The equation of motion for the transient response of the SDoF system is given by

$$x(t) = Xe^{-\zeta\omega_n t}\sin(\omega_d t + \phi) \tag{8.4}$$

In this equation, $x(t)$ is the displacement of the body as a function of time. X, which dictates the magnitude of the response, is a constant of integration that must be determined from the initial conditions and is given by

$$X = \frac{x_0}{\sin\varphi} \tag{8.5}$$

The damping ratio ζ is given by two equations above. The natural frequency ω_n is given by

$$\omega_n = \sqrt{\frac{k}{m}} \tag{8.6}$$

where k is the spring stiffness and m is the mass of the body. This term is always called the natural frequency that is scarcely logical, given that it is actually an angular velocity and is measured in rads/s. The 'real' natural frequency is given by

$$f_n = \frac{1}{2\pi}\sqrt{\frac{k}{m}}$$

The quantity ω_d is the damped natural frequency. The natural frequency ω_n relates to the undamped version of the system. When the effect of damping is included, the system vibrates more slowly because the damping opposes the motion all the time. ω_d is given by

$$\omega_d = \sqrt{1-\varsigma^2}\omega_n \tag{8.7}$$

The phase angle φ is a second constant of integration that must be determined for the initial conditions. It is necessary to have two constants because the initial displacement and initial velocity could take any value and a constant is needed for each one. The angle is given by

$$\Rightarrow \varphi = \tan^{-1}\left\{\frac{\omega_d x_0}{\dot{x}_0 + \tau\omega_n x_0}\right\} \tag{8.8}$$

In the formula for φ, the value x_0 is the initial displacement, the value $\dot{x}_0$ is the initial velocity, and the other quantities are as defined above.

Worked Example 1

A car is being driven along a road when it encounters a kerb. The kerb is 7 cm high. The car is travelling fast enough that the kerb may be approximated to an instantaneous step input. Prepare a spreadsheet that determines the time response for the movement of the body as a result of the kerb strike. Assume that the road surface is completely flat other than the kerb. Model the system as a single-degree-of-freedom system using a spreadsheet. The car has the following dynamic data:

Mass: 800 kg	Individual spring rate: 30 kN/m	Damping value: 14000 N s/m

Prepare a spreadsheet solution allowing the user to input the dynamic data for the car and then obtain the displacement of the body as a function of time. A screenshot from a working spreadsheet is provided below to allow you to test your solution by using the same input data and comparing the output data.

A spreadsheet solution for worked example one is shown above in Fig. 8.4. The spreadsheet works as follows. Input data are entered in cells B2/D10. The values for the mass, spring and damper values are

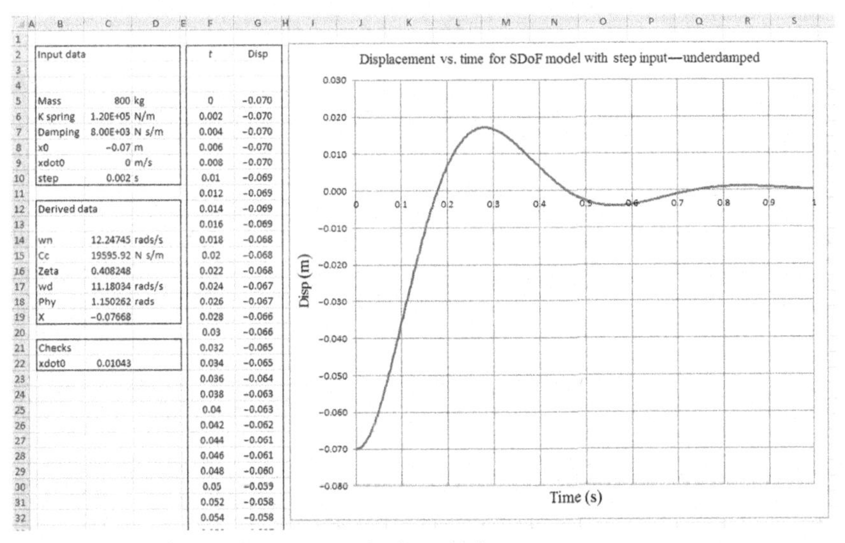

Fig. 8.4 Spreadsheet solution to worked example one.

self-explanatory. The value for x0, the initial displacement, is chosen to be −0.07 to model the fact that the road surface is 7 cm below the rad surface after the kerb has been struck. Xdot0 is chosen to be zero to reflect the fact that the car body has no vertical velocity when it first strikes the kerb. Parameters that are calculated from the input data are determined in cells B12/D19. This starts with the natural frequency denoted 'wn' in the spreadsheet determined by Eq. (8.3). Cc is the value of critical damping obtained from Eq. (8.2). Zeta is the damping ratio defined by Eq. (8.3). 'wd' denotes the damped natural frequency determined by Eq. (8.6). The constants determined from initial conditions are the phase angle φ, determined from Eq. (8.8), and the amplitude X, determined from Eq. (8.5). The quantity 'step' in cell C10 is used to advance the time domain in column F where each cell is determined by adding 'step' to the cell above. In column G, the displacement is determined for each time value using Eq. (8.4).

A graph is presented above to show the evolution of displacement with time. It can be seen from the graph that the car body is expected to reach its first overswing and maximum vertical displacement at about 0.28 s and to settle to its final value after about 0.75 s.

Critically Damped Transient Response

In the case where the damping ratio is one, the equation of motion for the single-degree-of-freedom system is given by

$$x = \left[x_0 + \left(\frac{v_0}{\omega_n} + x_0\right)\omega_n t\right]e^{-\omega_n t}$$

where x_0 is the initial displacement, v_0 is the initial velocity, and ω_n is as defined above. This response is plotted in the graph below for two different initial condition sets (Fig. 8.5).

Overdamped Transient Response

In the case of the overdamped system, the solution of the homogeneous differential equation of motion yields the following expression for the displacement of the body as a function of time:

$$x = Ae^{\left(-\varsigma + \sqrt{\varsigma^2 - 1}\right)\omega_n t} + Be^{\left(-\varsigma - \sqrt{\varsigma^2 - 1}\right)\omega_n t}$$

Again, there are initial constants A and B: these are given by

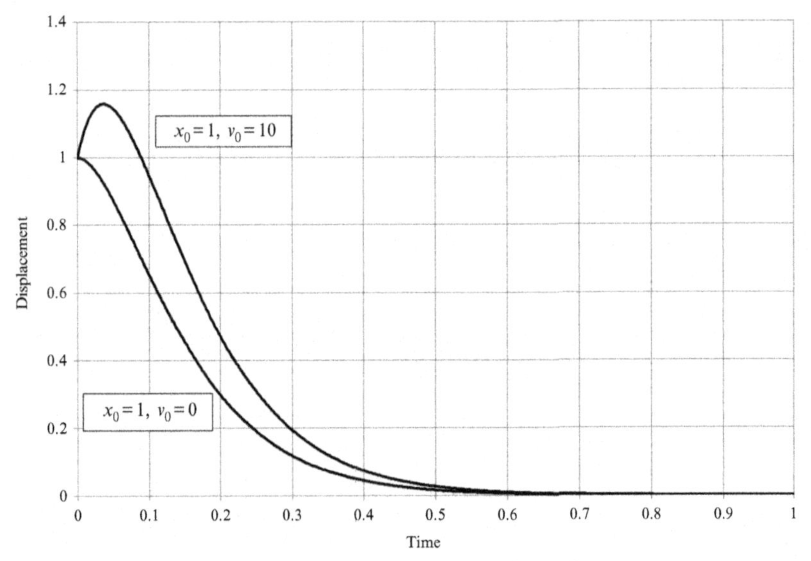

Fig. 8.5 Response of a critically damped single-degree-of-freedom system to a step input with zero and non-zero initial velocity.

$$A = \frac{x_0\omega_n\left(\varsigma + \sqrt{\varsigma^2 - 1} + \dot{x}_0\right)}{2\omega_n\sqrt{\varsigma^2 - 1}}$$

$$B = \frac{-x_0\omega_n\left(\varsigma + \sqrt{\varsigma^2 - 1} - \dot{x}_0\right)}{2\omega_n\sqrt{\varsigma^2 - 1}}$$

A spreadsheet may be prepared as shown in the Fig. 8.6 below which calculates the overdamped response. The spreadsheet shows the input data, the derived constants of natural frequency, the critical damping, the initial constants A and B and the response *disp* as a function of time.

A graph is also presented that clearly shows the exponentially decaying motion as a function of time.

Thus, the response of the single-degree-of-freedom system to a step input is defined by whether the damping ratio is below the critical damping value, equal to it or greater than it. In all cases, the response of the system decays away and after some period of time has become zero. For this reason, the response to the step input is called a *transient* response.

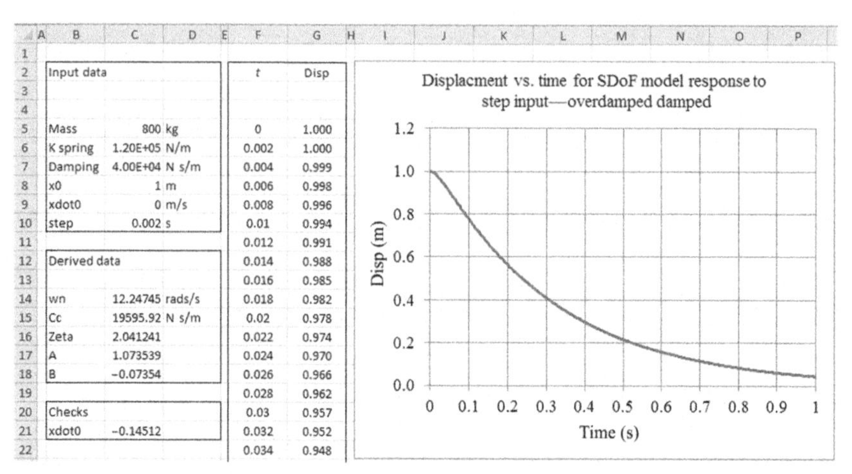

Fig. 8.6 Displacement versus time for an overdamped SDoF—step input.

Settling Time

Since this section is all about the transient response to a step input, we must answer the obvious question; when is the transient over? The mathematical functions developed above are all either asymptotic towards zero in the case of the underdamped and critically damped, or oscillatory about zero in the underdamped cases. In this sense, the transient is never over. However, it is clear that the separation between the current value of the function and its final value decreases with time, and so, the normal way to treat the problem is to choose an acceptably small error and define the transient as being over when the function in hand settles to within this error.

The graph below shows this approach. Two dotted lines have been inserted above and below zero. We define the function to be settled when it enters this region but does not come out of it again. So, for example, the fact that the curve for a zeta value of 0.1 enters the region at 0.14 s does not indicate that the function has settled because it soon leaves the region. Alternatively, the curve for a zeta value of 1.5 enters the region at around 0.85 s and does not re-emerge, and so, we would describe it as having settled at this time.

On this basis, we can see that the value of zeta that lets the function settle most quickly is 1.0, the critical value. If zeta is less than this, the function oscillatory takes longer to settle. (One could argue that for any acceptable error, there will always be a zeta value just below 1.0 that is indeed oscillatory but not sufficiently oscillatory that it ever leaves the acceptable error region. This value of damping therefore offers the quickest time to settle. In practice, such adjustments are generally too small to be accounted for.) If zeta is larger than 1.0, then clearly the function takes longer to settle too.

Summary

To summarise

- The response of a single-degree-of-freedom system to a step input depends critically on the damping in the system.
- There is a *critical damping* value for the system.
- Above *critical damping* value, the response is an exponential decay.
- Below *critical damping*, the response is an exponentially decaying sinusoid.
- Marking the border between these two regions is when the system has the critical damping.
- A system with the critical damping will settle in the shortest possible time.
- Since all the functions are asymptotic to the final value, an acceptable error must be specified to define when the system has settled.

8.2.2.2 SDoF Steady-State Response to an Harmonic Input

We have seen in the section above that the 'transient' response to a step input tends to zero. In most practical situations, damping is present, and before long, the transient response is over. Thus, if we are concerned with the steady-state response, we can ignore the step response. It will still have a role to play in finding the ideal damper values as we shall see, but in this section, concerned with the steady-state response, we can ignore it.

We must be clear, however, that the solution so obtained will only describe the vibratory behaviour of the system *after* any initial transient, as determined in the section above, has decayed away, and the system has settled down. The equation of motion describing the steady-state response to harmonic input of the SDoF system shown in Fig. 8.3 is as follows:

$$m\ddot{x} + c(\dot{x} - \dot{y}) + k(x - y) = 0$$

$$\Rightarrow m\ddot{x} + c\dot{x} + kx = c\dot{y} + ky$$

Here, the impressed motion takes the form

$$y = X\sin(\omega t)$$

where X is the amplitude of the impressed waveform. The development of a solution to this differential equation is a lengthy process not presented here. Instead, we shall quote the solution. Given that the input is sinusoidal, so too must the output be, and it takes the form

$$y = Y\sin(\omega t)$$

where Y is the amplitude of the response meaning here the amplitude of the mass. The amplitude response of the system is defined by the following equation:

$$\Rightarrow \frac{X}{Y} = \sqrt{\frac{1 + (2\varsigma r)^2}{[1 - r^2]^2 + [2\varsigma\, r]^2}} \tag{8.9}$$

In this equation, the amplitudes of the input and output appear as a ratio on the left-hand side as X and Y. The right-hand side is a function that depends on only two parameters; these are zeta, ζ and r. These are defined in the following way:

$$\zeta = \frac{C}{C_C}$$

where $C_C = 2\sqrt{kM}$.

Thus, zeta is the ratio of damping present in the system to the value of critical damping. The second parameter, r, is the frequency ratio and is a measure of how fast the system is being driven in terms of input frequency. When we dealt with the step input case above, we were concerned with obtaining the output over time. This is because the system responded with a monotonic output that decayed away with time, and so, time is the correct domain over which to present the output. In this case, the input has a given amplitude, X, and what we are interested in, is the response of the output as the frequency is increased. Determining the response in the frequency domain is the correct approach here. We could simply prepare response curves against input frequency, but this does not make comparison between systems possible. Instead, we use a frequency axis in units of natural frequency of the system. We call this idea *normalising* because it means that regardless of the system in hand, we shall always be looking at the response against the same domain. The frequency ratio r is defined as

$$r = \frac{\omega}{\omega_n}$$

where $\omega_n = \sqrt{\frac{k}{m}}$.

Thus, the output response of the mass of the system above is a sinusoidal wave whose amplitude is related to the amplitude of the input wave by Eq. (8.9) above. However, there is slightly more to the response than just this amplitude ratio. The output motion of the mass is not in phase with the input but instead lags behind it. If, for example, the spring and damper values are low, then not much force is communicated to the mass, and it will take time to respond to the motion. The phase angle φ by which the output lags the input is given by

$$\tan\varphi = \frac{2\varsigma r^3}{1 - r^2 + (2\varsigma r^2)}$$

where the symbols have the same meaning as above. The response of the amplitude ratio and phase angle has the following graphs:

8.2.2.3 Combined Response to Step and Steady-State Input

If a single-degree-of-freedom system is held stationary and suddenly starts to receive a harmonic input through the base that is at some offset displacement, then the motion of the mass will be the superposition of the response to step input and steady-state input. Such a situation would correspond to a car travelling on a smooth plane and then encountering an undulating surface whose mean altitude is displaced vertically downwards from the plan.

In the figure, the mass M is constrained to move vertically in the coordinate Y. The spring and damper, k and c, respectively, link the motion of the mass to the wheel that is constrained to move vertically in the coordinate X. A spreadsheet may easily be prepared using the knowledge gained above that determines the combined response of the system to the step input supplied and the simultaneous start of a harmonic input (Figs. 8.7–8.9).

The input data are as before, but the derived data have been added. Cells C24 and C25 contain the amplitude ratio 'X/Y' and the phase angle offset for the harmonic response term *Phy-harm*. Column 'G' contains the response to the step input as before, column 'H' contains the response to harmonic input determined as above, and column 'I' contains the total response that is the sum of 'G' and 'H'. The graph shows the total response of the system, and it is clear that the transient response to the step input fades away after a few cycles, and we are left with the steady-state response. Comparison of the graph to the step input depicted in Fig. 8.10 shows that the response is certainly believable; the body drops as a results of the step and being underdamped and bounces a few times before that motion decays away. After this, the motion is sinusoidal with a different amplitude to the input given by the amplitude ration and lags behind the input by a phase angle *phy*. The diagram below which shows the input profile and output profile on the same axis makes these observations abundantly clear.

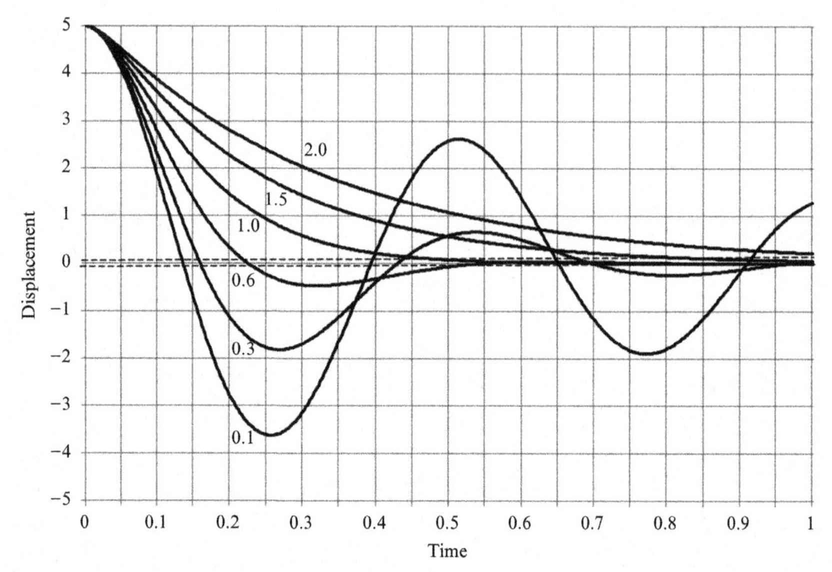

Fig. 8.7 Settling time for a range of damping values.

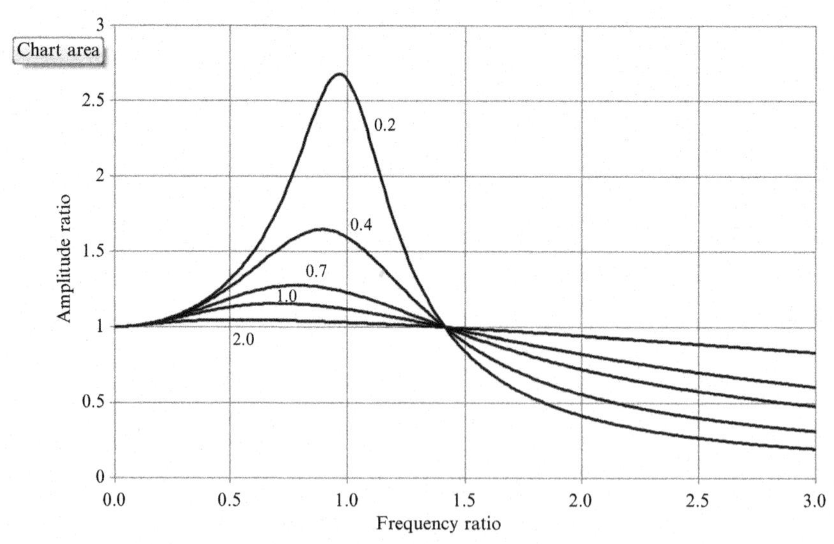

Fig. 8.8 Amplitude ratio versus frequency ratio with different zeta values for an SDoF.

We have now developed a sound ability to model the response of a single-degree-of-freedom system to a step input, a harmonic input and a combined input. We shall now turn our attention to seeing how this understanding may be used to optimise a suspension.

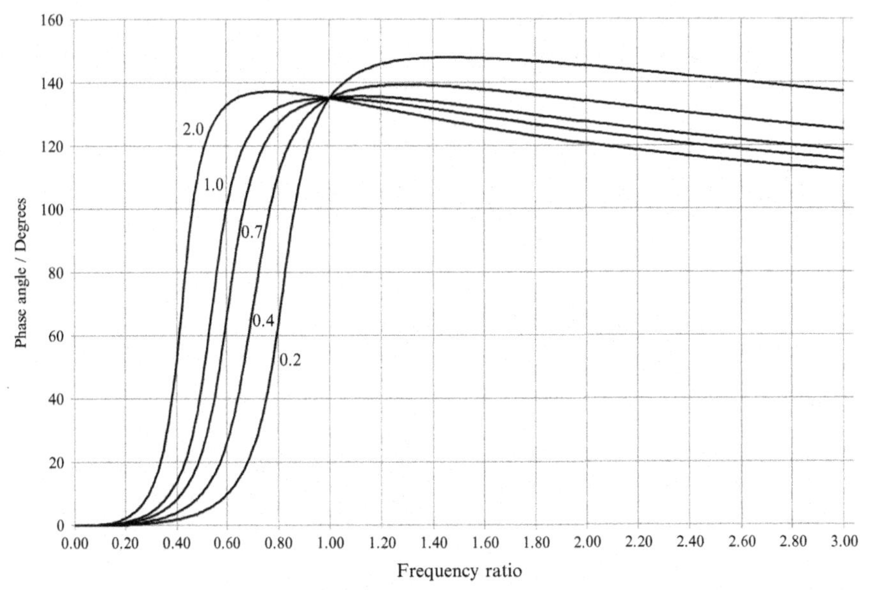

Fig. 8.9 Phase angle versus frequency ratio for a range of damping ratio values for a single-degree-of-freedom system.

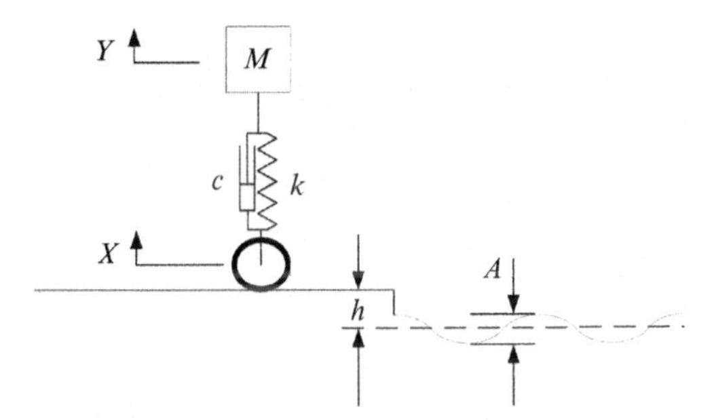

Fig. 8.10 SDoF system with combined harmonic and step input.

8.2.3 Optimising Using a Single Degree of Freedom Analysis

We have to start somewhere with an attempt to optimise a suspension using an SDoF, and the best start is a simple one. We shall examine the SDoF under the conditions of harmonic input of constant amplitude swept sine wave. This has the distinct advantage that we can solve the situation analytically, and it must at least in some way be representative of the suspension response. It certainly fails to deal with real road profiles and with bumps, and there are no uprights, pitch modes, etc., but it is a start. The rest will follow. Even with this highly simplified model, some very useful conclusions can be drawn.

8.2.3.1 SDoF Ideal Spring and Damper Values Swept Sin and Step Input

From the analysis above, we developed equations for the amplitude ratio of the body displacement to input displacement and also for the phase angle between the input waveform and body displacement. What it needed here is an equation for the force transmitted. This is given by

$$F = k(x - y) + c(\dot{x} - \dot{y})$$

Also,

$$F = m\ddot{x} = m\omega^2 X \sin(\omega t + \varphi) = F_T \sin(\omega t + \varphi)$$

where F_T is the amplitude of the force transmitted:

$$\begin{aligned} F_T &= m\omega^2 X \\ &= m\omega^2 Y \left[\frac{1 + (2\varsigma r)^2}{(1 - r^2)^2 + (2\varsigma r)^2} \right]^{\frac{1}{2}} \\ &= \frac{km\omega^2 Y}{k} \left[\frac{1 + (2\varsigma r)^2}{(1 - r^2)^2 + (2\varsigma r)^2} \right]^{\frac{1}{2}} \\ &= \frac{k\omega^2 Y}{\omega_n^2} \left[\frac{1 + (2\varsigma r)^2}{(1 - r^2)^2 + (2\varsigma r)^2} \right]^{\frac{1}{2}} \end{aligned}$$

It is convenient to arrange this equation for F_T in a normalised form to allow comparison between different systems. Thus, it is common to express this equation as

$$\frac{F_T}{kY} = r^2 \left[\frac{1 + (2\varsigma r)^2}{(1 - r^2)^2 + (2\varsigma r)^2} \right]^{\frac{1}{2}}$$

From this equation, we can conclude immediately that in order to minimise the contact patch force, we have three options: minimising Y, minimising k or minimising the bracketed term. The first option is not of interest since the terrain has its input amplitude and that is a given. We cannot alter that and instead seek to find the best performance for the terrain in hand. The second option involves minimising k. Certainly, from the point of view of minimising contact patch force variation, the smaller k the better. However, this will lead to other problems. The soft spring will mean that road inputs produce only small increases in spring force, and so, the body will rise by only a small amount. If the vehicle encounters long-wavelength inputs, the suspension travel will have to be sufficiently large to deal with them. Nevertheless, we can conclude that the contact patch force variation will be least when the spring rate is as small as possible. We shall simply have to use other criteria, such as *bottoming out* of the suspension to set its value. This leaves the possibility of minimising the bracketed term, and to start with, we consider its graph below.

With the graph in mind, we can see that selecting an optimum damping value is problematic. At high frequencies, meaning above a frequency ratio of $\sqrt{2}$, the lower the damping ratio, the lower the force transmitted (Figs. 8.11–8.13). Below $\sqrt{2}$, the higher the damping ratio the lower the transmitted force. Alternatively, we could select a range of input frequency to be important. For example, for simplicity, let us suppose that a frequency ratio range of 0–4 was determined to be important. We could then choose the value of damping coefficient that minimises the response over that range.

Fig. 8.14 shows the results of applying this process over the frequency range r=0–4. As can be seen, a damping ratio value of somewhere around 0.1 returns the lowest value of average contact force.

For the spring rate, it is trivial to see that the softer the spring, the less the spring force will vary, and we can conclude that for the minimisation of contact patch force variation, we require the lowest possible spring rate (we don't have a way of knowing the lower bound yet) and a damping ratio of around 0.1, one-tenth of critical.

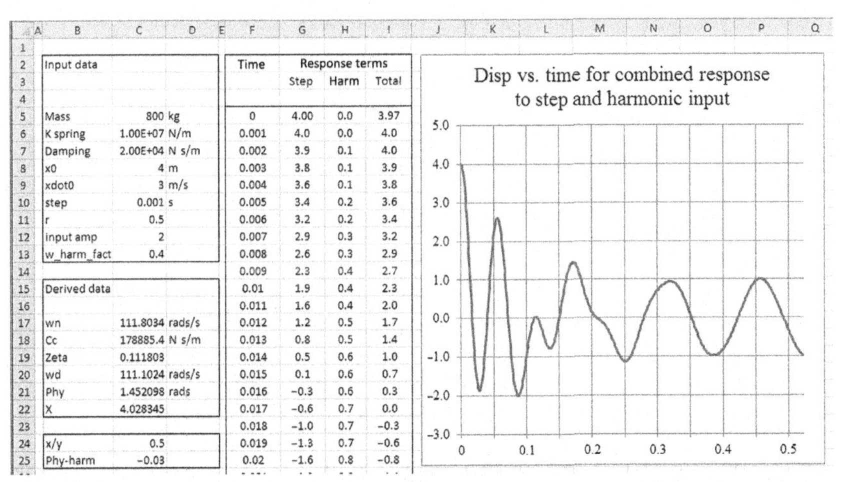

Input data		
Mass	800	kg
K spring	1.00E+07	N/m
Damping	2.00E+04	N s/m
x0	4	m
xdot0	3	m/s
step	0.001	s
r	0.5	
input amp	2	
w_harm_fact	0.4	

Derived data		
wn	111.8034	rads/s
Cc	178885.4	N s/m
Zeta	0.111803	
wd	111.1024	rads/s
Phy	1.452098	rads
X	4.028345	

x/y	0.5	
Phy-harm	-0.03	

Time	Response terms Step	Harm	Total
0	4.00	0.0	3.97
0.001	4.0	0.0	4.0
0.002	3.9	0.1	4.0
0.003	3.8	0.1	3.9
0.004	3.6	0.1	3.8
0.005	3.4	0.2	3.6
0.006	3.2	0.2	3.4
0.007	2.9	0.3	3.2
0.008	2.6	0.3	2.9
0.009	2.3	0.4	2.7
0.01	1.9	0.4	2.3
0.011	1.6	0.4	2.0
0.012	1.2	0.5	1.7
0.013	0.8	0.5	1.4
0.014	0.5	0.6	1.0
0.015	0.1	0.6	0.7
0.016	−0.3	0.6	0.3
0.017	−0.6	0.7	0.0
0.018	−1.0	0.7	−0.3
0.019	−1.3	0.7	−0.6
0.02	−1.6	0.8	−0.8

Fig. 8.11 Spreadsheet determination of the combined response of a single-degree-of-freedom system to step input and simultaneous start of harmonic input.

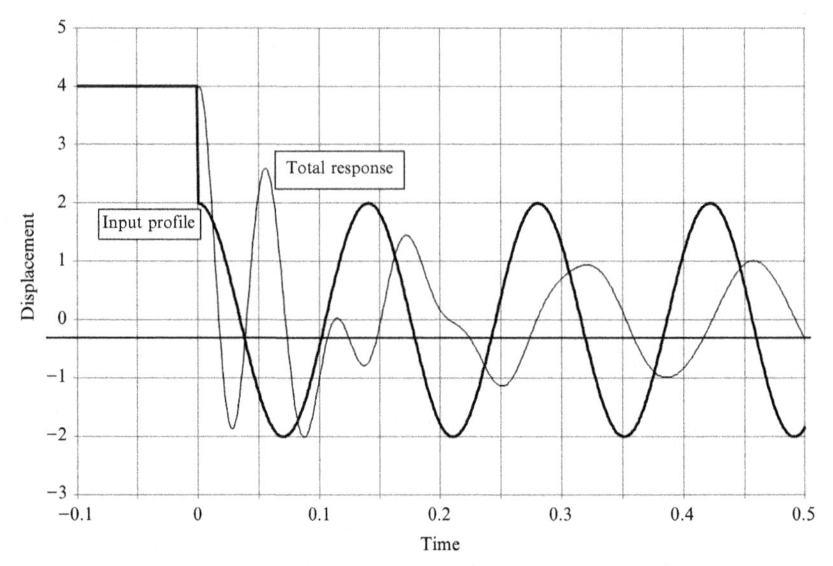

Fig. 8.12 Input and response for combined input.

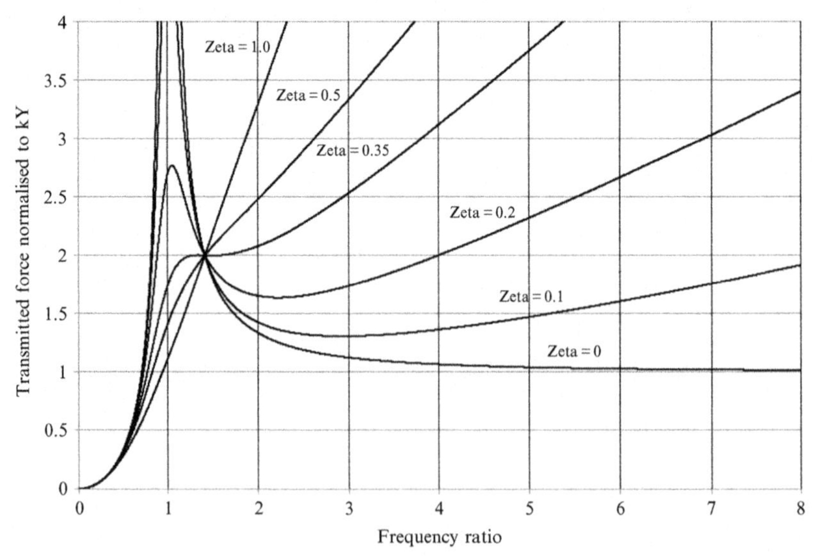

Fig. 8.13 Normalised transmitted force against frequency ratio for a range of damping values.

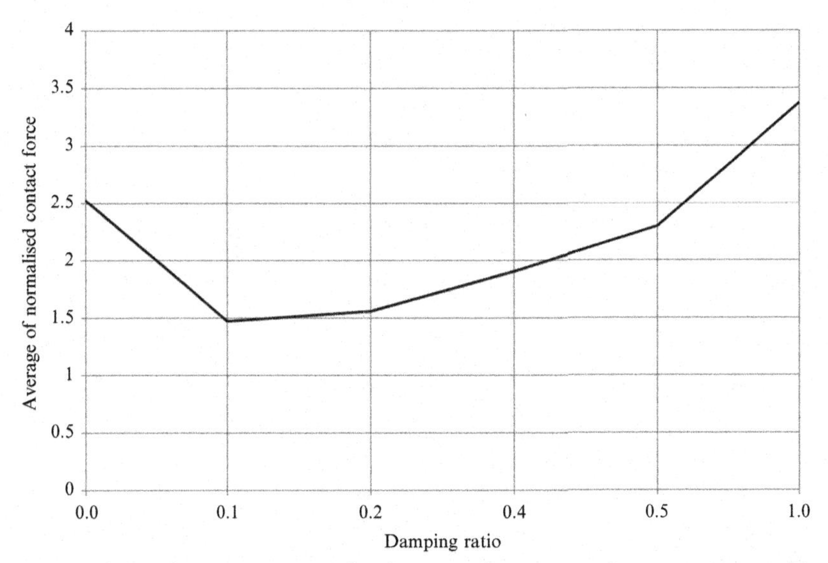

Fig. 8.14 Shows the average normalised contact force over a frequency range of 0–4 against damping value.

From the point of view of minimising the spring force for a transient step input, the material above shows that a damping value of 1.0 is best, and again, the lower the value of spring, the less the spring force varies. We may summarise the findings for the SDoF system with either constant amplitude, swept sine input or step input as follows:

Input	Ideal damper value	Ideal spring value
Step	1.0	As low as possible
Harmonic below $r=\sqrt{2}$	As high as possible	As low as possible
Harmonic above $r=\sqrt{2}$	As low as possible	As low as possible
Harmonic over the range $0 \leq r \leq 8$	0.1	As low as possible

8.2.3.2 Real Inputs, Road Surface Profiles and Pot-Holes

The analysis above is presented for two main reasons. Firstly, it is an excellent starting point that any engineering student with a knowledge of dynamics would rightly pick as a place to get going. Secondly, it offers some real insight into dynamic system behaviour. The response to step input, the harmonic response and the combined response are all going to be important as our analysis of suspensions advances.

However, there are many shortcomings in the treatment above, and we need to put them right before proceeding. The main problems stem from the inputs provided to the system. The input to the system was assumed to be either a step input or a harmonic one of swept frequency with constant amplitude.

In the first case, the step input, the problem is that the step response gives us what happens after the step, not during it. For example, if a car drives over a steep rise, the wheel will rise very rapidly indeed. The step response will show what happens in response but not tell us, for example, what the force in the suspension spring was during the bump. The second problem is that the input is assumed to be a swept sine wave of constant amplitude. This is far from true as we shall see. We must therefore turn our attention to more realistic ways of representing the input to the single-degree-of-freedom system before proceeding to use it further.

For the time being, whilst considering the SDoF, we shall limit the range of road inputs considered to two sources. The first is the harmonic profile describing the undulations from the road surface as the car proceeds. The second is the profile experienced when the car negotiates excursions in the road surface such as potholes. These are considered in turn.

8.2.3.3 Harmonic Road Profile

In the previous section, we approximated the road profile to be a constant amplitude sine wave of increasing frequency. Clearly, this must be incorrect. One would not be surprised to find a road surface had components with amplitudes of a metre or more if the wavelength of these components were in the hundreds of metres. If, however, the wavelength was just a few metres, then an amplitude of a metre would be more of a low wall than a bump. Thus, it is clear that real roads have amplitudes that depend on their wavelengths. Long are large, short are small. But what does a real road profile actually look like? We need to answer this first and then develop a profile that is representative. It would be unwieldy to have to use input profiles lasting as long as an entire journey so we are seeking a representative wave of acceptably short duration to use as an input to the SDOF. This process starts with the profile of real roads.

The earliest known contracts involving the *levelness* of a road date back to the Roman Empire where sticks were used to measure the local road surface height relative to a long planks on the road surface. This is the idea behind the early wheeled profilometer shown in Fig. 8.15.

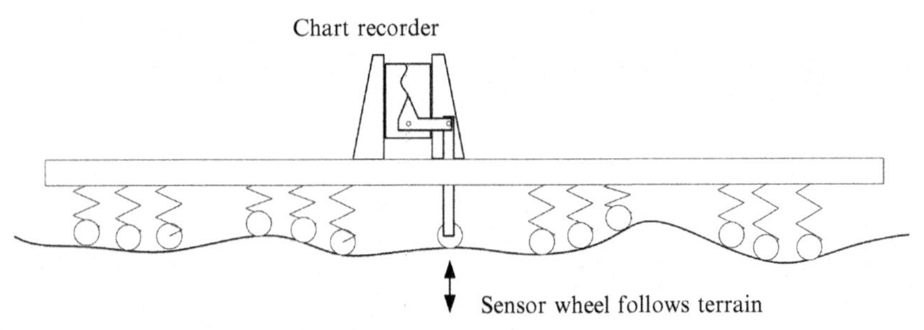

Fig. 8.15 An early design of profilometer.

Clearly, methods such as this have limitations; the whole vehicle will rise and fall a little as is moves forwards. It is also restricted to recording vibrations of wavelengths rather less than the vehicle itself. The modern approach is to measure vertical acceleration at the upright of a test vehicle and then, by using a validated dynamic model of the suspension, determine what road profile must have been experienced in order to produce the acceleration log recorded.

A possible road profile is shown in Fig. 8.16 above, in which the vertical axis is in millimetres and the horizontal one in metres. For the purposes of our analysis, we seek a wave that accurately represents the one shown but is practical and simple function. Ideally, it should have an analytical wave form of acceptably short duration that it can be used for practical tests on a four-post test rig where it is possible compare theoretical estimates of expected suspension performance with actual measured values.

A function is needed that is swept through the frequency domain. A car travelling over a road surface receives inputs that simulate its suspension in

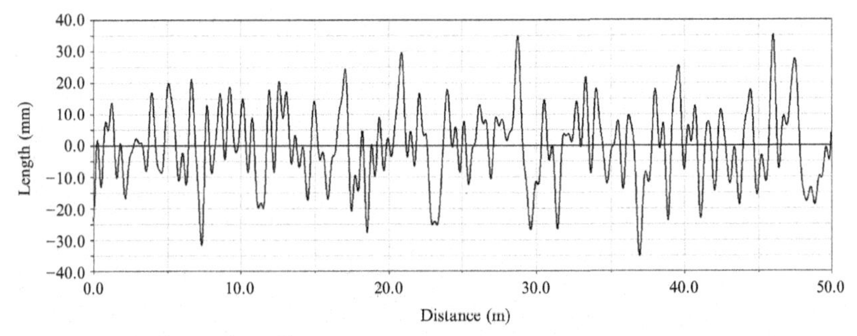

Fig. 8.16 Example road profile.

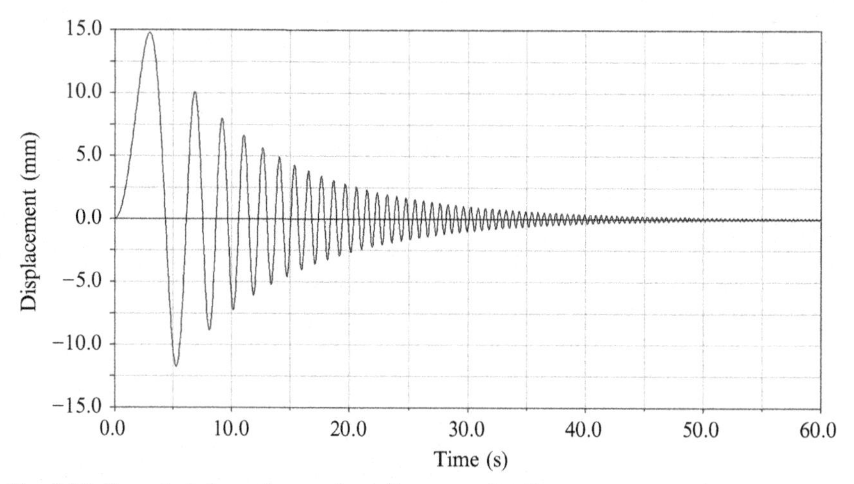

Fig. 8.17 Expected shape for road profile approximation.

the range 0.5–40 Hz. Much below this range and the body simply moves up and down with the road surface with negligible movement in the suspension. Much above this range and the tyre simply accommodates all the movement without communicating it to the upright at all.

Given the earlier comments on how the longer wavelengths must have larger amplitudes, we should expect our function that represents the road surface to look something like

The function shown in Fig. 8.17 is of a sinusoidal wave in which the amplitude decreases exponentially and the frequency increases quadratically with advancing time. It is a guess for a waveform that represents the road. To evaluate whether the waveform does indeed represent the road, we shall make use of the 'power spectral density' concept.

Power Spectral Density

One approach familiar to many engineers is that of Fourier analysis. In this technique, one can decompose a waveform into its component sinusoidal components. For example, if we have a waveform consisting of just three sinusoidal components and three different frequencies, then in the time domain, we would see a wavy graph containing the three components. In the frequency domain, after using a Fourier transform, we would have an output line that was zero except at the three constituent frequencies, where there would be a peak with a height corresponding to the amplitude of that wave in the original waveform at that frequency.

We could use this technique to assess whether a proposed road profile is representative of the real road by taking the Fourier transform of the road and of the approximation, and if the resulting transforms are very similar, then the approximation must be valid since they both contain equal quantities of each component, even if their overall shapes are very different.

This idea works very well for waveforms that consist of a large number of oscillations at any given frequency. However, we are seeking a waveform that of its very nature will have a rapid frequency sweep; we don't want the wave to be any longer than necessary.

Decomposing a swept frequency wave into its frequency components is problematic mathematically. Consider a function such as that in Fig. 8.17. If we now try to estimate the amplitude of the component that has a frequency of exactly 0.3 Hz, there clearly must be a component at this frequency since at times between 0 and 10 s just one wave is completed (much less than 0.3 Hz), yet between 40 and 50 s, around 10 cycles are completed (much more than 0.3 Hz). However, what would be the amplitude of the frequency component at *exactly* 0.3 Hz? If we were to find the point at which the waveform was at exactly 0.3 Hz, there would still be a problem because within just the one cycle, the frequency would have increased. Each individual cycle is faster at the end than the start so perhaps in theory at least, the amplitude of any given precise frequency is zero! Instead, we need to be able to consider frequencies that lie within a given range and not at a given value. To do this, the method of power spectral density is used.

Fig. 8.18 above shows the power spectral density for two roads, 'A' and 'B'. We interpret the graph in the following way. We start by choosing a frequency interval, for example, between 1 and 2 Hz, as shown. The amplitude of the wavelengths between these two frequencies is given by the square root of the area of the rectangle shown. This result comes from lengthy treatments of frequency analysis not reproduced here. Instead, it is quoted here because of its importance. It allows us to move to a waveform that faithfully represents the road without actually being the road but instead a much simpler and easy to manipulate function.

In Fig. 8.19 below, a waveform that offers a very good approximation to a typical road surface is shown. It is a good approximation in the sense that when its PSD is compared with that of a real road, they agree well.

The input amplitude waveform consists of a sinusoidal wave of constant peak velocity. Such a signal is easy to generate analytically. A simple version of the velocity waveform is given by a sinusoidal waveform of the following form:

$$\dot{x}(t) = \sin\left\{k_1 \frac{\left(k_2^T - 1\right)}{Ln(k_2)}\right\}$$

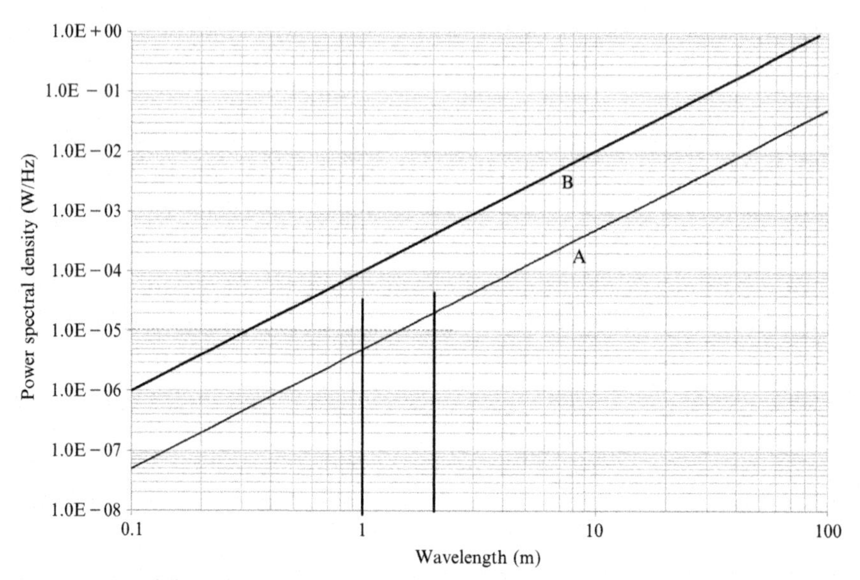

Fig. 8.18 Graph based on ISO 0608, road PSD versus wave number.

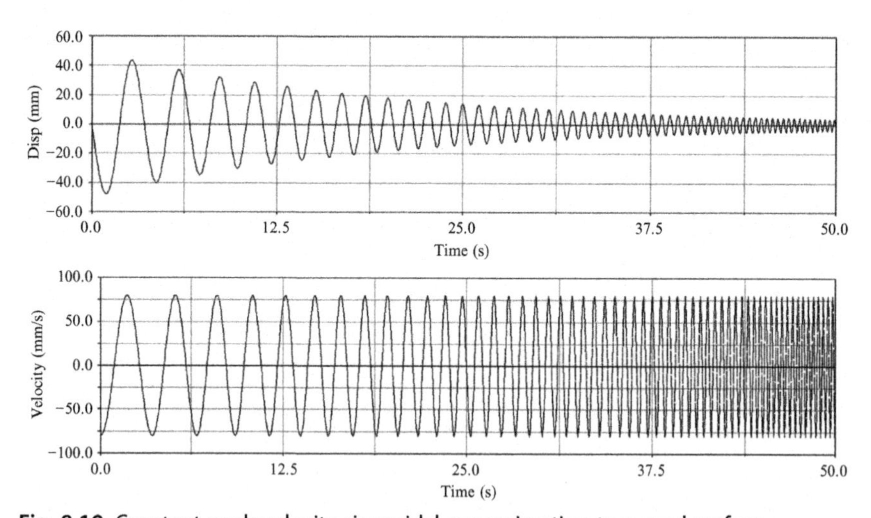

Fig. 8.19 Constant peak velocity sinusoidal approximation to a road surface.

where

$$k_1 = 2\pi f_0$$
$$k_2 = 1.087629$$

and f_0 is the frequency at which the sweep starts.

A very marked difference between this waveform and the swept sine one used earlier is that in this case, the velocity has a constant peak value. Thus, the damper will be stroked at the same maximum value at *every* frequency. In the case of the swept sine wave of constant amplitude, the peak velocity raises with frequency. Here, it is constant and takes a value of around 80 mm/s. The corresponding input amplitude starts at around 40 mm.

We may finally modify the wave a little to make it more practical, particularly for use with a four-post test rig.

Fig. 8.20 shows a well-prepared four-post rig input waveform. The wave starts with a dwell at zero, and the oscillatory component builds up rapidly from this. The first couple of cycles are the same amplitude, and this is enough to get the vehicle through the transient associated with the rapid initial evolution of the amplitude. After this, two important factors are at work. The first is that the amplitude decays. This decay is a specially selected function above and permits a wave whose duration is acceptably short, in this case, just 60 s. The second affect at work is that the frequency increases as described above. At around 58 s, the function is seen to have a fade out, so that the rig actuators can be guaranteed to be at rest when the input stops.

For a single-degree-of-freedom system, we need only model the vertical input applied directly to the 'tyre'. In a real vehicle, the situation is much more complex. Each wheel could move independently, and the vehicle

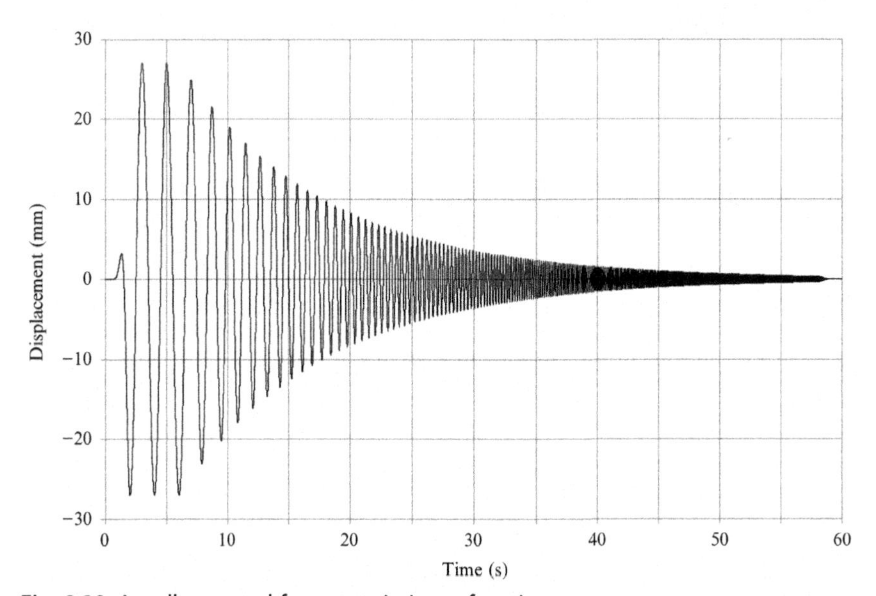

Fig. 8.20 A well-prepared four-post rig input function.

be stimulated to vibrate in four modes, heave, pitch, roll and warp. We shall deal with these cases in due course.

Such a waveform cannot be fed into an SDoF model and then analysed *analytically*, as the resulting maths is simply insoluble. Instead, we shall have to make use of numerical computing to solve the suspension modelling problems from now on. This was an inevitable occurrence as we move towards two degrees of freedom and above, the equations are bound to become too unwieldy to use at some point. Computer packages able to do this sort of analysis easily include the multibody codes such as Adams, the symbolic modellers such as MatLab's Simulink or even one's own spreadsheets using a numerical approach rather than an analytical one.

8.2.3.4 Excursive Style Step Inputs Such as Pot-Holes Etc

We now consider the second type of input we want to use; that of the pot-hole (Fig. 8.21). Various standards have been presented on road data and working from them a good approximation to a pothole is given by the following dimensions.

If traversed at 50 mph, this feature would be passed in

$$v = \frac{s}{t} \Rightarrow t = \frac{s}{v} = \frac{10 \times 10^{-3}}{50 \times 0.447} = 0.0004\,\text{s}$$

The upward velocity would be

$$\begin{aligned}\frac{50 \times 10^{-3}}{0.0004} &= 125\ \text{m/s}\\ &= 125{,}000\,\text{mm/s}\end{aligned}$$

The peak value of velocity here is well over a thousand times faster than for the harmonic road input. Given that the force transmitted through the damper is proportional to the stroking velocity, it is no surprise that hitting a pothole at 50 mph causes a very noticeable jolt to the chassis.

In reality, the tyre would not communicate such a large velocity to the upright. Using a 2DoF model of the system such as that prepared below, it is

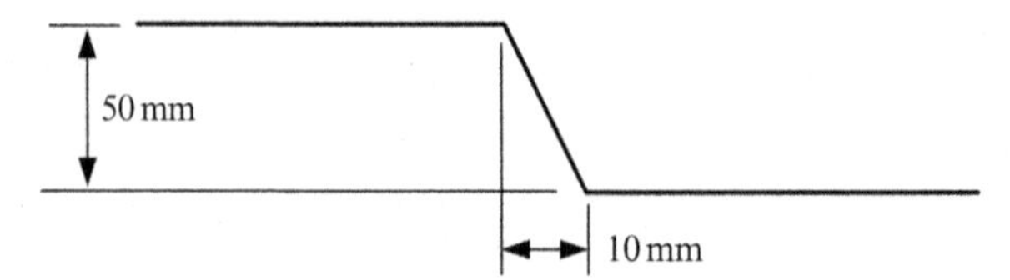

Fig. 8.21 Pothole standard dimensions.

possible to determine the peak damper speed that can result from the negotiation of a step bump of this kind, and it is fact in around 270 mm/s. In the interests of avoiding the rather spurious results that would come from using inputs as fast as 125,000 mm/s, we shall make use of the sneak preview result of the 2DoF system and approximate a step input to be 50 mm high bump but being traversed in 0.185 s. This forward speed gives a vertical velocity of approximately 270 mm/s, and this is representative of a pothole excursion for modelling an SDoF. Clearly, something of an approximation but it is at least a logical one.

We therefore conclude that bumps of this kind impart velocities well above those that result from the road profile inputs. It is tempting to think that these pothole style inputs only operate in the upward direction, but this is not the case. The precompression in the suspension spring is able to accelerate the upright downwards at very high speed in rebound if the pothole permits. The situation is different for the occupants of the vehicle, they can be accelerated upwards at whatever the chassis and seats supplies but only downwards at roughly 1 g, and this is indeed important for comfort, as we shall see.

Thus, there are two inputs that need consideration for the SDoF:

- Harmonic road profile of constant peak velocity and exponentially increasing frequency (peak damper velocity 0–85 mm/s) as given above.
- Step input consisting of a haversine displacement input of the base of 50 mm competed in 0.185 s (damper velocity 0–270 mm/s).

(Note a 'haversine' function is a function used to approximate a step in multibody code modelling. One cannot have an actual step since this involves discontinuities and has undefined gradients at the step. Instead, two half sine waves are joined to from a 'sinusoidal' step. Such a function can be differentiated as many times as desired and will still be continuous.)

8.2.4 Optimisation of Damper Function Using SDoF With Real Inputs

We shall now optimise the SDoF in terms of the vertical acceleration and its dependence on damping for each of the two inputs above. We start by briefly reviewing the numerical model SDoF system that shall be used for the analysis.

The model used was created in the multibody code Adams and is shown in Fig. 8.22. In the figure, the upper block models the vehicle body and is supported by a spring damper above the lower block. This lower block models the road. The motion of each block has been set up in the package to be displayed in the insert graphs. It can be seen that the road input follows

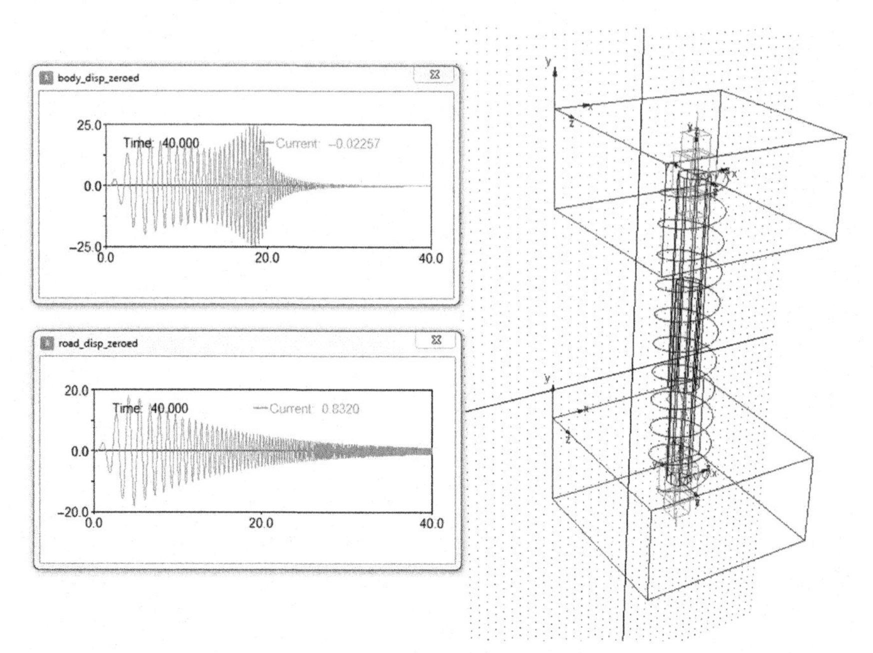

Fig. 8.22 Adams model of a single-degree-of-freedom system.

the form of the constant peak velocity input function developed above. The motion of the body shows a sinusoidal form whose amplitude at the very start rises as the input function rises. It then starts to reduce following the reduction in input amplitude. Shortly before 20 s, a resonant region is reached, and the amplitude increases to a peak. After this, it again reduces rapidly for two reasons. The first is that the response basic SDoF response function decreases after resonance (as shown in Fig. 8.8) and secondly because the amplitude of the input function itself is also decreasing.

To illustrate this, consider Fig. 8.23. This shows the response firstly of an SDoF system to a swept sine input and takes exactly the form shown in Fig. 8.8 above.

This is the dotted line with the higher response at resonance. The second, solid lower line shows the amplitude response of the system when provided with the road profile. Initially, at low frequency, there is little effect since the initial amplitudes are the same and the response function is unity. As the frequency is increased, the input amplitude of the road profile has reduced so much by resonance that the peak scarcely exceeds the initial height. At high-frequency values, the two curves are again similar because both the road input and the amplitude ratio tend to zero.

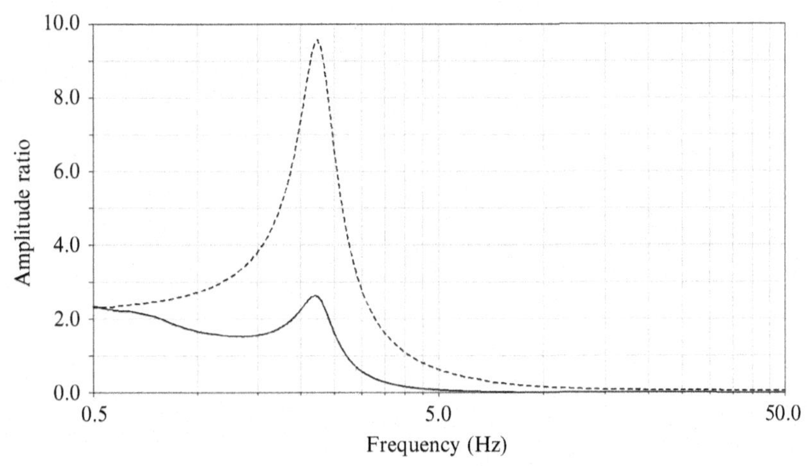

Fig. 8.23 Amplitude response of an SDoF system for a constant amplitude swept sine and road profile inputs.

8.2.4.1 Analysis Using Harmonic Road Profile Input

The SDoF system can be analysed by using a sweep of damping ratio and graphing the average of the body displacement and acceleration (Figs. 8.24 and 8.26). This analysis is shown below, starting with a graph of body displacement as a function of time for a low and high value of damping ratio.

The low damping, zeta = 0.1, is shown as a dotted line and has a peak at around 16 s. At this time, the frequency was just below 2 Hz, and this is the natural frequency of the system. The second solid line shows the time response for a damping ratio of just over 1.0. It can be seen that as the

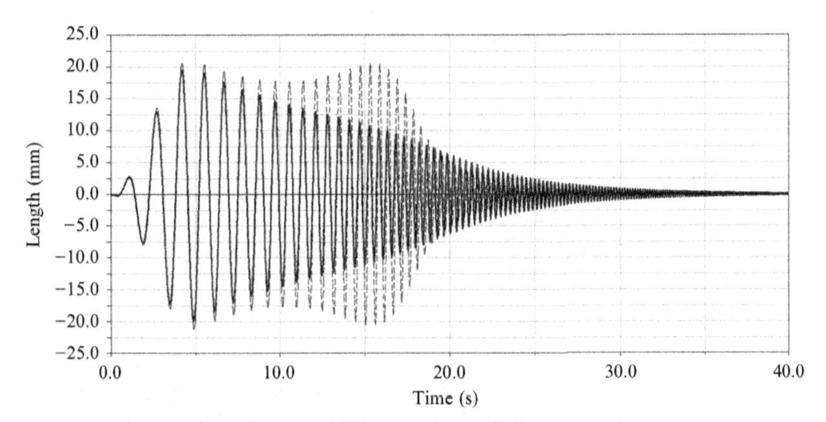

Fig. 8.24 Body response for two different values of damping.

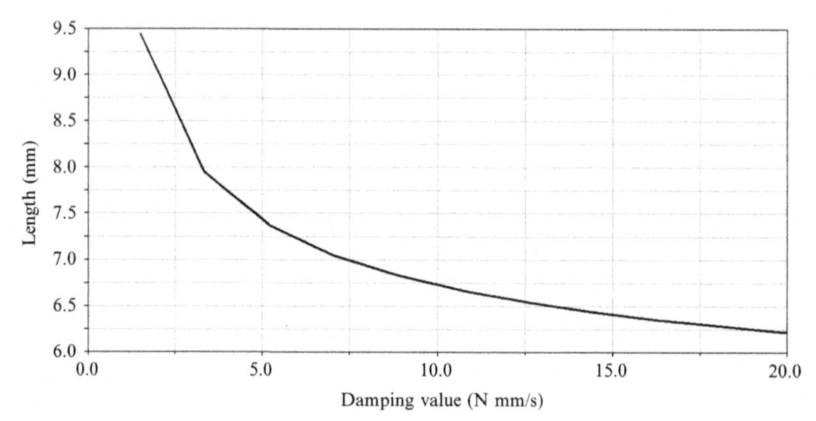

Fig. 8.25 RMS body displacement versus damping.

damping is increased, the RMS of the response also decreases. This is shown more clearly in the graph below where the RMS of a series of analyses with different damping values were prepared and the RMS for each plotted against damping value (Fig. 8.25).

We conclude that the larger the damping, the smaller the body vibration RMS becomes. This is because the larger damping progressively restricts the size of the resonance, and in the end, the displacement will follow the input displacement. We now turn attention to the acceleration of the body as shown in Fig. 8.26. The time domain response is shown again for damping values of 0.1, dotted line, and just over 1.0, solid line. It can be seen that as the damping increases, the peak at resonance gets smaller but the steady-state

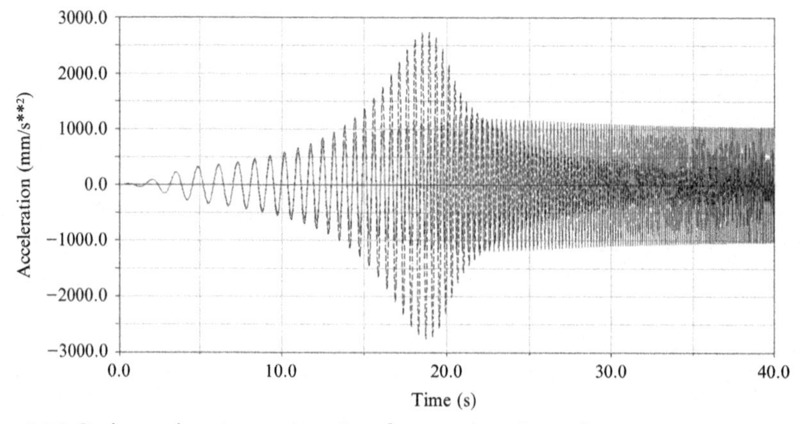

Fig. 8.26 Body acceleration against time for two damping values.

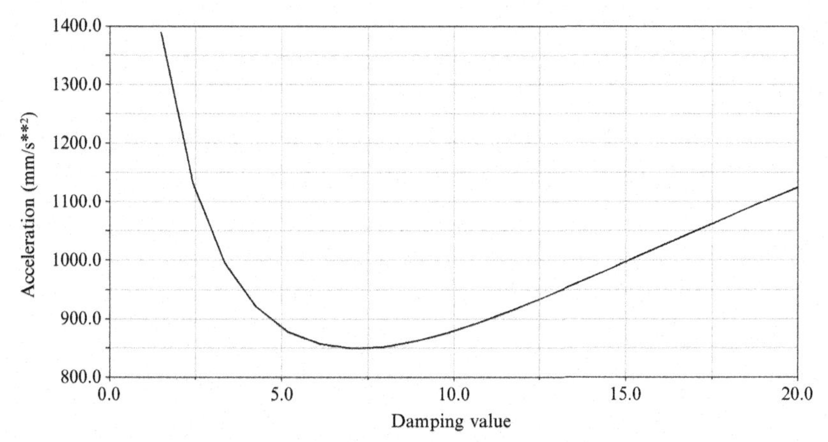

Fig. 8.27 RMS body acceleration versus damping value.

response at high frequencies gets larger. This is because as the frequency of the input waveform increases, its peak velocity stays constant but the acceleration increases. This is transmitted to the body more if the damping is higher.

One should expect a minima to occur between these extremes, and indeed, it does as Fig. 8.27 shows.

In this figure, we see that a minima occurs at a damping value of just over 7 N m/s. With a critical damping value of 14, this corresponds to a damping ratio of just over 0.5. The minima occurs because if the damping value is lower, then the RMS increases due to a larger resonance peak and if larger, then the body is too rigidly connected to the road at higher frequencies where accelerations are high because the frequency is high. Since, in the SDoF system, the body acceleration is determined directly by the suspension spring force, we conclude that the value of damping determined here is ideal not only from the point of view of body acceleration but also from that of the contact patch force too.

8.2.4.2 Analysis Using Step Input to Simulate Pot-Hole Type Transients

Earlier, we developed a step input to simulate a pothole type excursion. The displacement input provided at the 'road' to the SDoF system above is shown in Fig. 8.28.

Using the same SDoF system as above, Fig. 8.29 can be prepared showing the body vertical displacement response to the haversine input as a function of time for different damping ratios.

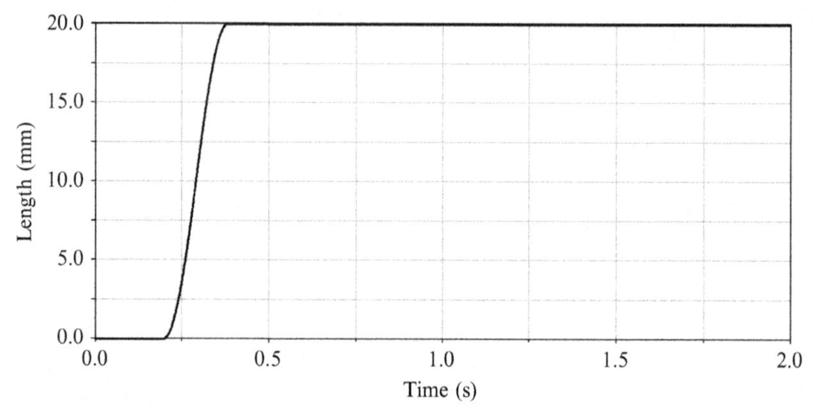

Fig. 8.28 Displacement input for pothole simulation SDoF.

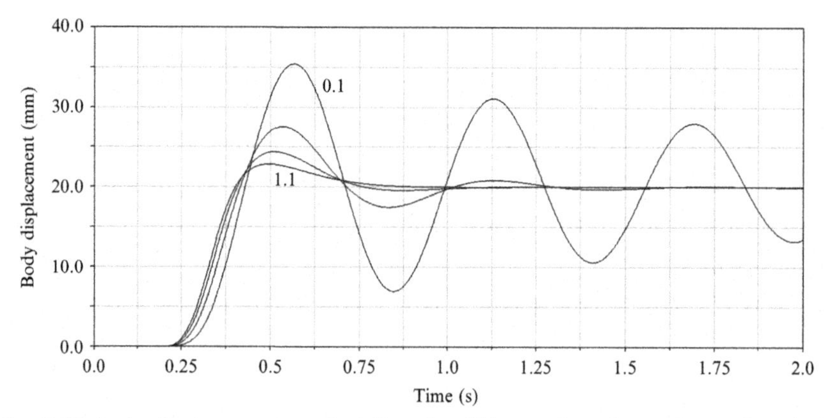

Fig. 8.29 Body displacement against time for different damping ratios and pothole input.

We can see that as the damping is increased from a low value of 0.1 to a high one of around 2.0, the amount of overswing reduces. Fig. 8.30 shows the RMS of body displacement given in Fig. 8.29, and it can be seen that as damping increases, the RMS of the body displacement continues to get smaller. We conclude that from this point of view, the larger the damping, the better.

Fig. 8.31 below shows the body acceleration for the same range of damping values. As damping increases from the low to a high value, we move from the line with most overswing through to the line that settles the quickest. The behaviour is complex. The impulse displacement is applied between 0.2 and 0.385 s and during this period, the body is initially

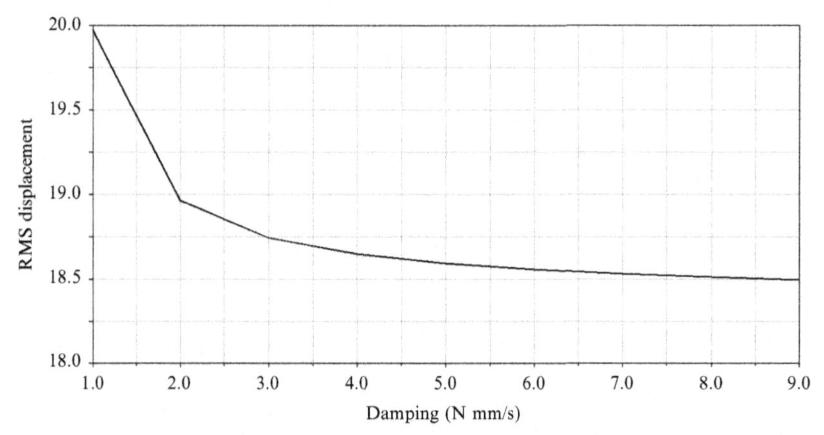

Fig. 8.30 RMS body displacement against damping value.

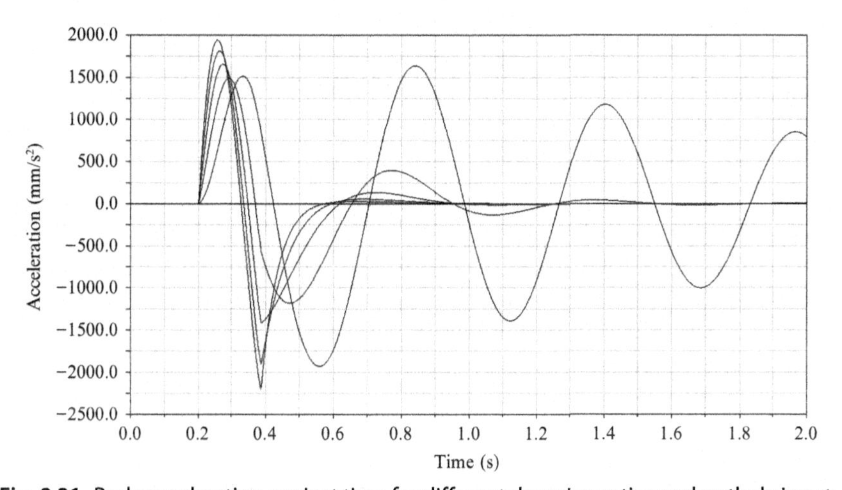

Fig. 8.31 Body acceleration against time for different damping ratios and pothole input.

accelerated upwards, but then, as the rising road slows down, it decelerates. There is a relatively abrupt point shown in the high damper curves at the end of the impulse displacement. This is because when the damping is high, it can communicate high forces leading to larger accelerations. When low, this can't happen, and the acceleration graph shows no such abrupt changes.

In general as damping increases, the curve becomes closer to the axis, and its integral decreases. However, at high damping, the initial peak increases, and so, we should expect that there is a minima in the RMS of the acceleration graph. Fig. 8.32 below shows that there is indeed such a minima, and it occurs at a damping value of 5 N s/m.

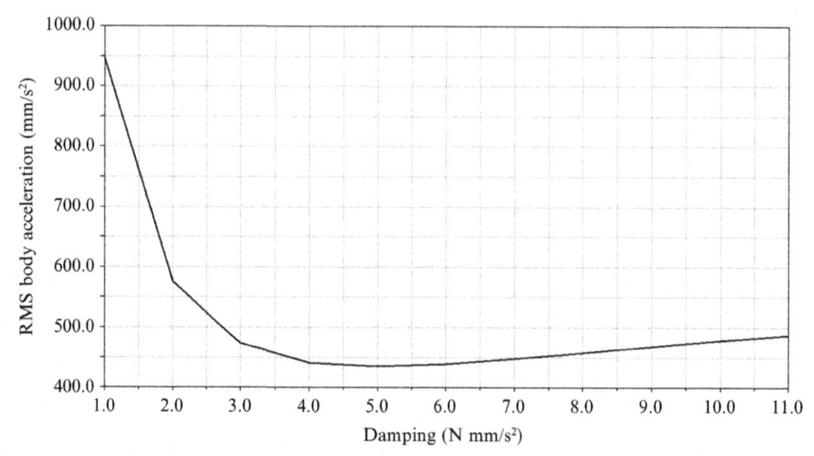

Fig. 8.32 RMS body acceleration against damping value pothole input.

This corresponds to a damping ratio of 0.18, and we conclude that this value of damping ratio is optimal for this kind of input.

8.2.4.3 SDoF Idealised Damper Function, Combined Inputs, High and Low Speed

In the sections, above we have examined the behaviour of a single-degree-of-freedom system to a range of three inputs. The response has been determined for differing bump and rebound values of damping, and we are now in a position to draw conclusions and suggest a 'best' damping function based on the SDoF model. The results are summarised:

Input profile	Ideal damping ratio	Max damper velocity mm/s
Harmonic road profile	0.5	80
Step input for pothole type excursions	0.18	270

The ideal damping value proposed for the harmonic road profile is a very persuasive value. It applies for the majority of the time since under normal driving conditions the tyre is in good contact with the road and vibrating with it. The damping value found to be ideal for a step input is rather lower, and we have the problem that the damper clearly can't have two values at once. However, we notice that the damper velocity associated with the harmonic inputs are always below around 80 mm/s, whilst the damper speed associated with the pothole are higher, around 270 mm/s.

This allows for the provision of an optimal damper for both inputs simultaneously. All that is required is to make the damper value depend on stroking speed. Using these two inputs, we conclude that for speeds up to 80 mm/s, a value of damping ratio of 0.5 should be used and above this 0.18. Further to this, there is the opportunity to make the damper function different in bump than rebound. For the case of the harmonic road input, this is not advantageous since the tyre is in constant contact with the road and there is no dynamic difference from bump to rebound. However, with the pothole, there is the possibility that the wheel will actually lose contact with the road surface. If this happens, then the suspension spring, sized to support the body will be accelerating the very light unspring mass downwards, and the resulting acceleration will be very high. If the damping is increased in rebound, the effect will be mitigated against, and a value ratio of around three is common. Thus, the bump-to-rebound ratio should be 1:3 for high speeds but 1:1 for low speeds. Such a function would look as follows for the car in the example where the critical damping value was 26.8 N s/m:

In conclusion, we can see that despite being a very simple system, the SDoF can, with care, be used to make complex and remarkably good predictions for an idealised damper function.

8.2.5 Optimisation of Spring Function Using SDoF With Real Inputs

Having been able to determine an ideal damper value based on the analysis of an SDoF system, we now need to do the same for a spring value. Earlier, we showed that the smaller the spring value, the better from the point of the minimisation of contact patch force variation for a steady-state input. For a step input, the same is true. However, if the vehicle encounters a section of uphill gradient, we would require that the spring was sufficiently stiff that it would accelerate the car body upwards sufficiently quickly to avoid grounding. In addition to this, we have the observation that most vehicles will have a variable payload, even in racing this happens as fuel is consumed. For a road car, the difference between one driver and five occupants with luggage is very significant especially for a small car.

Elementary consideration of the spring value shows that it cannot be considered in isolation from the available travel for the suspension. A soft spring will clearly result in more suspension movement than a stiff one—so how much would be too much? This depends on the layout of the car. A circuit racing car with high aerodynamic downforce will have little ground clearance, perhaps as little as 30 mm, and a desired suspension movement as little as 5 mm for aerodynamic reasons. An off-road 4 × 4 is clearly going to have rather more; as high as 400 mm. We can introduce the suspension travel limit to the SDoF schematic we met earlier as shown below.

In practice, the limit to the suspension travel in bump must not be a hard one, and is generally a 'bump stop' consisting of a conical rubber stop mounted rigidly on the chassis and making contact with a stiff and structurally sound moving part of the suspension. When road excursions such as potholes are met, loads can be very high and the bump stop arrangements must communicate wheel loads straight into very stiff parts of the chassis. In rebound, the requirement is that there is a limit to the suspension maximum extension. When a major excursion is met in rebound, the wheel simply leaves the road surface, and the limit stop must resist the load in the spring at full extension.

One approach to choosing the spring value is therefore to start with the acceptable suspension travel; analyse the input waveform from the road, including its occasional excursions; and select a spring value such that the bump stop is only engaged for an acceptably small percentage of the time, say 2%.

In practice, suspension travel that results from heavy braking that compresses the front suspension and extends the rear, or hard cornering, which compresses one side and extends the other, provides a lower limit for suspension stiffness. This limit is a separate consideration from the ideal value from the point of view of ride optimisation which, as we have seen, always requires the lowest possible value. Once the spring rate is chosen, the damper value can be optimised.

This process allows the determination of a particular spring rate, but significant improvements can be achieved by allowing the spring function to be non-linear. It is clear that the lowest spring rate practical is best and also that we need to limit the travel of the suspension. If the spring rate is made to increase as the bump displacement progresses than the fraction of time spent on the bump stops will be decreased, and a softer suspension at the design point can be used, making for better grip optimisation and a soft ride. A suspension with this characteristic is called a *rising rate* suspension and can be achieved through kinematic design using bell cranks, as is explained in Chapter seven. Alternatively, springs with increasing coil diameters that come into play as the spring becomes progressively more coil bound can be used.

Even in the case of a car with large aerodynamic downforce, the softest spring possible returns the best performance. It's just that in order to control a ground clearance and angle of attack, very stiff springs are needed. This brings with it a compromise in performance but more is gained by the downforce than is lost by suspension compromise. If aerodynamic loads could be passed directly into the uprights the best of both worlds could be achieved, but this is almost universally banned.

8.2.6 Worked Example—SDoF Optimisation

Using the method above for the optimisation, determine the ideal spring and damper function for a vehicle with the following dynamic data:

Vehicle mass	700 kg
Wheelbase	2.6 m
Front axle to CoG distance	1.1 m
Maximum suspension movement	±40 mm
Nonbump stop at max deceleration	0.8 g
Centre of gravity height	1.2 m

Compare these with the values in Figs. 8.33 and 8.35.

We start by estimating the lowest acceptable spring stiffness from the point of view of clearance at maximum deceleration (Figs. 8.34 and 8.35).

By balancing moments about the rear axle, we have

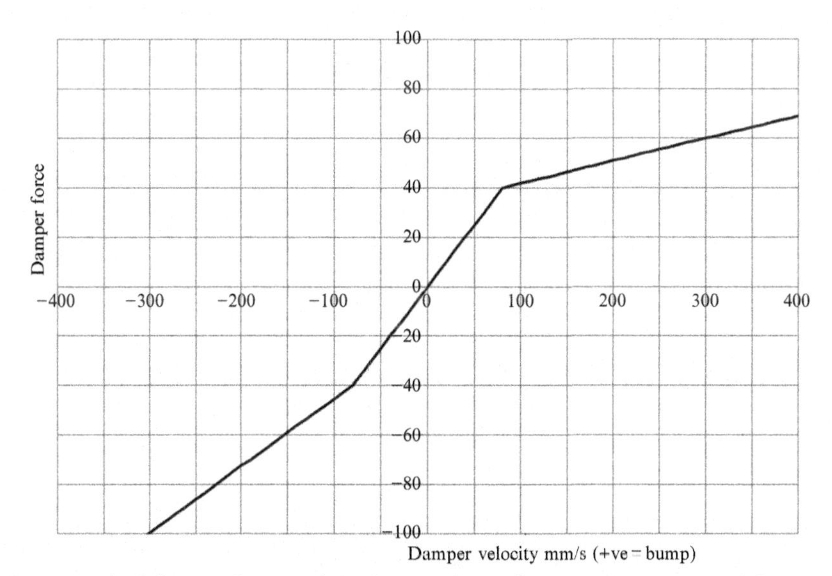

Fig. 8.33 Ideal damper function based on analysis of SDoF suspension model.

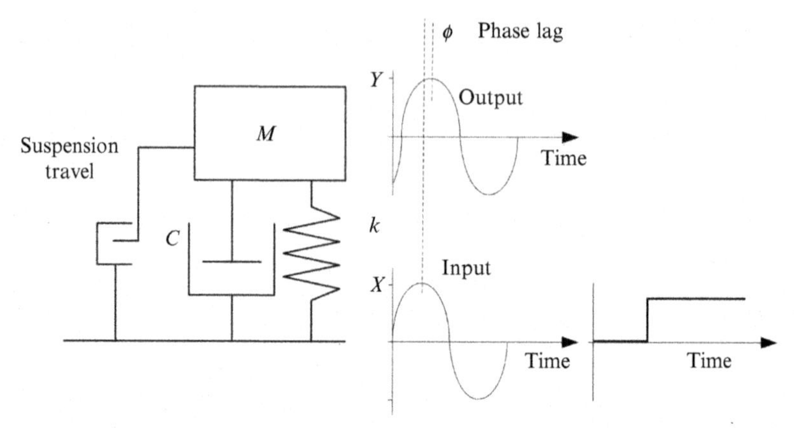

Fig. 8.34 Ideal spring function must include consideration of bump stops.

$$F_{ZF} = F_{SZF} + \frac{h}{l} m a_X$$

also $F_{SZF} = \frac{1}{l} mg \times b$

$$\begin{aligned} F_{SZF} &= \frac{1}{2.6}(700 \times 9.81)(2.6 - 1.1) \\ &= 3961 \text{ N} \end{aligned}$$

Hence,

$$\begin{aligned} F_{ZF} &= 3961 + \frac{1.2}{2.6} \times 700 \times (0.9 \times 9.81) \\ &= 3961 + 2852 \\ &= 6813 \text{ N} \end{aligned}$$

Thus, when the car is at rest on the ground or at uniform velocity, the load in both the front springs combined will be 3961 N (Fig. 8.36). Under braking, this will rise to 7447 N and must not result in a compression of more than 30 mm. Thus,

$$\begin{aligned} 6813 - 3961 &= k \times 40 \times 10^{-3} \\ \Rightarrow k &= \frac{2852}{40 \times 10^{-3}} \\ &= 71300 \\ &= 71.3 \text{ N/mm} \end{aligned}$$

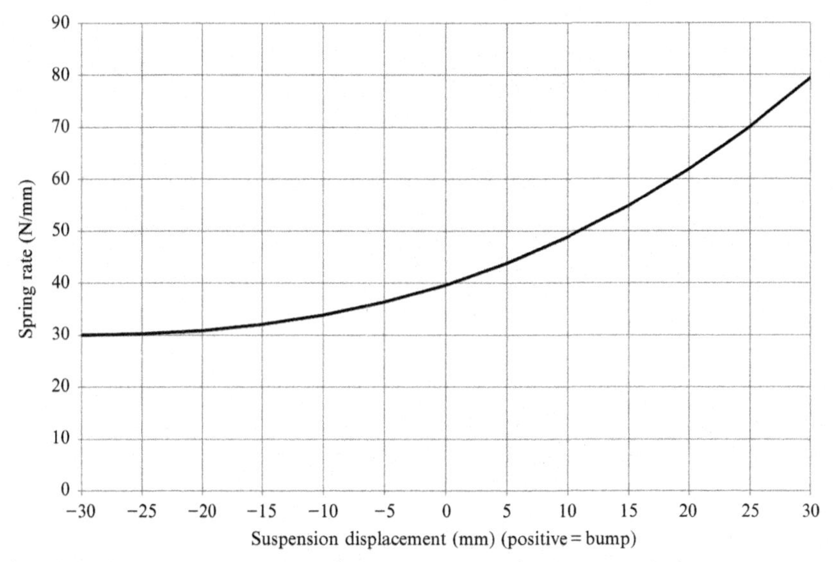

Fig. 8.35 A rising rate suspension has a spring rate that increases in bump.

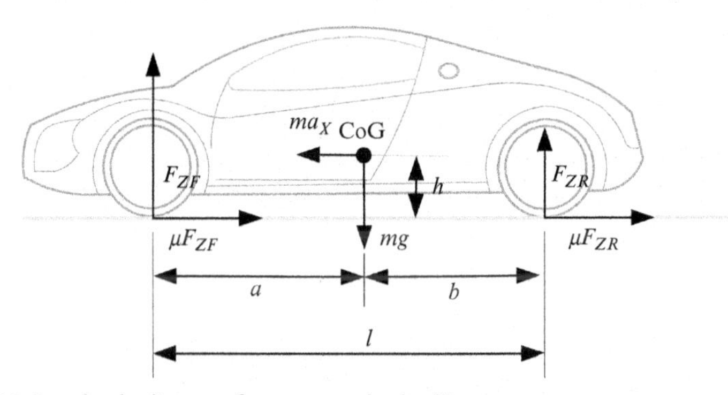

Fig. 8.36 Free body diagram for a car under braking.

The total spring value for both axles will be twice this, 142.6 N/mm. From this, the natural frequency can be determined:

$$\begin{aligned}\omega_N &= \sqrt{\frac{k}{m}} = \sqrt{\frac{142600}{700}} \\ &= 14.3 \text{ Rads/s} \\ &= 2.2 \text{ Hz}\end{aligned}$$

From this, we may determine the critical damping value C_C:

$$\begin{aligned} C_C &= 2\sqrt{km} \\ &= 2 \times \sqrt{142600 \times 700} \\ &= 19982 \text{ N s/m} \end{aligned}$$

As an initial estimate for the system damping, we shall assume a damping ration of 0.4 giving a value of

$$\begin{aligned} C &= 0.4 \times 19982 \\ &= 7993 \text{ N s/m} \end{aligned}$$

8.3 2DoF MODEL OF SUSPENSION

We shall now extend the analysis of the SDoF system above to a two-degree-of-freedom system, 2DoF.

8.3.1 2DoF Equations of Motion, Adams and Simulink Modelling

In the analysis of an SDoF system, we started with analytical equations that describe the dynamics of the system when simulated with a sinusoidal input. It is possible to produce analytical equations for a 2DoF in a similar way, but they are much more complex.

We start with the diagram in Fig. 8.37 below. The upper mass, M, represents the body mass. The Z coordinate is used for the direction of motion of the body, upright and road. The suspension stiffness and damping are represented by k_S and C_S, respectively and the unsprung mass by m and the tyre stiffness by k_T.

Equilibrium of body M requires

$$M\ddot{Z}_b + k_s(Z_b - Z_u) + C_s(\dot{Z}_b - \dot{Z}_u) = 0$$

$$\Rightarrow M\ddot{Z}_b + C_s\dot{Z}_b + k_s Z_b = k_s Z_u + C_s\dot{Z}_u$$

Equilibrium of upright M requires

$$m\ddot{Z}_u + C_s(\dot{Z}_u - \dot{Z}_b) + k_s(Z_u - Z_b) + k_t(Z_u - Z_r) = 0$$

Using the method of complex numbers, we start with

$$\begin{aligned} Z_r &= Re^{j\omega t} & \Rightarrow \dot{Z}_r &= Rj\omega e^{j\omega t} & \Rightarrow \ddot{Z}_r &= -R\omega^2 e^{j\omega t} \\ Z_u &= Ue^{j\varphi} e^{j\omega t} & \Rightarrow \dot{Z}_u &= Ue^{j\varphi} j\omega e^{j\omega t} & \Rightarrow \ddot{Z}_u &= -U\omega^2 e^{j\varphi} e^{j\omega t} \\ Z_b &= Be^{j\psi} e^{j\omega t} & \Rightarrow \dot{Z}_b &= Be^{j\psi} j\omega e^{j\omega t} & \Rightarrow \ddot{Z}_b &= -B\omega^2 e^{j\psi} e^{j\omega t} \end{aligned}$$

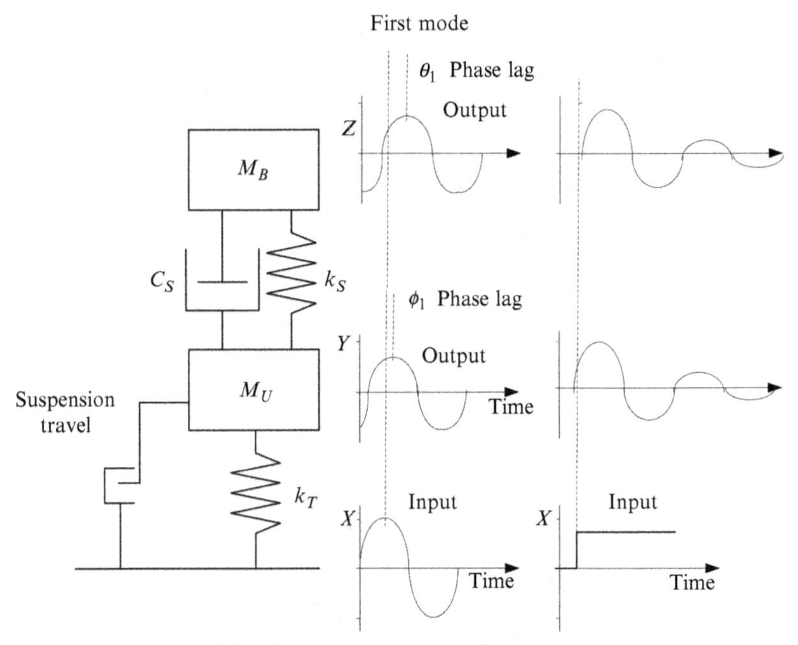

Fig. 8.37 Two-degree-of-freedom representation of racing car suspension.

After some very lengthy and complex manipulation, one can produce expressions for the amplitude ratios of the motion of the upright with respect to the road and also for the motion of the body with respect to the road. One example, that of the equation describing the motion of the upright with respect to the road surface, is given below:

$$\left|\frac{U}{R}\right| = \sqrt{\frac{k_t^2\left(M^2\omega^4 + \omega^2 C_s^2 + k_s^2 - 2k_s M\omega^2\right)}{\begin{array}{l} m^2\omega^6 C_s^2 + 2m\omega^6 C_s^2 M - 2m\omega^4 C_s^2 k_t + C_s^2\omega^6 M^2 \\ -2C_s^2\omega^4 M k_t + k_t^2\omega^2 C_s^2 - 2m^2\omega^6 k_s M - 2m^2\omega^6 k_s M \\ -2k_t^2 k_s M\omega^2 - 2k_t^2 k_s M\omega^2 - 2k_s^2 M\omega^2 k_t + 2k_s M^2\omega^4 k_t \\ -2m\omega^6 M^2 k_t - 2m\omega^6 M^2 k_s + 2m\omega^4 k_s^2 M + 4m\omega^4 k_s k_t M \\ +k_t^2 M^2\omega^4 + m^2\omega^8 M^2 + m^2\omega^4 k_s^2 + k_t^2 k_s^2 - 2m\omega^2 k_s^2 k_t \end{array}}}$$

Such equations are too unwieldy to be practical, and whilst we might be able to cope with a 2DoF, we shall soon need to deal with model involving much larger numbers for degrees of freedom, and the approach will become totally impractical. For this reason, we shall, from now on, deal only with numerical simulations. A Simulink model for the 2DoF system is shown below (Fig. 8.38) and after that, an Adams model (Fig. 8.39).

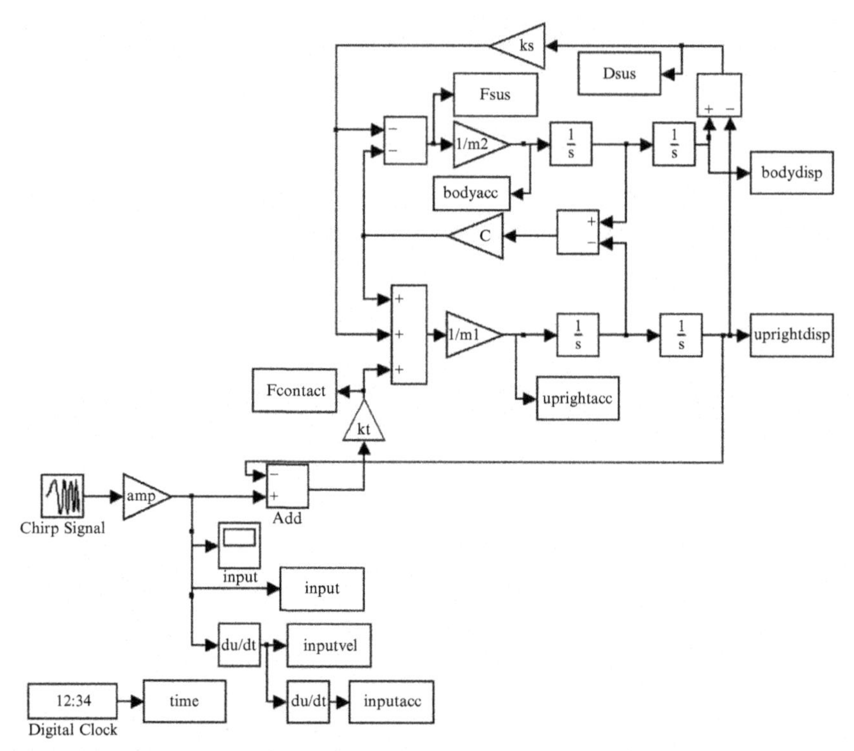

Fig. 8.38 MatLab model of a 2DoF in Simulink.

In the model, the road is represented by the block seen lowest in the model. It is made to vibrate with the road input displacement graph shown lowest in the group of three graphs. The next block up represents the upright, and its motion in response to the input from the road is shown in the middle graph. The upper block and graph correspond to the body in the same way.

8.3.2 Classical Response of 2DoF to Step and Harmonic Inputs

A 2DoF system presents two separate modes of vibration. These are illustrated in Fig. 8.40 below. In the first, the input waveform results in motion of the upright and body in the same sense. Broadly, they move up and down together, in phase. In reality, there will be some phase lag in both the response of the upright and the body as shown with φ_1 and θ_1.

In the second mode of vibration, the upright and body are broadly in antiphase, and as one moves upwards, the other moves downwards. Again, there are phase lags, shown this time as φ_2 and θ_2.

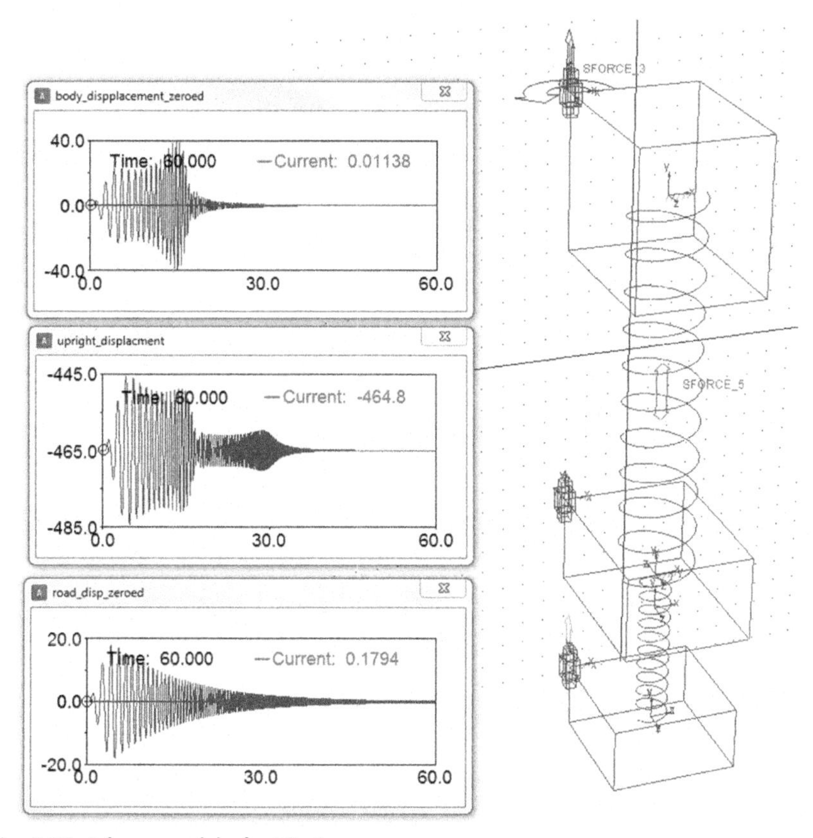

Fig. 8.39 Adams model of a 2DoF.

If we consider the effect of damping on the system, we can see that at very large values of suspension damping, we should expect a third kind of output vibration. This is because when the suspension damping is so high, the body and upright are locked together and must follow the same displacement. In this case, the system has become an SDoF system with the mass consisting of the sum of the body and upright bouncing together on the tyre spring.

In the first case, the body and upright move in-phase together but have different amplitudes. This mode is associated with the car body mass vibrating on the suspension springs. As the heavy body vibrates slowly on the soft springs, the upright is taking up an equilibrium point in between the car body and the ground. The natural frequency for the properties given above may be estimated as

$$\omega_{n1} = \sqrt{\frac{k_s}{M}}$$

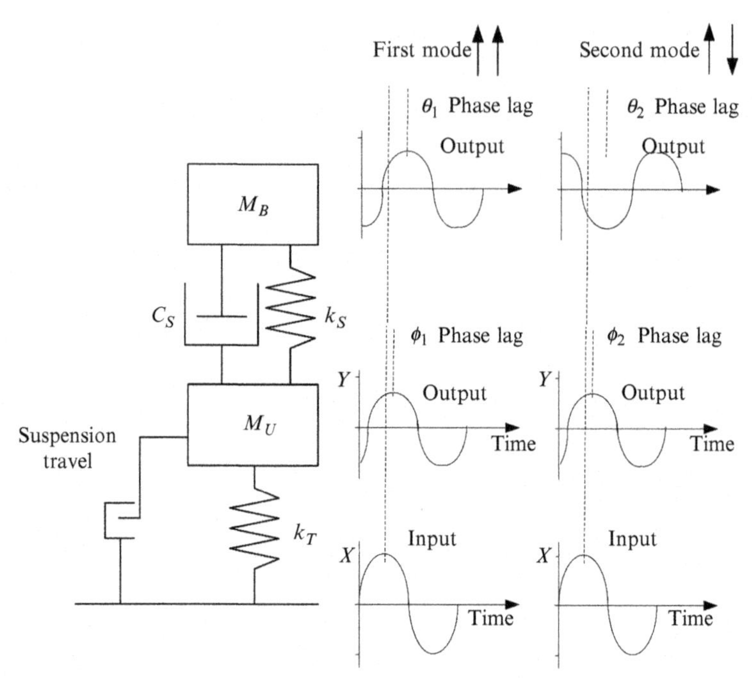

Fig. 8.40 Three modes of vibration for a 2DoF suspension.

In the second mode, the upright vibrates in antiphase with the body. This mode of vibration is associated with the upright vibrating up and down against the combined springing of the suspension and the tyres. Since the mass involved, the upright, is low and the spring rate is high, the natural frequency will be rather above the previous two:

$$\omega_{n2} = \sqrt{\frac{k_s + k_t}{M_U}}$$

In the final mode of vibration, the whole car, body and uprights, is bouncing on the tyres. This mode is associated with a high value of damping in the suspension, so high that the body and upright move as one. Since many practical dampers are nowhere near this high, the mode may not even be observed:

$$\omega_{n3} = \sqrt{\frac{k_t}{M + m}}$$

Thus, we expect the 2DoF system to exhibit two resonances, the *body mode* and *upright mode* when damping is low and a third *tyre mode* when damping is higher. Near these resonances, the amplitudes of vibration will be large. Their size will vary with the damping ratio. In addition, we would expect a phase shift across the frequency range. We can illustrate this by using some actual values and obtaining the response.

Using the Adams multibody code or MatLab using Simulink and making a 2DoF model with the parameters shown in the table below, the graph of amplitude of vibration of the upright over road vibrational amplitude shown in Fig. 8.41 can be prepared.

Body mass, M/kg	700
Suspension spring rate, k_s/N/m	100×10^3
Unsprung mass M_U/kg	120
Tyre spring rate k_T/N/m	250×10^3
Damping ratio	0.12, 0.24, 0.48, 1.0 and 3.0

For this system, the value of critical damping determined from $C_C = 2\sqrt{km}$ will not have the same direct correspondence to that value of damping that ensures the system settles as quickly as possible for the SDoF. However, it continues to be a useful guide to approximately where the damping value lies. For the system above, it takes the value 16.7 N mm/s. In the Fig. 8.41 below, the 2DoF amplitude ratio of the body to the road is shown for a range of damping values.

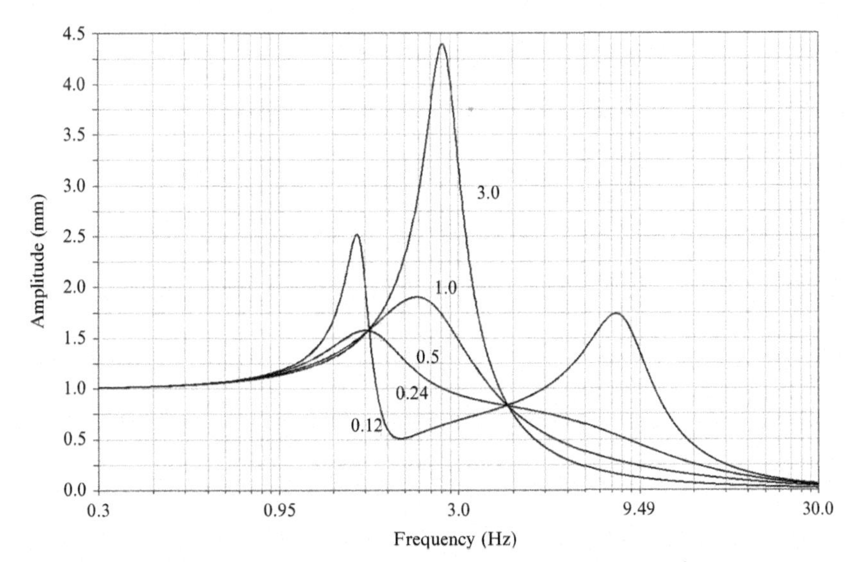

Fig. 8.41 Amplitude ratio of the upright displacement versus input frequency.

The graph for the damping ratio of 0.12 shows two peaks. The first at around 0.7 Hz is associated with resonance of the body and is called the body mode. The second is near 5 Hz and is associated with resonance of the upright and so is called the hub mode. As the damping is increased, so the displacement between the upright and the body progressively locks up, and by the time a damping ratio of 3.0 is reached, the body and hub mode are no longer visible, and instead, the system has become an SDoF vibrating on the tyres.

The phase response of the upright with respect to the road is shown below in Fig. 8.42. It can be seen that for the damping ratio of 3.0 in which the system behaves as an SDoF system, the phase transition takes the same shape as seen above. As the damping ratio is reduced, the 2DoF system behaviour is revealed, and in the 0.12 case, two distinct regions of phase transition are clear.

The first occurs at around 0.8 Hz when the upright temporarily has a phase lag whose maximum value is nearly 90°. After this, it returns to a small value before again growing at around 5.0 Hz. After this second transition, the phase remains shifted.

In Fig. 8.43 below, the amplitude ratio for the body to road input of the 2DoF system above is presented for the same range of damping values.

It can be seen that for the high damping ratio of 3.0, there is one resonant peak corresponding to the body and upright vibrating together on the tyre spring. As the damping ratio decreases, the motion changes with a significant resonance at the lower body mode near 1.1 Hz, and at the same time, a very small peak can be discerned near 8 Hz. This corresponds to the hub mode.

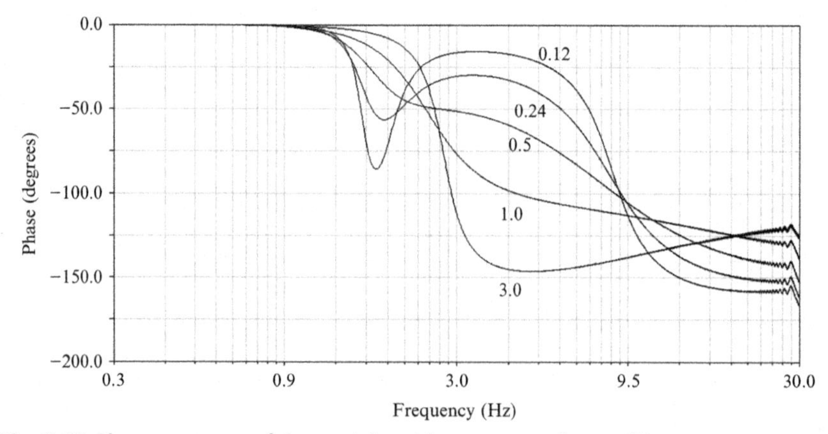

Fig. 8.42 Phase response of the upright with respect to the road input.

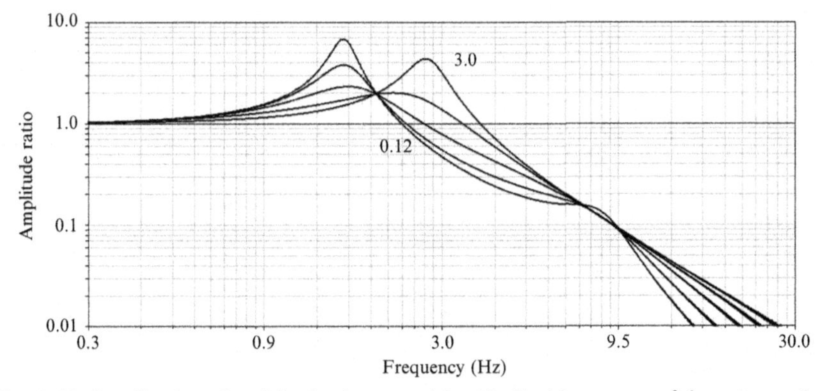

Fig. 8.43 Amplitude ratio of the body to road for 2DoF with a range of damping ratios.

As the upright vibrates at this mode with a significant amplitude, the body is urged to follow it, but the suspension spring is very effective at isolating the vibration and body movement resulting from this cause small. Indeed, the line for the 0.12 zeta value is a very modest peak at this frequency.

The pothole excursion can also be modelled, and we shall now examine a pothole profile that consists of a 40 mm haversine step completed within 0.0001 s, an input that corresponds to passing over the pothole at 50 mph. This gives a length of the pothole entry as

$$50 \times 0.447 \times 0.0001 = 2.2\text{mm}$$

which is a very abrupt step (Fig. 8.44). A graph of body displacement against time for the same range of damping as used above is shown below:

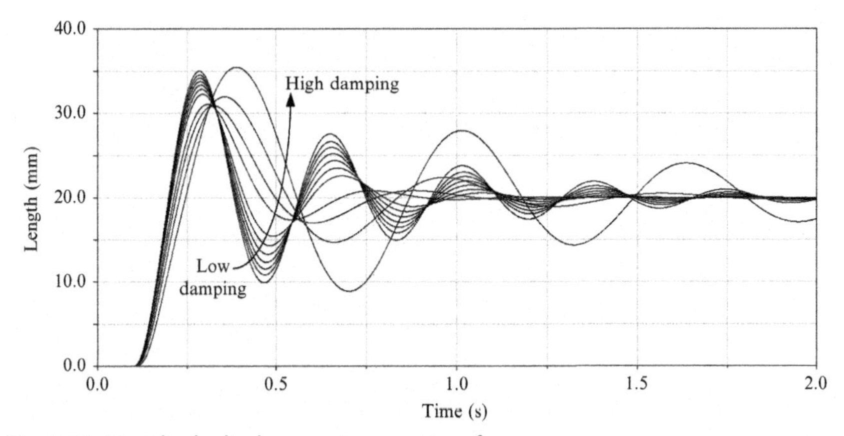

Fig. 8.44 2DoF body displacement versus time for.

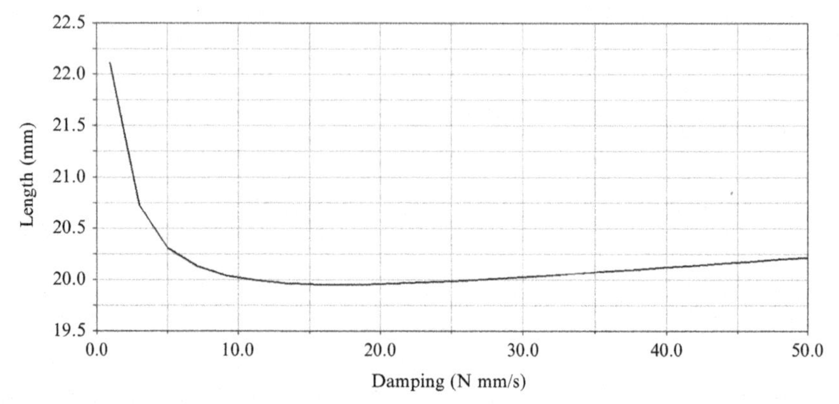

Fig. 8.45 Ideal damper value for 2DoF with pothole input.

The move from body mode, low-frequency red line, to tyre mode, higher frequency mode, green line, is clear.

Fig. 8.45 shows the RMS of the body displacement against damping with a minimum at a value of around 18, for this system corresponding to a zeta value of 1.0

8.3.3 Characterisation of 2DoF System

We shall now proceed to analyse the 2DoF using the two inputs developed above. Since the 2DoF offers increased modelling fidelity, we are able extract a greater range of output parameters. We can now examine the system from the point of view of contact patch force and body acceleration separately. These measurements are the important ones available for the optimisation of ride and handling. It is however, informative to examine the response of the body displacement and we shall examine this too. Specifically, we shall analyse the following cases:

Input	Measure
Harmonic road input	Contact patch force
	Body acceleration
Pothole excursion	Contact patch force
	Body acceleration

These are considered in turn below.

8.3.3.1 2DoF Analysis with Harmonic Road Profile Input

We can now use the model developed to analyse the response to harmonic road input. The two main output parameters are of interest with the 2DoF system are the contact patch force and the body acceleration.

Contact Patch Force

We shall now analyse the 2DoF system with a road profile and optimise it from the point of view of the minimisation of contact patch force variation. Fig. 8.46 shows an example of the contact patch force with time. Fig. 8.47 shows how the RMS of the contact patch force varies with damping value. The parameters used are given above. Fig. 8.47 was obtained by repeatedly preparing versions of Fig. 8.46 for a range of damping values and then plotting the RMS of each one against the damping value used.

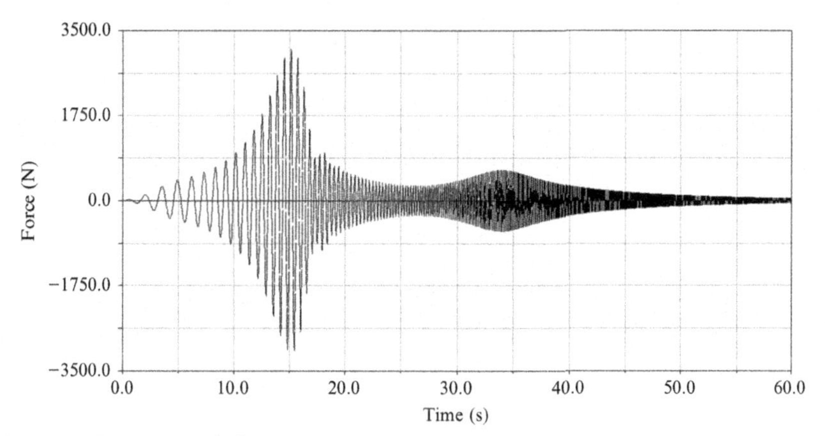

Fig. 8.46 Contact patch force against time 2DoF.

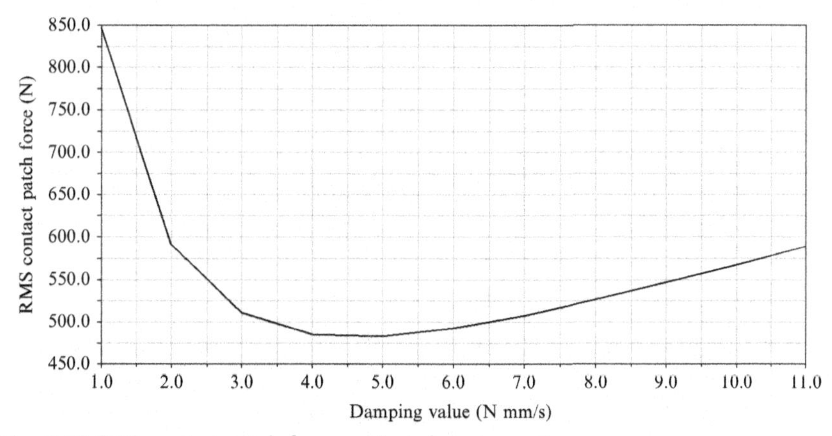

Fig. 8.47 RMS contact patch force against damping ratio.

It can be seen that the minimum occurs at a damping value of around 4.9, corresponding to a damping ratio of 0.3, slightly less than was suggested by the SDoF. The graph below shows the same analysis but for the body acceleration.

The process was then repeated but this time varying the suspension spring rate between 25 and 300 N/mm, and Fig. 8.48, RMS of contact patch force against spring rate, was produced.

Body Acceleration

We shall now analyse the 2DoF system with a road profile and optimise it from the point of view of the minimisation of body acceleration. Fig. 8.49 on the left shows an example of the body acceleration with time. It is the case

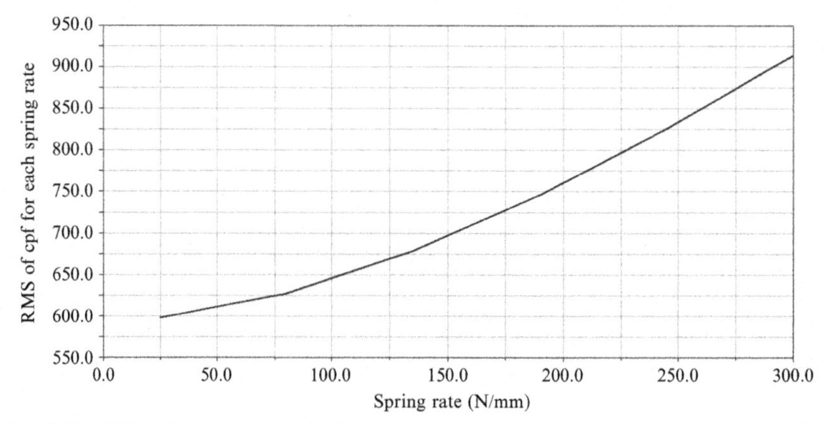

Fig. 8.48 RMS of contact patch force over road input sweep against suspension spring rate.

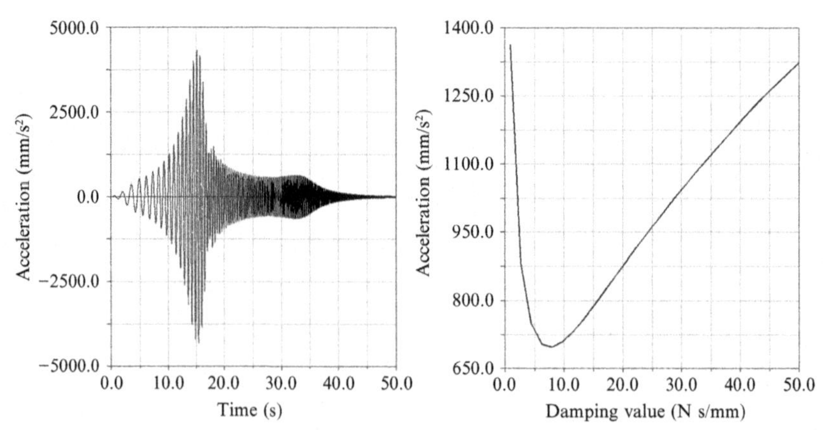

Fig. 8.49 Body acceleration vs time (left) and RMS body ace versus damping ratio (right).

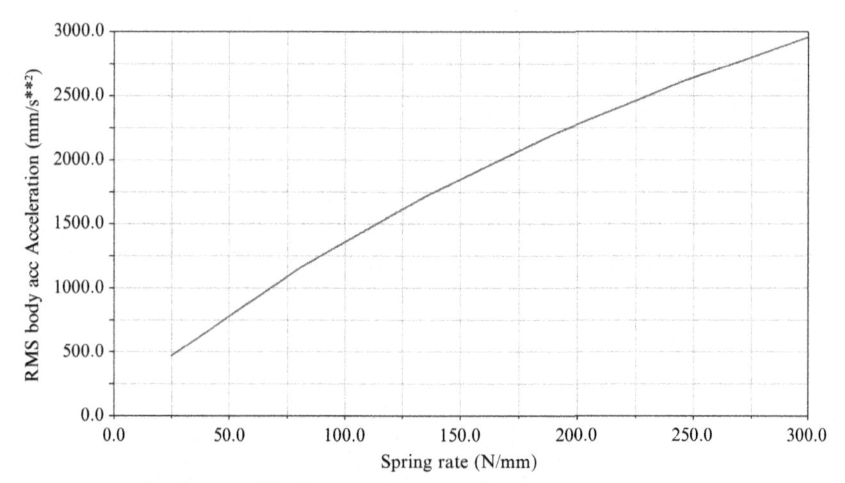

Fig. 8.50 Graph of RMS of body acceleration versus spring rate.

where the damping was 2.75 and had an RMS value of 880, from which its position in the graph on the right can be found. The graph on the right shows how the RMS of the body acceleration force varies with damping value. The parameters used are given above. The graph on the right was obtained by repeatedly preparing versions of the graph on the left for a range of damping values and then plotting the RMS of each one against the damping value used.

In this case, the minimum is slightly higher at about 7, corresponding to a value of 0.33 for the damping ratio.

Fig. 8.50 shows that as the suspension spring rate is reduced, so the RMS of the body acceleration also reduces. We conclude that from this point of view, the softer the suspension spring, the better the suspension.

8.3.3.2 2DoF Analysis Using Step Input to Simulate Pot-Hole Type Transients

We now analyse the 2DoF system for a pothole type input described above. Specifically, the input will consist of a haversine displacement input of the base of 40 mm completed in 0.001 s to simulate pothole-type transients.

Contact Patch force

The graph below, Fig. 8.51 shows the contact patch force over at the same range of damping values used above but for a system with a small amount of tyre damping.

It can be seen that as suspension damping is increased from 0.1 to 0.4, the RMS of the contact patch force will go through a minimum, and this is

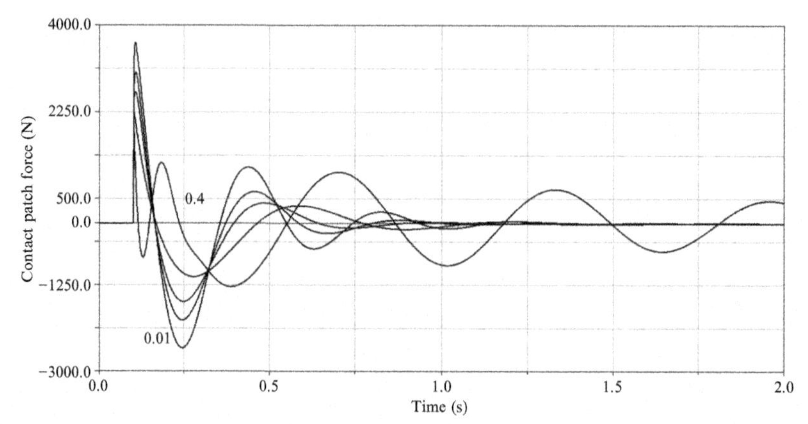

Fig. 8.51 2DoF Contact patch force against time for a pothole type input.

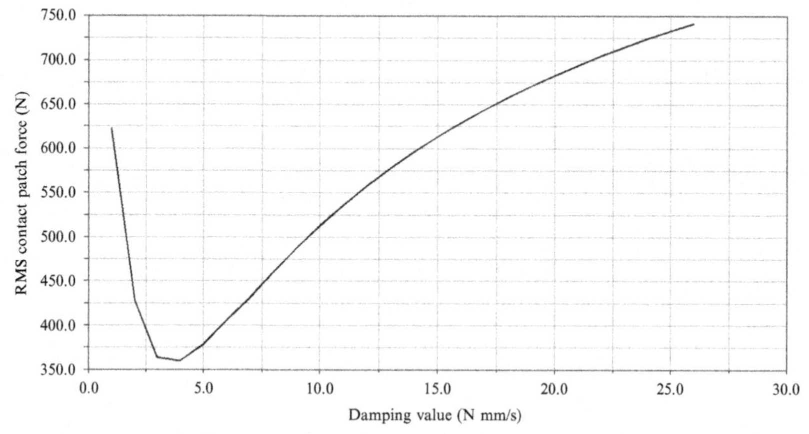

Fig. 8.52 2DoF RMS of contact patch force versus damping.

confirmed below in the plot of RMS against damping (Fig. 8.52). However, we also see that as the suspension damping is increased, the peak force experienced at the start of the bump is larger, and this is a rationale for having the minimum value of damping for this input.

This time, Fig. 8.52 shows that a value of around 3 is optimal, corresponding to a damping ratio of 0.17. The RMS of the contact patch force is shown against suspension spring rate in the graph below and we see that the lower the spring rate, the better this is (Fig. 8.53).

Body Acceleration

The graph below shows body acceleration for the same range of damping values as used before.

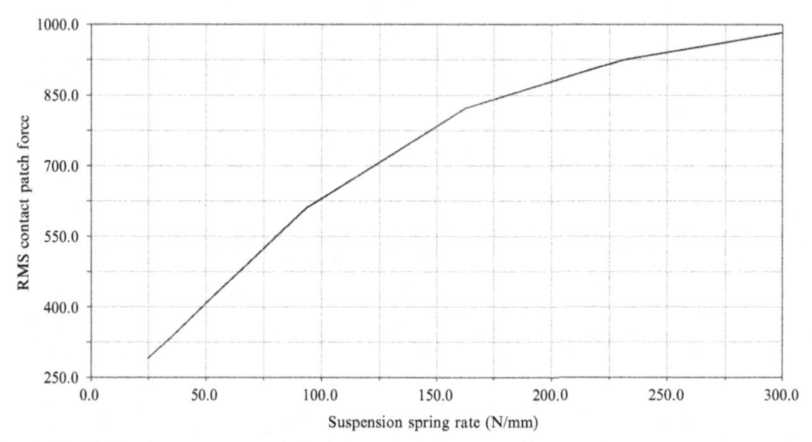

Fig. 8.53 RMS of contact patch force against spring value.

Starting with the graph on the left, high damping results in a large initial spike followed by some oscillations. As we move towards lower damping, the amount of oscillation reduces, but if damping decreases sufficiently, the size of the oscillation increases again. We therefore expect a minimum in the curve of RMS of body acceleration against damping vales, and this is shown in Fig. 8.54. This graph shows that the lowest value is returned for a damping value of around 7 corresponding to an optimal damping ratio of around 0.17. However, scrutiny of the peak acceleration experienced as a result of the bump, shown in the graph on the right in Fig. 8.55, shows that the lower the damping the better this is.

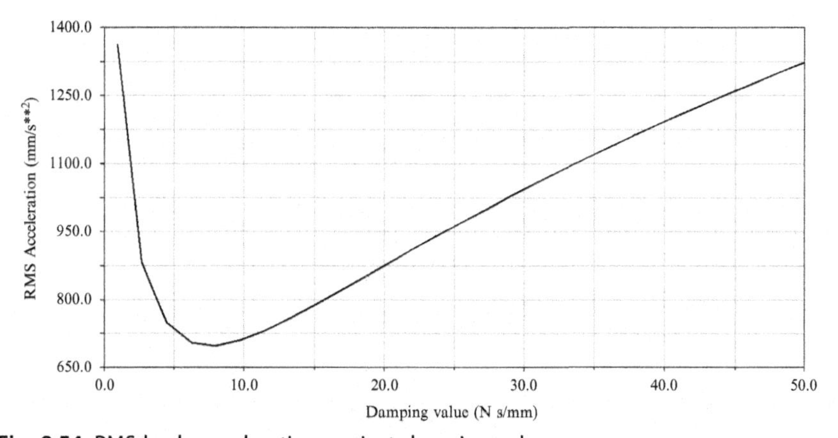

Fig. 8.54 RMS body acceleration against damping value.

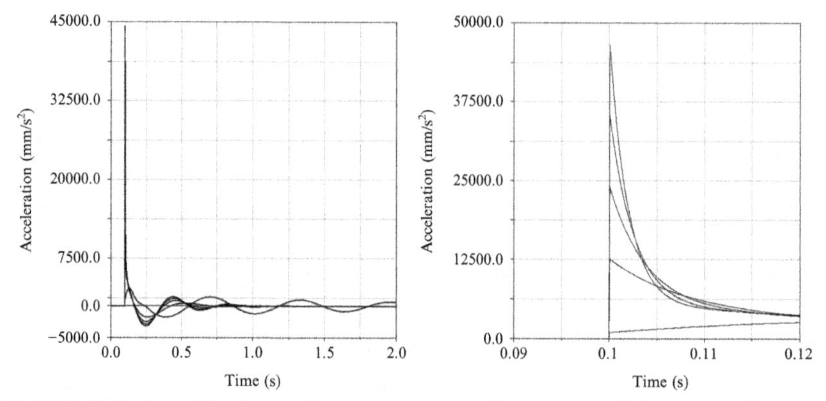

Fig. 8.55 Variation of body acceleration over time for a range of damping values.

The peak acceleration of the body is significant both from a comfort point of view and from a chassis response point of view. We conclude that from this point of view, the lower the value of damping, the better the suspension will deal with a pothole.

In Fig. 8.56, we see the RMS of the body acceleration against suspension spring rate. The lower the spring rate, the lower the RMS. We conclude that the lower the spring rate, the better.

8.3.3.3 2DoF Idealised Spring and Damper Function for Combined Input

A summary of the analysis using the 2DoF is given below:

We conclude that to deal with all the regimes covered and for up to suspension velocities of around 100 mm/s, a damping ratio of 0.3 is best and that above this speed, the lower, the better.

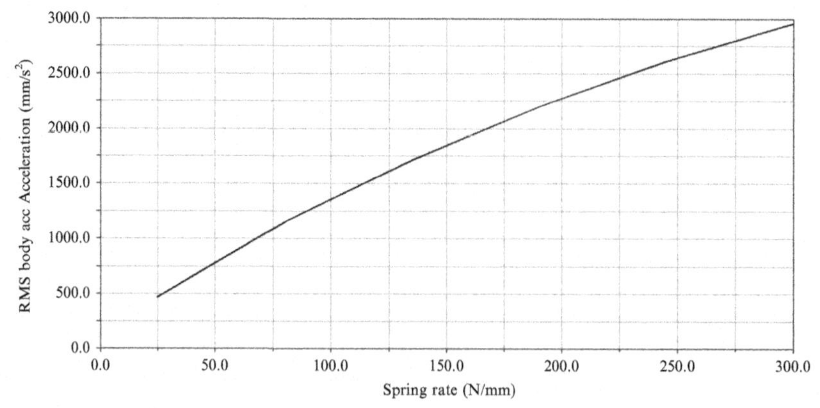

Fig. 8.56 RMS of the body acceleration against spring rate.

Input profile		Ideal damping ratio	Ideal spring rate	Damper velocity mm/s
Harmonic road profile	Minimisation of RMS of contact patch force	0.3	Low as possible	80
	Minimisation of RMS of body acceleration	0.33	Low as possible	
Pothole style step Input	Minimisation of RMS of contact patch force	0.17	Low as possible	>150
	Minimisation of RMS of body acceleration	0.17	Low as possible	
	Minimisation of peak body acceleration	Low as possible	No advantageous ratio found	

The ideal spring rate remains as low as possible, and so far, the process of analysis is not proving a means of determining the best value.

8.4 4DoF MODELS OF SUSPENSION

We now consider the next step in the escalation of modelling fidelity that we are pursuing, the 4DoF system.

In this model, the uprights are free to move vertically as before. However, we now have two uprights instead of just one, as above. The body is free to move both vertically and rotationally about its centre of gravity. The 4DoF model could equally well be applied to the front view, to model roll or the side view and to model pitch. To aid the text in this section, we shall always describe the model in the roll mode but be aware that everything that applies in roll equally be applied in pitch.

In addition to these extra outputs from the model, we have new possible inputs. We clearly can still supply a road profile heave input to both the left and right tyres. Indeed, this would give exactly the same results as obtained for the 2DoF with the equivalent input. Thus, there are some inputs that don't need to be modelled since they are already covered in earlier models.

One new input would be that of roll, in which the input on one side is in antiphase to that on the other. This could happen for both road and pothole style inputs. There is however, an entirely new input that can be modelled; that of body roll under cornering (pitch under acceleration or braking). This input cannot be modelled in the 2DoF system because it requires the rolling of the body to be included and this depends on the body's polar moment of inertia, not just its mass. The input is called the *roll transient* meaning the movement cause by roll that soon decays away. The input consists of a force applied to the body at the centre of gravity height and of a magnitude consistent with that generated by the vehicle in question under cornering conditions.

The model also allows for the modelling of the dynamic effects of a new feature, the anti-roll bar. An anti-roll bar serves to impart increased roll stiffness without affecting heave stiffness.The inputs that can be modelled and that we shall consider are listed below, together with the outputs that are of interest in the optimisation of the suspension:

Input	Measure
Harmonic road input, heave	Contact patch forces
	Body acceleration vertically
	Body roll displacement
	Body roll acceleration
Harmonic road input, roll	Contact patch forces
	Body acceleration vertically
	Body roll displacement
	Body roll acceleration
Pothole excursion, heave	Contact patch forces
	Body acceleration vertically
	Body roll displacement
	Body roll acceleration
Pothole excursion roll	Contact patch forces
	Body acceleration vertically
	Body roll displacement
	Body roll acceleration
Roll transient	Contact patch forces
	Body acceleration vertically
	Body roll displacement
	Body roll acceleration

The first comment to make in the table is that it is so much larger than the equivalent table for the 2DoF, and consequently, there is going to be a very large amount of modelling required by comparison. The second observation is that some measures seem counter-intuitive. For example, the table lists body roll displacement under a harmonic road input in heave. Since this input is symmetric left to right, one would expect the output to be symmetric too, and so, there would be no body roll displacement. This is indeed true, but only when the centre of gravity is in the geometric centre of the body. In general, in the front view, the centre of gravity is normally so close to the geometric centre that we can assume it is symmetric. However, in the side view, this is not normally the case. The centre of gravity is normally some way away from the geometric centre and an input in heave causes outputs in both heave and pitch. The amount can be defined by the 'heave-pitch couple', a constant used to define the ratio of output in roll per unit input of heave across the frequency range.

In addition to the table of inputs needed for the 4DoF, we need to consider which parameters we might adjust in order to optimise the suspension. These are as shown in the table below:

Parameter	Symbol and units
Suspension spring stiffness	k_{SUS} N/mm
Suspension spring rate of rise	k_{SUS}/mm
Left-to-right spring stiffness ratio	k_L/k_R
Damper value—low speed bump	N s/mm
Damper value—high speed bump	N s/mm
Damper value—low speed rebound	N s/mm
Damper value—high speed rebound	N s/mm
Spring to unsprung mass ratio	—
Polar moment to mass ratio	mm^2
Body mass	kg
Centre of gravity location	mm
Anti-roll bar stiffness	

From the above, it is clear that an exhaustive approach considering every possible case of the 4DoF would be a lengthy undertaking; one that is certainly too long to include in this work (Fig. 8.57). However, our aim in this whole chapter is to develop the understanding necessary for the reader to be able to undertake his or her own analysis, and so, presenting every case for the 4DoF model is not in fact necessary. Instead, we shall examine just one case, that of the contact patch load under roll transient, since this represents a new kind of input and allows for the determination of an ideal bump-to-rebound ratio that is not possible with the 2DoF model.

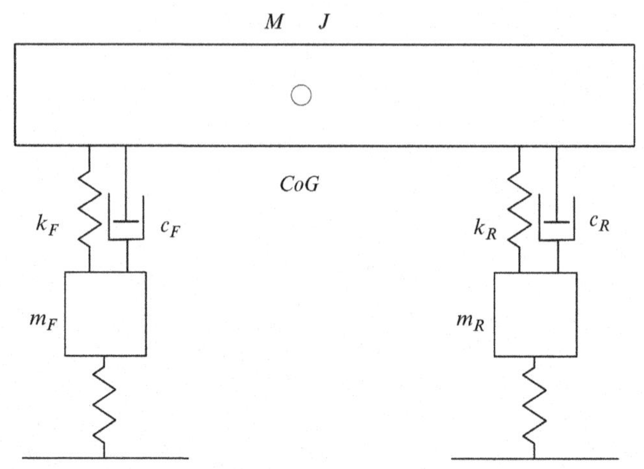

Fig. 8.57 The four-degree-of-freedom model of vehicle suspension.

8.4.1 Inclusion of Roll and Pitch Transients in SDoF Analysis and Numerical Modelling

In addition to the two inputs for harmonic road input and pothole type inputs used above we have a new input for use with the 4DoF. This is the roll (or pitch) transient. When a car enters a corner the steering is applied from the straight ahead position through to the steady-state cornering position. The application of this steering input can take several tenths of a second. It can be much more for a corner that tightens on entry, but it is never normally less than around 0.05 s. The Fig. 8.58 shows the suspension velocity in a car damper caused by a roll transient. The lateral force was applied

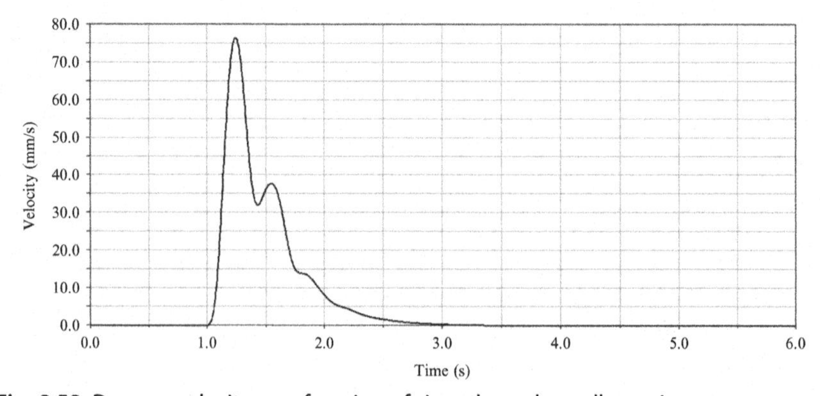

Fig. 8.58 Damper velocity as a function of time through a roll transient.

over 0.2 s, and it can be seen that the suspension stroking velocity is never more than around 100 mm/s but has an average value of around 40 mm/s.

We therefore conclude that suspension stroking velocities associated with roll and pitch transients are similar but smaller than the stroking speeds associated with the road profile. Thus, there are two inputs that need consideration for the SDoF:

- Harmonic road profile of constant peak velocity and exponentially increasing frequency (damper velocity 0–85 mm/s)
- Step input consisting of a haversine displacement input of the base of 50 mm competed in 0.185 s (damper velocity >270 mm/s)
- Step input consisting of a force applied to the body at the centre of gravity, representative of a roll transient event (damper velocity 40 mm/s)

8.4.2 4DoF Equation of Motion, Adams and Simulink Modelling

Modelling of the 4DoF suspension system can be achieved through either Simulink solution of the equations or multibody code, such as Adams simulation.

The figure above shows an Adams model of the 4DoF mode that includes an anti-roll bar. The *road* pads left and right can be assigned motions, remaining still for a roll transient input with a step force applied at the centre of gravity or a displacement function for the road input and can be set to move either in phase or antiphase. The uprights left and right are constrained to move vertically. Drop links connect the uprights to the anti-roll bar and a torsional spring resists torsion caused by vertical movement in opposite directions of the uprights. Each side of the anti-roll bar is connected to the body by a hinge joint. In this way, the torsion spring resists roll but leaves heave unaffected. The suspension springs connect the uprights to the body that is free to move in heave or roll.

8.4.2.1 The Anti-Roll Bar Optimisation

The anti-roll bar is an important element of the suspension since it allows the designer to control the roll stiffness of the vehicle independently from the bump stiffness. We saw in Chapter 5 that it can also be used to adjust the understeer gradient. To understand how it works, we start by considering a suspension without an anti-roll bar. The suspension springs will offer a given vertical stiffness. They will also offer some roll stiffness since when the car rolls, one suspension spring is compressed and the other extended. The values for the suspension springs will be determined first as described in this chapter. Once selected, the vehicle's response to roll may be determined. The amount of roll that is acceptable can be determined from a consideration of the suspension geometry. For example, if static camber is being used, then there will be a desirable amount of roll that will bring the tyres to the ideal camber angle in a corner. Generally, the suspension springs selected will be too soft to restrict the roll to this desired amount, and the remaining roll stiffness is proved by the anti-roll bar. An Adams model as shown in Fig. 8.59 can be used to determine the torsional stiffness required at the point in the anti-roll bar where the elastic compliance is provided. It is then a simple enough design problem to convert this into the dimensions of, for example, a torsion bar. The roll bar stiffness front and rear can then be adjusted from this point to control the understeer gradient. An analysis of the effect of the modification to weight transfer for the case in hand can be used to inform the design of the amount of adjustment provided by the anti-roll bar such that it can provide the required amount of understeer control.

The increase in roll stiffness will raise the natural frequency in roll and with it the value of critical damping in roll. If the vehicle is underdamped in roll, the increase in critical damping will make the response more underdamped. Ideally, the level of roll damping would correspond to the critical value, and if it is too small, then roll damping can be introduced directly if the geometry allows for it. Alternatively, the suspension damping can be increased and a compromise between roll and heave damping used.

Full vehicle simulation and transient lap-time simulators can also be used to examine this problem, as described in Chapter 9, though this is a much more lengthy undertaking.

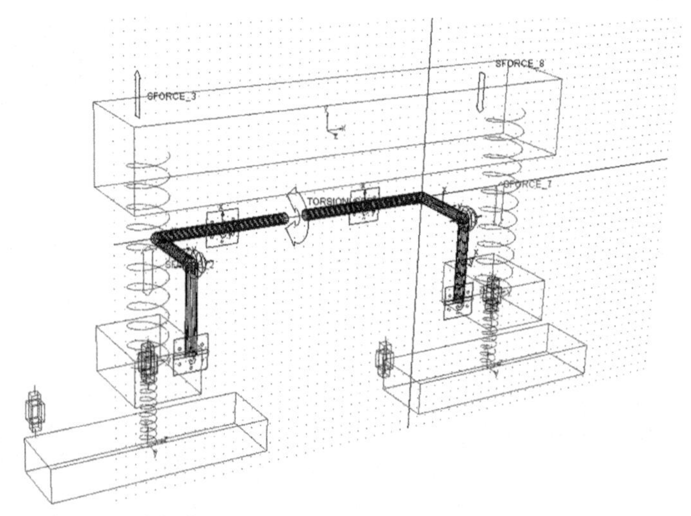

Fig. 8.59 Adams model of a 4DoF suspension system.

8.4.3 4DoF Analysis Contact Patch Force Under Roll Transients

Using the 4DoF system in Adams shown above and using the same dynamics parameters as determined from the 2DoF analysis and the new ones needed for a 4DF as listed in the table below produce the analysis below:

Body Mass: 1200 kg
Body polar moment in roll: 3E8 kg-mo2
Suspension springs: 75 N/mm each
Upright mass: 55 kg each
Tyre springs: 300 N/mm each
Central damping value: 15 N s/m
Track: 1.8 m
Roll transient input: torque of 9 kN m over 0.2 s

An important point to note here is that we are optimising this 4DoF system using many parameters that were previously determined from the point of view of the 2DoF. For example, the suspension damping values have already been chosen to optimise the system against road and pothole inputs in heave, and so, we are not now free to simply optimise them against all of the 4DoF system inputs.

The graph above (Fig. 8.60) shows the roll angle as a function of time through the application of the roll transient force input.

From the graph, we conclude that the roll transient is underdamped since it overswings its final value before settling. In Fig. 8.61 below, we can see the transient behaviour of the 4DoF system under a roll transient. The lower line shows the displacement of the right-hand side of the body as the roll torque is applied. Naturally, it descended. As a result, the upright starts to move downwards as is shown by the upper line. However, as the upright descends, its movement is resisted by the tyre stiffness. This is larger

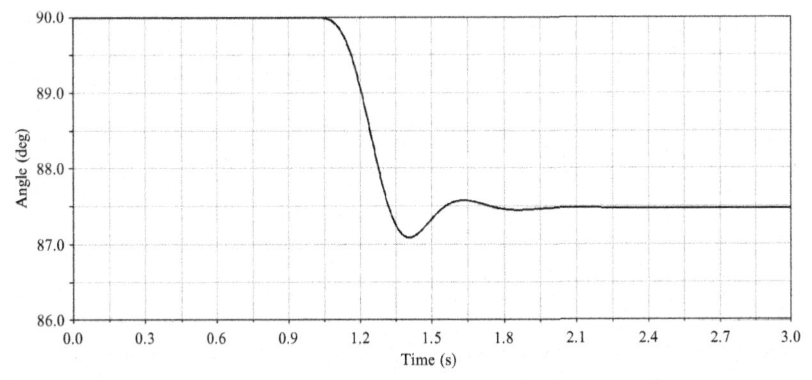

Fig. 8.60 Roll transient for 4DoF system with previously determined parameters.

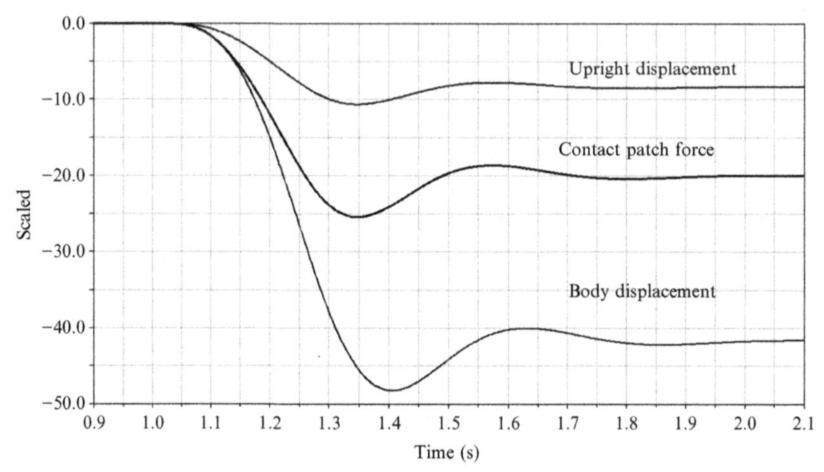

Fig. 8.61 Transient behaviour of the 'outside' wheel under roll transient.

than the suspension stiffness, and so the upright reaches its maximum decent before the body. The contact patch force, shown in the middle line and scaled to fit on the axes, copies the profile of the upright. It can be seen that the upright overshoots its steady-state position. There is a benefit to this since it means that, temporarily during the roll transient, the load on the outside wheel is more than expected, and consequently, the lateral force that tyre can develop is greater, and so the cornering response will be improved. Before long, around 2 s in this case, the contact patch force must of course return to is equilibrium value and the transient is over. However, during this transient improved performance can be gained by optimising this effect.

We would not wish to modify the average damping value used because this has already been set to optimise the 2DoF system performance. However, we can reduce the damping in bump and increase it in rebound and so keep the overall value as desired. The effect of reducing damping in bump is that the body overswings more. In doing so, the spring is further compressed by the time significant upright velocity has developed. This in turn causes the upright to reach a lower maximum extent and produces a larger peak in contact patch force. The bump-to-rebound ratio was varied from 1:1 to 1:4 and the following graph of contact patch force on the 'outside' tyre shown in Fig. 8.62 below obtained.

The '1:1' line shows the starting point. Moving to the '1:2' line, we see that the rise is slightly slower but that in the region 1.26–1.5 s, the values are

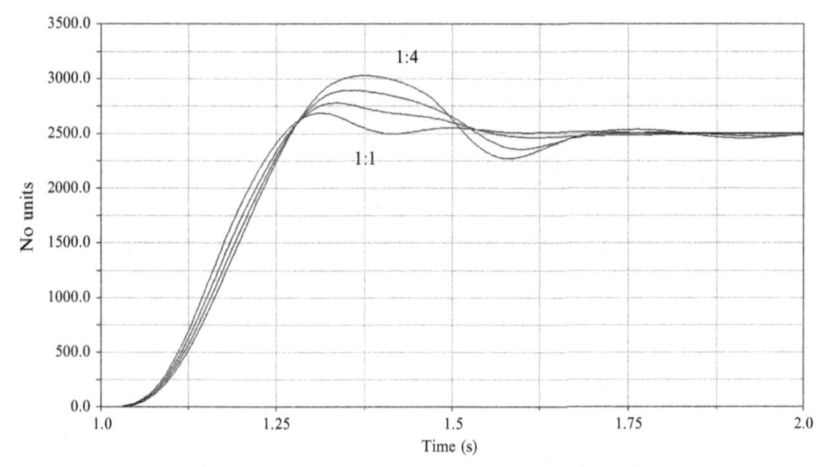

Fig. 8.62 Variation of contact patch force with bump-to-rebound ratio.

higher, and this is an advantage. If the car is, for example, cornering at the limit, almost all the lateral force will be coming from the outside wheel and more vertical load on it is good. Moving to the pink line, 1:3 the trend continues, but examination of the RMS shows that returns are diminishing. Further to this, we see that in the region 1.55–1.7 the vertical load is starting to decrease making things worse during this period. Moving to the black line 1:4 this process continues. This is contrary to the common view that stiffening in bump would cause the car to lean harder on the outside wheel as might be concluded from the consideration of the car as two SDoFs side by side. In an SDoF, the increase of bump damping will indeed cause the contact patch load to be higher transiently. In addition, the increase in rebound damping keeps the inside edge from rising, and so, the body pivots about that outer point under the action of the rolling moment and so lowering the outside to compress the tyre more. This can only be modelled in the 4DoF because the roll inertia and mass need to be present in the model to give a separate response for roll and heave.

There is no clear point at which the bump-to-rebound ratio is most advantageous here, but one can see that somewhere around 1:3 is a good choice. One can only take this selection further by undertaking a full transient lap-time simulation to quantify the effects for a given vehicle under a given set of conditions on a given track. However, the material presented does make clear the effects that are at work. We conclude that for low-speed damping, a bump-to-rebound ratio around 1:3 is ideal.

8.4.4 General Comments on the Optimisation of a 4DoF

We can see from the previous section that the optimisation of a 4DoF system is a complex affair that can only be performed for a particular case in hand. In the case of the SDoF system and to a lesser extent the 2DoF system, it was possible to produce parametric analysis. For example, stiffening the spring of an SDoF system, will always raise natural frequency. Once we have the complexity of the 4DoF, such analysis is not really possible, and we must treat each examination individually, not only mindful of the dynamic fundamentals but also attentive to how they affect the more complex system.

8.4.5 Optimisation Using Cost Functions

Once a working 4DoF system has been prepared and the vehicle dynamicist is ready to begin the process of optimisation, attention must turn to exactly what parameters are to be optimised and how this can be achieved. A list of parameters that one might want to optimise is listed below:

Parameter	Optimisation
Contact patch force	Reduce the variation to a minimum to reduce as much as possible grip loss
Suspension travel	Imposing a limit on suspension spring travel enables a lower limit on spring stiffness to be determined
Pitch displacement	Reduce to a minimum to enhance aerodynamic performance
Body vertical acceleration	Minimise to improve comfort
Body roll acceleration	Reduce to a minimum to improve comfort

Clearly, there are potential conflicts here. Minimising the RMS of the contact patch force will urge the designer towards softer springs and low damping, whilst minimising the pitch displacement will offer design pressure in the opposite direction. Further to this, some parameters will apply much more to one kind of vehicle than another. Ride comfort is simply not a consideration for a high-performance racing car but is absolutely essential for a luxury saloon. A method is needed to resolve this conflict logically.

To do this, we make use of the cost function approach. The metrics of interest are chosen, and from these, a cost function is produced that should be minimised. For example, imagine if we wish to optimise a car from the point of view of pitch angle variation, heave acceleration and contact patch force variation, we might prepare a cost function of the following form:

$$CF = C_1 \frac{(RMS(Pitch\_acc))}{Amp(Pitch\_acc)} + C_2 \frac{(RMS(heave\_acc))}{Amp(heave\_acc)} + C_3 \frac{RMS(CPF\_variation)}{Amp(CPF\_variation)}$$

The first step in preparing this cost function is to ensure that the variable of interest is made to vary around an equal magnitude, usually one. If one were simply to take pitch angle and RMS of heave acceleration and attempt to minimise these, there would be a problem. In all probability, one of these will be a much larger quantity than the other. Let's suppose the RMS of the heave acceleration is measured in hundreds of mm/s, whilst the pitch angle is measured at fractions of a degree, values in tenths. If we then set about minimising the sum, any algorithm will make much large changes to the heave acceleration than the pitch angle, since at best halving the pitch angle will produce roughly one-thousandth the improvement that halving the RMS of acceleration will. For this reason, we normalise each term and shown in the equation above.

The next step is scale each term by a factor, C_1, C_2, etc., that reflects the relative importance of each term. The normalisation can be performed any way desired, and practice varies from one organisation to another. Common starting suggestions are included in the equation above. One can find that as the optimisation process advances, the relative importance of each term changes and renormalising part way through is needed.

The preparation of these cost function coefficients can be a very lengthy process. For example, in a racing team, lap-time simulation can be used to determine the sensitivity of the car to changes in pitch compared with changes in the RMS of the contact patch force. This might well involve hundreds of simulations of a particular circuit to produce a map of the sensitivity of the car to each term in the cost function. In the case of a performance road car, the same approach can be taken, using a full transient vehicle simulation approach to determine a cost function that includes terms involving comfort and ride performance.

This is a rather unscientific process, but in the end, we have a cost function consisting of a series of non-dimensional terms all taking a value close to one and each being representative of the parameter it has been selected for and each term has a coefficient reflecting its importance.

The vehicle dynamist can then set about modelling the car as 4DoF system (or higher) and find the set of parameters that will return the minimum value of the cost function. As with the preparation of the cost function coefficients, this is a lengthy and computational-based process. The end result is the best parameter collection that can be determined. These are the parameters that should be used.

This process can be something of a computational challenge. One is searching for a minima in a multidimensional parameter space and being sure that a global minimum has been found and not just a local one is a matter of concern. Diligent cross-sections of the entire parameter space are at the lengthy but simple end of the range of possibilities. The use of genetic algorithms is extremely fast, but a mathematically very complex approach at the quick and difficult end of the spectrum.

8.5 7DoF AND LARGER MODELS

In principal, the approach taken in the above section on 4DoF models can be extended without limit. One can imagine a seven-degree-of-freedom system such as that in Fig. 8.2 being used with a cost function to optimise not only contact patch variation but also parameters only accessible with a seven-degree-of-freedom system such as anti-roll bar stiffness front and rear or sensitivity to 'warp' input from the road.

8.5.1 Input Parameters

The input table for a seven-degree-of-freedom model would extend considerably that of the 4DoF. The three fundamental inputs of road profile, pothole and lateral transient all still need to be modelled, but the modes of input need to be extended to cover not just heave and roll but pitch and warp as well. The track will not necessarily be the same front to rear meaning that a roll input will result in some warp output.

	Heave	Pitch	Roll	Warp
Heave	—	Heave/pitch	Heave/roll	Heave/warp
Pitch		—	Pitch/roll	Pitch/warp
Roll			—	Roll/warp
Warp				—

The centre of gravity can be placed asymmetrically on the track and wheelbase, and so, there will be coupling constants between all inputs as shown in the table below:

8.5.2 A 'Perfect' Suspension

In Fig. 8.63 below, we see a system that can be described as a 'perfect' suspension. The system works in the following way. Each wheel is connected

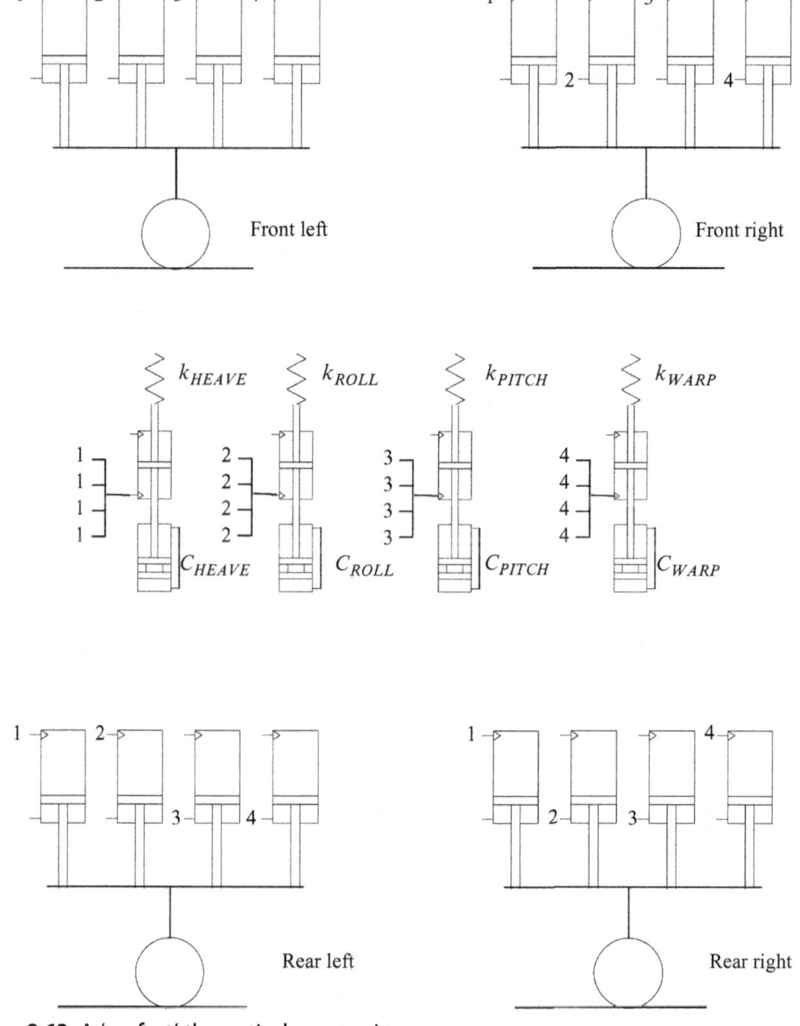

Fig. 8.63 A 'perfect' theoretical suspension.

to four separate hydraulic actuators. As the wheel rises and falls, fluid is pumped in and out.

Each terminal at each wheel (numbers 1–4), is connected via pipework, not shown, to the set of four actuators in the middle. Scrutiny of the side of each wheel actuator to which connection is made will show that in pure heave, only the heave spring damper will be in motion, similarly so for pitch, roll and warp. Each actuator in the middle consists of an independently adjustable spring damper. In this way, it is possible to independently control the stiffness and damping of each input entirely independently. In practice, such a system is not really practicable. The tiny movements are being communicated significant distances, the inertia even of the fluid involved and the inevitable weight and complexity compared with conventional approaches make it unworkable in reality. The system is presented here because it represents an ideal that all simpler systems strive to emulate in function.

8.6 LIMITATIONS OF SINUSOIDAL-BASED SUSPENSION ANALYSIS

Throughout this chapter on the dynamic analysis of suspensions we have made use of sinusoidal waveforms. They have been modified to approximate the road profile, but nevertheless, the wave is sinusoidal. This is a very sensible place to start; after all Fourier analysis shows that any waveform can be approximated by a collection of sinusoidal components and so being able to analyse a sinusoid makes possible the analysis of any waveform at all. In addition to this use of sinusoidal wave, we have made use of the linear second-order differential equation to model the system. Again, this is a very sensible place to start, it offers a good approximation, and the maths involved in its manipulation is not too onerous. However, the real world is not this simple; here are some observations concerning problems that result from the application of the above analysis using normal laboratory techniques and mathematical analysis:

- Optimising body acceleration is not as good as paying attention to its differential, called *jerk*. Passengers perceive jerk as more significant than acceleration, and it should be included in the optimisation process.
- The use of different bump and rebound functions in the damper is a much more complex process than covered above. The difference can

be used to generate non-sinusoidal waveforms offering new opportunities for optimisation.

- Stiction in the suspension bearings and hysteresis in the bushes makes for a modified response. In the case of stiction, the wishbones can come to rest on every up and down stroke, then when made to move again as the waveform advances, high-frequency components appear in the output waveform.
- As one sweeps the frequency, any experimental set-up will normally include data sampling at a constant clock rate. This results in waveforms at low frequencies having a very large number of data points compared with the high-frequency waveforms. If mathematical analysis is performed, for example, a Fourier transform, the low-frequency components will be unduly weighted. What is needed is a sampling regime where every frequency has an equal number of data points.
- Accuracy is an issue when one integrates a displacement waveform; for example, to obtain a velocity, it will be necessary to perform the integration over thousands of cycles, and errors are collative.
- Noise is a very significant issue. At Oxford Brookes, around a third of the thousands of lines of code used to perform four-post rig analysis are dedicated to filtering.
- Chassis rigidity is a significant issue. It is one thing to design a chassis to be stiff under steady-state or very slow moving forces such as those that result from weight transfer. However, when inputs are even as low as 5 Hz, the resulting body displacements can be surprisingly large.

All these observations make it sound as if the analysis presented is too flawed to be of much use, but this is far from the case. Even the SDoF analysis produced meaningful and useful results. The point is that the analysis presented above is a starting point rather than the whole story.

8.7 4-POST RIG TESTING

To perform practical testing of a vehicle suspension, a four-post test rig is an essential piece of equipment. Such a rig consists of hydraulically actuated rams that can supply vibrational input to the car in exactly the way that the road does. The vehicle then vibrates just as it would in the road but in the lab instead. This allows the car to be instrumented and to make measurements of important dynamic parameters such as wheel displacement and body accelerations. In addition, it is possible to supply the car with specified input profiles that represent the road but are much quicker to input that any actual road profile over a journey would be. The harmonic road input profile developed above is a good example. This wave, only 60 s in duration, contains sufficient harmonic components at each frequency band to faithfully represent the road profile.

A view of the actual rig at Oxford Brookes University is shown in Figs. 8.64 and 8.65. The rig is connected to a computer and drive equipment to control the actuators and record the test data from sensors such as accelerometers, displacement transducers and string pots.

Fig. 8.64 Four-post test rig at Oxford Brookes University.

Fig. 8.65 View underneath the 'road' surface.

8.7.1 Basic Function Check

With a car being vibrated on a four-post test rig, it is possible to get up close to all the moving parts of the suspension and simply observe them first hand at close quarters in a way that one simply can't do when the car is moving. A surprising amount can be learnt this way especially if the car has problems. One can observe first hand all the natural frequencies occurring during the frequency sweep. Any asymmetry stands out clearly and leads one visually to the source of the problem. With time and experience, one even gets to hear good and bad acoustic responses. Further to this, it is possible to view basic output data very quickly, for example, the suspension displacements at each wheel, and, using a computer, overlay the amplitude responses to see the bigger picture very quickly.

8.7.2 Acquisition of Parameters

One of the first purposes of the rig is to allow the determination of vehicle dynamic parameters such as spring stiffness, damper functions and unsprung mass. To do this, the rig is used to supply known inputs, such as the swept road input described above, and the resulting signals from the instrumented car are recorded. In Fig. 8.66 below, we can see that parameters such as the

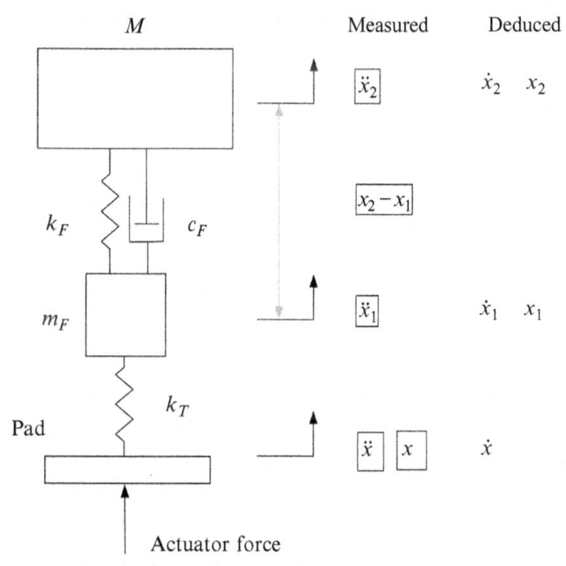

Fig. 8.66 Some system parameters are measured; from these, the others are worked out.

body acceleration, damper displacement and upright acceleration are measured.

From a knowledge of these parameters, the remaining ones can be deduced. For example, consider the unsprung mass. By finding a point in the data set when the body is at rest, we can work out the unsprung mass because all the force at the pad must be expended in either accelerating the upright, stroking the damper or displacing the upright against the suspension and tyre spring. It is by no means easy to do such experimental work. Practical issues associated with noise make methods that work easily and quickly in simulation impossible to implement in practice. However, it is indeed possible to do although organisations that have successfully achieved this tend not to publish their methods and instead keep it as propriety knowledge.

8.7.3 Validation of Modelling

The second main usage of a four-post rig is the validation of suspension models of the kind developed above against real data from the car in question. For example, imagine a 7DoF model of a car in a specific race series has been developed. If the real car is tested on the four-post test rig and data taken are shown to agree well with data produced in simulation, then it is a reasonable step to use the model in simulation to optimise the parameters for an ideal suspension. Once determined this set of parameters can be confirmed on the rig again before being used on the track. When one considers how many parameters there are to adjust and the range of values they can take, it is clear that simply trying these alternatives out on the track is impractical, not to say prohibitively expensive. By using a computer, however, literally millions of parameter sets can be searched to find an ideal set.

8.7.4 Overview of Four-Post Rig Testing

A complete analysis of a car using all the functionality available from a four-post rig is a much more lengthy affair. It involves running a series of frequency sweeps at different amplitudes for each of the four input modes of heave, pitch, roll and warp. These tests can be completed within a few hours but the subsequent analysis is a very lengthy undertaking. At Oxford Brookes University, the rig makes use of analysis software running to tens of thousands of lines of code, all written 'in-house' since it can't be bought, and, once prepared, its use and interpretation is a skilled and complex process. For these reasons, once an organisation has reached a point where the investment in a rig is economically justified, they tend to develop in house proprietary skills and experience not for sale outside. If an organisation or team wish to have the benefits of a test but do not have the resources for their own rig, they can make use of contract test companies or universities such as Oxford Brookes that have the equipment and know-how.

8.8 DAMPERS AND INERTERS

In the treatment of suspension dynamics above, we have dealt with dampers as a dynamic element, a device that produces a force proportional to the velocity at which the two terminals of the device move with respect to each other. The values of the force developed are given by a coefficient multiplied by the velocity:

$$F_D = C(\dot{x}_1 - \dot{x}_2)$$

In practice, such a force is normally produced by moving a fluid through an orifice within a telescopic, oil-filled cylinder. This general approach is shown Fig. 8.67. The oil in the oil-filled chamber is forced to flow through the orifices, and the flow is restricted by the shims. Adjustment of the shims offers control over the force function. A second piston with a gas spring is necessary since as the rod enters the chamber the volume available for the oil is reduced because of the pushrod, and accommodation must be made for it. This basic approach can be developed to produce a vast number of different design realisations of a damper unit. Commercially available units have low- and high-speed adjustability separately in bump and rebound. There are units available with separate gas reservoirs and through rod damping. Often, the unit is pressurised to increase the stroking speed at which cavitation starts to develop across the orifice. The way in which the valves operate to control the flow of oil is a complex art and restricted proprietary knowledge for each manufacturer. In our analysis, we have assumed that the damping force is a

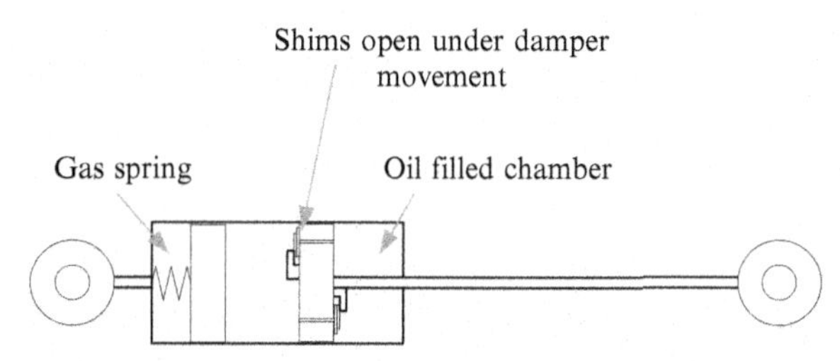

Fig. 8.67 Damper internal layout.

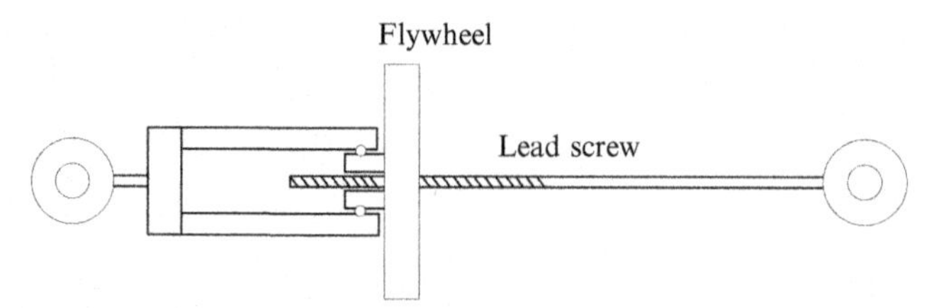

Fig. 8.68 Schematic layout of an inerter.

function of velocity, but it does not have to be. One could produce a device that has any function at all, perhaps quadratic with velocity and with a small term of acceleration dependence. An inerter, for example, produces a force dependent on the mutual acceleration between the two terminals. A schematic showing how this is usually done is shown in Fig. 8.68. Once one considers this possibility, suspension optimisation enters a whole new infinity of possibilities making their optimisation a major work.

In Fig. 8.68, we see the internal layout of an inerter. The lead screw forces the flywheel to rotate when a mutual movement between the terminals occurs. Elementary consideration shows that when at rest or constant velocity, the force developed will be zero but when a mutual acceleration occurs that will indeed be a force proportional to the acceleration

$$F_{INERT} = J(\ddot{x}_1 - \ddot{x}_2)$$

where J is a quantity known as the *inertance*, with units of kg, and represents the equivalent mass that the inerter *appears* to be. The lead screw may be sized to provide a force much greater than would result from accelerating an object of the same weight as the inerter and an inerter weighting a few kilograms can have an inertance value of 80 kg.

To analyse the effect of an inerter on a suspension, one replaces the force developed by the damper with one having a force modelled by the equation above. Inerters are often thought to be a relatively recent invention, but an early appearance was in an American patent No. 893,680, lodged in 1908.

In order to characterise the performance of a damper, vehicle dynamicists make use of a piece of test equipment called a damper dynamometer. Such an instrument, belonging to Oxford Brookes University, is shown in Fig. 8.69. The device has a motor driving an eccentric cam to stroke the damper over a range of speeds and a load cell to produce a force output.

The unit produces graphs that characterise the performance of the damper under test of the kind shown in Fig. 8.70.

Fig. 8.69 Damper dynamometer Oxford Brookes University.

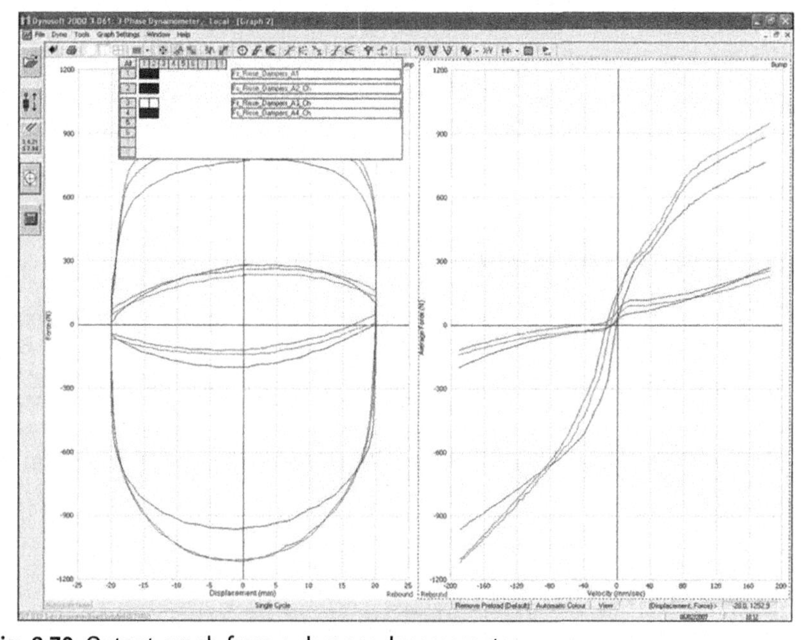

Fig. 8.70 Output graph from a damper dynamometer.

8.9 QUESTIONS

1. A dynamic system has a mass of 100 kg, is supported on a spring of stiffness 100 N/mm and has a damper with a damping coefficient of 3000 N s/m. What is the damped and undamped natural frequency?
2. Prove that the units of damping are N s/m.
3. If the system in (1) is provided with a step input of 10 mm, produce an expression for the displacement as a function of time and determine how long it will take to settle to within 2% of its final value.
4. A car is be approximated to a single-degree-of-freedom system. The car has a mass of 750 kg, a spring of 1.1E+5 N/m and a damper with a coefficient of 4.5E+3 N s/m. The car encounters a step of 5 cm. Determine the response as a function of time. Numerically or analytically, differentiate the function twice to obtain an expression for acceleration and determine the largest value taken during the event.
5. Adjust the damping in (4) above to produce a zeta value of 1.5 and repeat the analysis.
6. A vehicle with the same parameters as (4) is subjected to a sinusoidal input of 4 cm of wavelength 20 m whilst travelling at 10 m/s. Determine the response amplitude. What can be done to reduce it?
7. A vehicle has a body mass of 970 kg, a suspension spring stiffness of 120 N/mm, a total unsprung mass of 130 kg and a tyre stiffness of 290 N/m. Estimate the natural frequency when the damping is high and the frequencies of the hub and body mode when the damping is low.

8.10 DIRECTED READING

Milliken and Milliken [9] in their text deal with single- and two-degree-of-freedom modelling.

Gillespie [11] again provides a good coverage and deals effectively with ride.

Blundel, M., Harty D. [17] 'The multibody systems approach to vehicle dynamics'.

8.11 LEARNING PROJECTS

- For a vehicle you have access to, estimate the undamped and damped natural frequencies. You should estimate the spring rate by sitting on one corner and measuring the deflection. The corner mass should be determined by estimating the centre of gravity location and balancing moments. The damping ratio can be estimated by suddenly applying and removing yourself from the corner and estimating the displacement function (accuracy could be significantly improved here by filming the event). The damping ratio can then be determined by comparing it with the curves in Fig. 8.7. You should do the same for a selection of other vehicles and form a database of parameter values.

- Produce your own input function based on the curves in Fig. 8.20 using a spreadsheet.
- Produce your own set of SDoF, 2DoF and 4DoF models as shown above and analyse them with step and harmonic inputs. Use the graphs to validate your models.
- Research ISO standards for pothole dimensions and road surface profiles characterisation.
- Produce a 2DoF suspension model featuring an inerter and examine the parameter space. Can an inerter be used to offer improvements over an optimised conventional damper?

8.12 INTERNET RESEARCH AND SEARCH SUGGESTIONS

- Authors have no common consensus on the reasoning for having different damping values in bump compared with rebound. Undertake an internet investigation of the topic and write up a literature review bringing in differing views and the reasoning used.
- Use the internet to find organisations offering vehicle dynamics consultancy and testing. What kind of four-post rig analysis do they provide? Horiba MIRA is a good starting place.
- Using Google Scholar, perform a literature review of papers offering suspension models with more than 7DoF.

CHAPTER 9

Lap-Time, Manoeuvre and Full-Vehicle Simulation

In designing a high-performance car, there are clearly some design aims that stem from common sense. We need an engine with plenty of power, tyres with good grip and a low overall weight. If we have a knowledge of the competition, we might be able to list some targets that will give our car an advantage. This is all well and good, but there is a distinct limit to how much one can improve a car this way. We will never know all the relevant details about the opposition, and if we simply try to emulate them, we shall always be behind either commercially or on the track. In addition, there is always the problem of resources. There is only ever a finite length of time before the next deadline, whether it's a race or a launch and always a limit on expenditure too.

If we are to be rational about designing a car, we need to put numbers on these quantities. When we say we would like to improve the design of a racing car, what we really mean is that we want to know which activities will produce the greatest improvements in performance per unit effort in making them happen. We need to know, for example, whether spending half our time and budget on improving the engine will yield as much improvement as spending it on improving the chassis dynamics. In a racing situation, it is lap-time simulation that is used to answer these questions, and in this chapter, we will develop an understanding of how it is done. However, for performance road cars, we still need the same ability to predict the whole vehicle performance, and full-vehicle simulation is the process by which this is done. In a lap-time simulator, one enters all the relevant vehicle dynamic data including the track, and the package will determine the lap time. Changes can then be made and their effect determined. In full-vehicle simulation, again, all the vehicle dynamic data are entered, and the package is then used to simulate manoeuvres. The performance of the vehicle in terms of ride quality and comfort can be determined as well as dynamic performance. The effect of changes in the design can be studied and improvements made. In both cases, validation is an important step, meaning comparison of

Performance Vehicle Dynamics
http://dx.doi.org/10.1016/B978-0-12-812693-6.00009-2

the simulation with test data to confirm that the simulation is acceptably accurate before using it for optimisation.

9.1 SIMPLE J-TURN SIMULATION

A 'J-turn' is a very simple 'track' consisting of a straight section followed by a constant radius corner. It can be thought of as a kind of building block; once a J-turn can be simulated, several can be joined together to form a general track. We can simulate the sector time for a J-turn relatively, and here, we shall assume that

- the engine has all the power we need and so forward acceleration is limited only by friction on the two driven wheels,
- the brakes provide enough torque to lock up the wheels and so deceleration is limited only by friction on the four wheels,
- we will ignore all the chassis dynamics such as weight transfer,
- aerodynamics is irrelevant.

We start by looking at the track and how the driver will drive the car around it.

In Fig. 9.1 above, we can see the J-turn has a constant 20 m radius at the end preceded by a 50 m straight. The corner clearly has a maximum speed; if this is exceeded, the tyres will not be able provide the lateral acceleration necessary to keep the car on a 20 m radius corner, and the car will spin out. Our first step is therefore to determine the maximum cornering speed. We then consider the straight. When the car leaves the start line, we know it to have zero velocity, and its maximum longitudinal acceleration is simply $0.5 \times \mu$; assuming only half the wheels have engine torque applied, we can

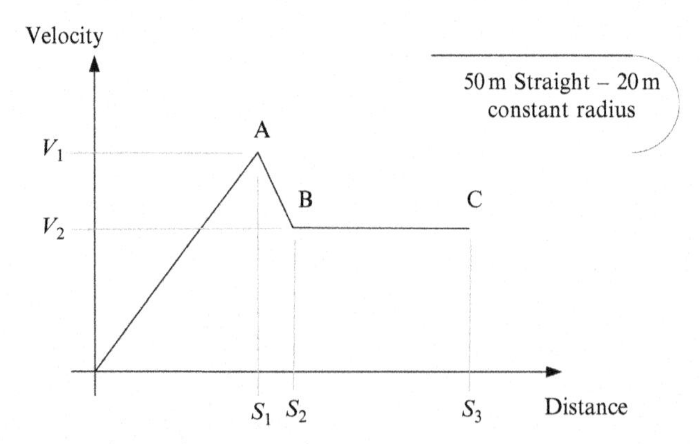

Fig. 9.1 A J-turn velocity diagram.

produce the line from the origin to point 'A'. We also know the maximum deceleration, $1.0 \times \mu$, and therefore the gradient of the deceleration line from S_1 to S_2. Since we know the speed with which the car must enter the corner, we can draw in the two lines, and the point in time at which they cross is the point at which the driver must go from wide open throttle to full braking, normally called the 'braking point'.

First determine V_2:

$$V_2 = \sqrt{(L\hat{at}_{acc} R)}$$

Since the maximum lateral acceleration is given by μg and the maximum braking acceleration is also given by μg, we can write

$$V_2 = \sqrt{(\mu g R)}$$

The numerical values for points on the diagram can then easily be determined using equations of uniform acceleration. This model is very simple; the only parameter to input is the friction coefficient of the tyres, since we haven't specified the weight, the tyre size, the engine power, etc.

9.2 EXTENSION OF THE J-TURN SIMULATION TO A GENERAL TRACK

The above analysis can be extended to deal with a general track by considering the track as a collection of J-turns.

Fig. 9.2 above shows that any circuit can be thought of as being composed of several J-turns one after another. In this way, the analysis above can be used for a general circuit.

9.3 CORNERS OF VARYING RADIUS

If more accuracy is required, one can subdivide each corner into a series of J-turns allowing a simulation of corners with varying radii. Fig. 9.3 shows this.

The approach taken here is much extended in a paper by Dominy and Dominy [18].

9.4 FULL VEHICLE SIMULATION IN ADAMS CAR

At various points in this textbook, reference has been made to the Adams multibody code. So far, this has always been to the code known as Adams View. In fact, Adams is a suite of packages designed to complete a range of

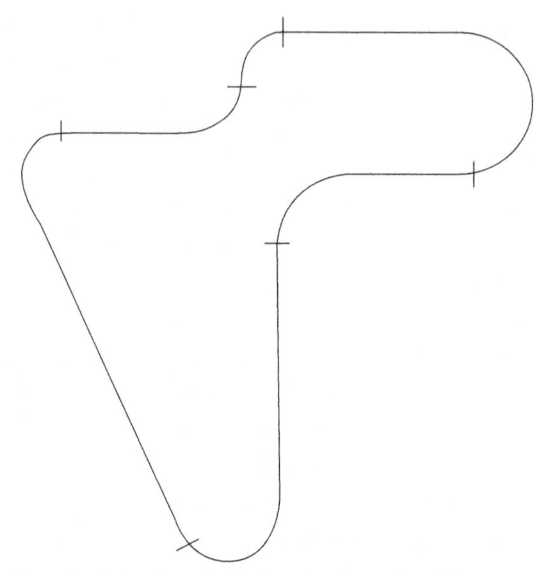

Fig. 9.2 Any racing circuit can be divided up into a series of J-turns.

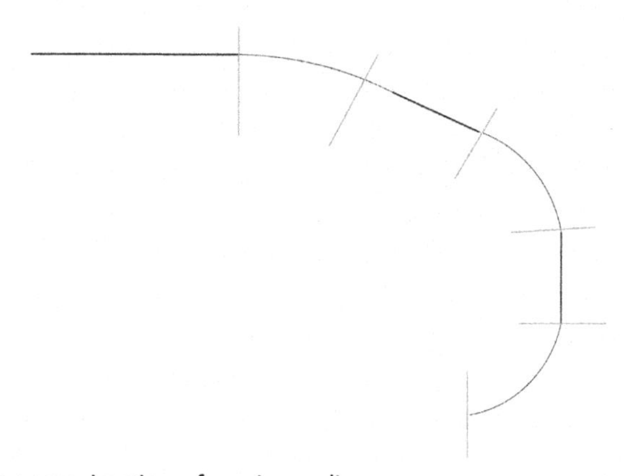

Fig. 9.3 J-turn approximation of varying radius corner.

complex dynamic simulations. The internet-based research will guide you to web-based material about the package and its supplier.

One of the members of the suite is Adams Car and, as it name implies, it allows the user to simulate vehicle performance. Learning such a package and becoming proficient in its use is a considerable undertaking and not within the scope of this text. However, packages like these offer tremendous potential for the simulation of vehicle performance. One can never replace all the knowledge we gained in Chapters 1–8 because one needs to have that level of understanding to know why any given vehicle behaves as it does. The limit on turn in response is a good example; you can't exceed the theoretical limit. However, armed with all this understanding, the availability of packages such as Adams Car makes for rich simulation possibilities and no doubt goes a long way towards explaining why current levels of refinement on road cars are so good and why racing cars are so fast.

Adams Car is divided up into two modes of use, the 'template builder' and the 'template user'. As a template user, you can call up different subassemblies from the library such as the one in Fig. 9.4 and link them together to build an entire vehicle.

Fig. 9.5 shows a full-vehicle assembly in Adams Car. It is made by linking a collection of subassemblies together. Once 'assembled', the vehicle can be put through a whole range of manoeuvres and events. It can be made to perform a step steer, for example, or a lane change or follow a prescribed path. A user enters the details for the manoeuvre and the package then determines the vehicle response, and this can be viewed as a movie or studied in detail with graphs of any parameter one might care to track. If you become a template builder, then it is possible to produce new vehicles of whatever design you wish to analyse. Fig. 9.1 shows a vehicle on a rough road and is a very advanced simulation.

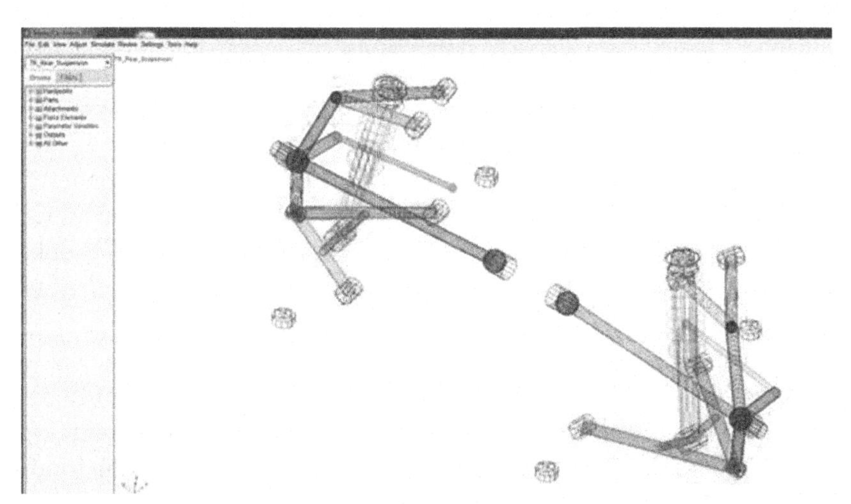

Fig. 9.4 A front wishbone assembly in Adams Car. *Adams, Adams Car, Adams View and other MSC product names are trademarks or registered trademarks of MSC Software Corporation and/or its subsidiaries in the United States and/or other countries. Image provided courtesy of MSC Software. 2017 MSC Software Corporation.*

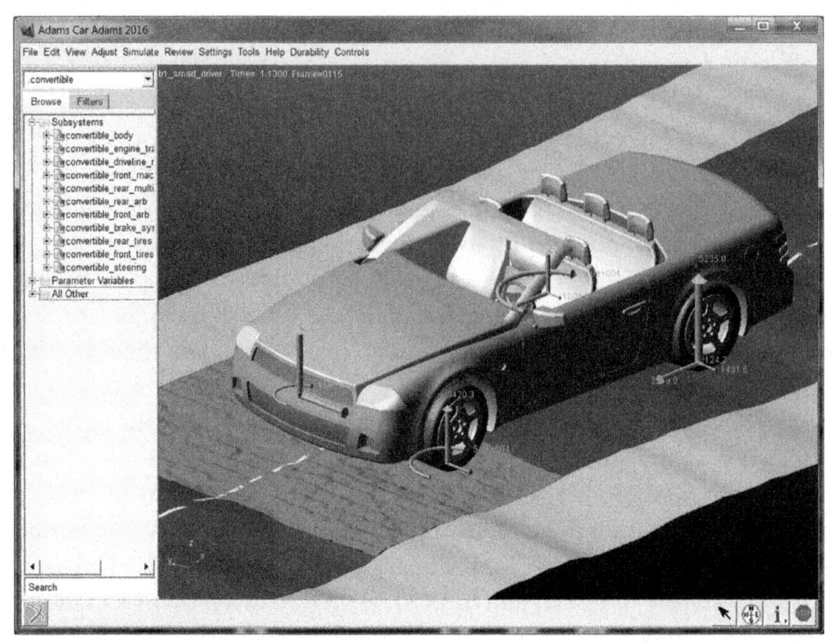

Fig. 9.5 A full-vehicle assembly in Adams Car. *Adams, Adams Car, Adams View and other MSC product names are trademarks or registered trademarks of MSC Software Corporation and/or its subsidiaries in the United States and/or other countries. Image provided courtesy of MSC Software. 2017 MSC Software Corporation.*

In addition to manoeuvre simulations, Adams Car can perform other design optimisations, for example, it has a 4-post rig emulator and suspension kinematics analysis built-in. There is also a very sophisticated optimisation functionality that allows multiple variables to be solved at once.

In Fig. 9.6, we see an off-road vehicle whose development was assisted by two of my MSc students as part of their dissertations at Oxford Brookes University. The vehicle is extremely rugged, but the development of refined handling is still important.

Fig. 9.7 shows a simulation developed in Adams Car, and in Fig. 9.8, we see one of the parameters of interest in a graph showing experimental data and computer simulation on the same axes. The preparation of computer-simulated data that agrees well with real-life telemetry from the car in hand is the final step in the process of being a successful road car vehicle dynamicist. Once simulation of this quality is produced, it can be used to make improvements to the car by changing its design and finding a parameter set that gives the desired performance. Once this is done, it is a reasonable expectation that

Fig. 9.6 Six-wheel off-road vehicle manufactured by Arctic Trucks. *Reproduced by kind permission of Nolan McCann and Arctic Trucks.*

Fig. 9.7 Simulation in Adams Car of the truck in Fig. 9.8. *Reproduced by kind permission of Nolan McCann and Arctic Trucks.*

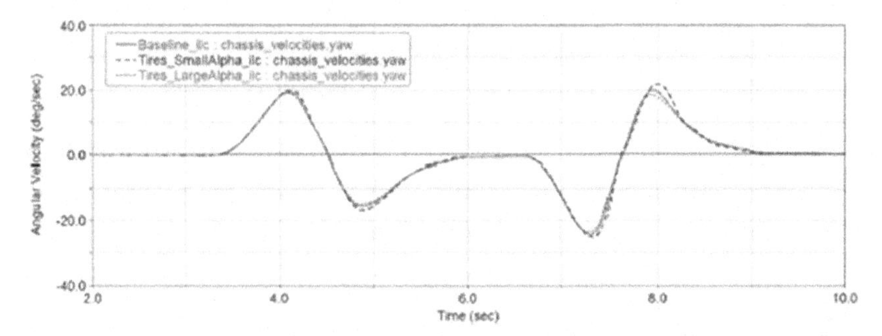

Fig. 9.8 Correlation between test and simulation for truck yaw velocity. *Reproduced by kind permission of Nolan McCann and Arctic Trucks.*

when the changes are also made on the real car, it too will improve as the simulation did. This process is applied equally to racing as we see in the next section. The two huge advantages offered are that firstly a vast number of parameter setups can be examined, far more than practical by testing alone, and secondly it is much cheaper than extensive testing. With simulations of this quality made possible, testing should really be a process of confirmation rather than experiment.

9.5 LAP-TIME SIMULATION IN CHASSISSIM

In racing, a very important use of vehicle dynamics is to produce lap-time simulations. For example, consider a circuit consisting of an oval with 2 mi straights connecting just two corners, each of 180°. Clearly, here, the terminal velocity will dominate, and the winner will simply be the car with the highest top speed. However, if the circuit were changed to include far more corners so that the cars are never at top speed, things would be very different. Here, a team would do better to swap to a lighter engine with less power and produce a chassis that corners better. The only way to answer the all-important question of what combination of dynamics parameters will make for the quickest lap of the track in hand is to undertake lap-time simulation. We saw above an introduction to lap-time simulation using a J-turn. Simple lap-time simulators follow a very similar approach and use static equilibria at each point on the track to estimate the speed over a very small local sector. Coding is required to determine whether the throttle should be fully open or the brakes fully on, depending on the distance to the next corner entry and current speed, nothing in between these two settings is used. More advanced lap-time simulators determine the transient conditions too. For example, the transient roll of the chassis can give an overswing to the roll displacement and then a consequent change in lateral force at each tyre in response to the vertical load on it. For this reason, such simulators are called 'transient' lap-time simulators.

ChassisSim is an example of a transient lap-time simulator. The internet-based research below will guide you to material about the package and its supplier. In essence, the package is simple, one enters all the vehicle dynamics parameters from very major parameters such as vehicle weight, cornering stiffness, polar moment, engine power and torque curves right through to effects such as camber gain, suspension geometry and aerodynamic map. The package comes with a library of parameters that can be used and modified, in a similar way to Adams Car above.

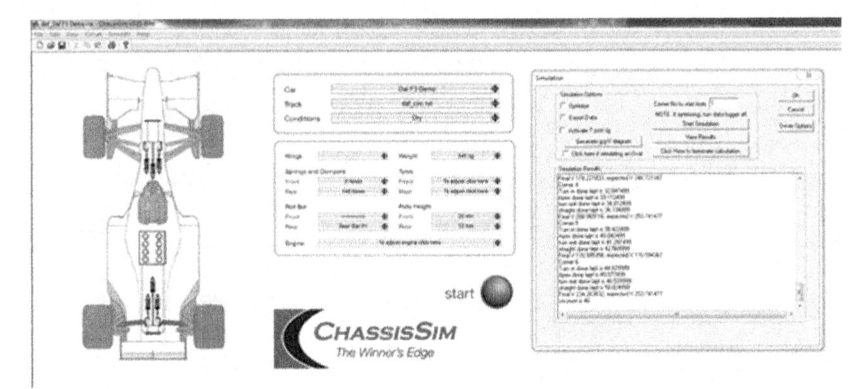

Fig. 9.9 The front end GUI for the lap-time simulator ChassisSim. *Reproduced by kind permission of Danny Nowlan—ChassisSim.*

The graphical user interface shown in Fig. 9.9 allows the user to call up existing library circuits, cars, etc. and edit them to suit the case in hand. After this, the control panel on the right is used to run simulations from which the results can be studied. Fig. 9.10 shows the track map for a sample simulation.

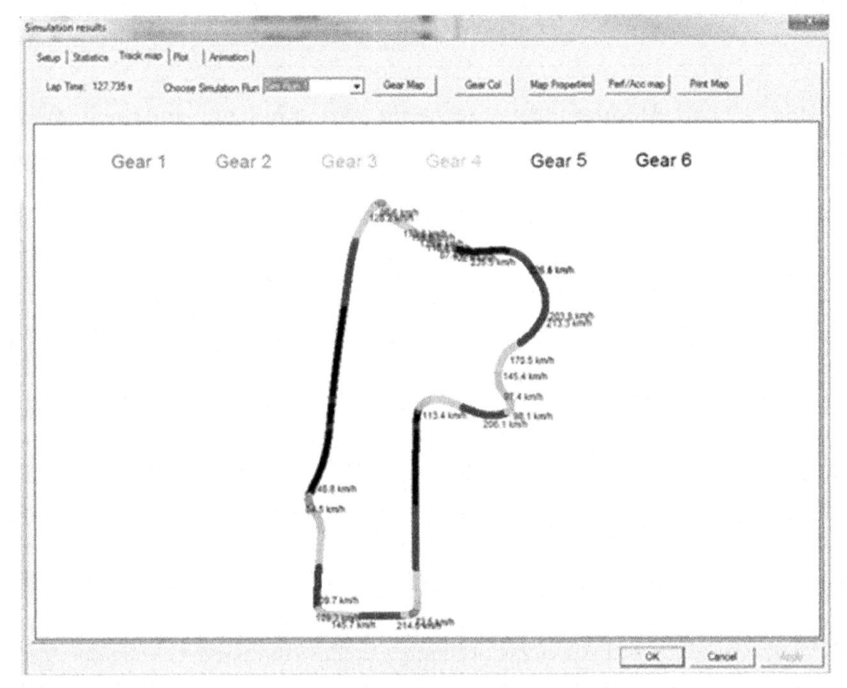

Fig. 9.10 Track map—ChassisSim. *Reproduced by kind permission of Danny Nowlan—ChassisSim.*

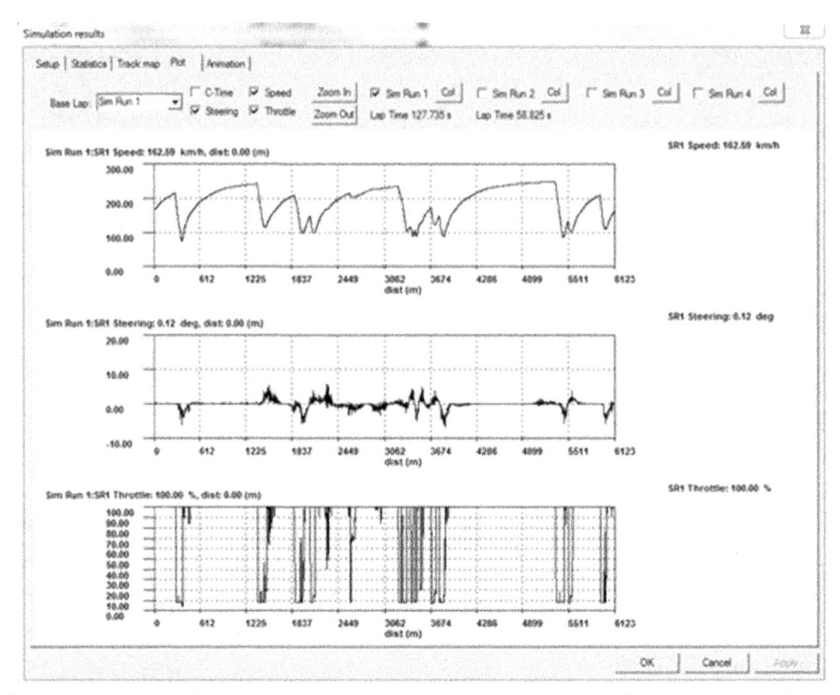

Fig. 9.11 Plotted data for the simulation. *Reproduced by kind permission of Danny Nowlan—ChassisSim.*

Fig. 9.11 shows output data for the simulation in hand, and the distance along the bottom axis refers to the distance from the start of the above circuit. In this case, speed, steering input and throttle are shown, but all dynamics parameters of interest can be plotted.

In the top line, we see the speed trace for the racing car, and it is clear that the car is never in equilibrium, accelerating all the time, and that it spends a minority of the time at terminal velocity. Using tools such as ChassisSim, vehicle dynamicists are able to determine exactly what set of vehicle parameters will result in the shortest possible lap times, something that is not otherwise easily answered. Issues such as damper settings bring advantages in one situation but not in another; exactly how all these play out for a particular circuit is not easy to anticipate. The existence of transient lap-time simulators such as ChassisSim makes for excellent design opportunities in racing and goes some way to explaining why racing packs are generally so close in performance.

The final step in the process of being a successful racing vehicle dynamicist is to overlap computer-simulated data with real-life telemetry from the car in hand, just as above from road car simulation. This is shown using

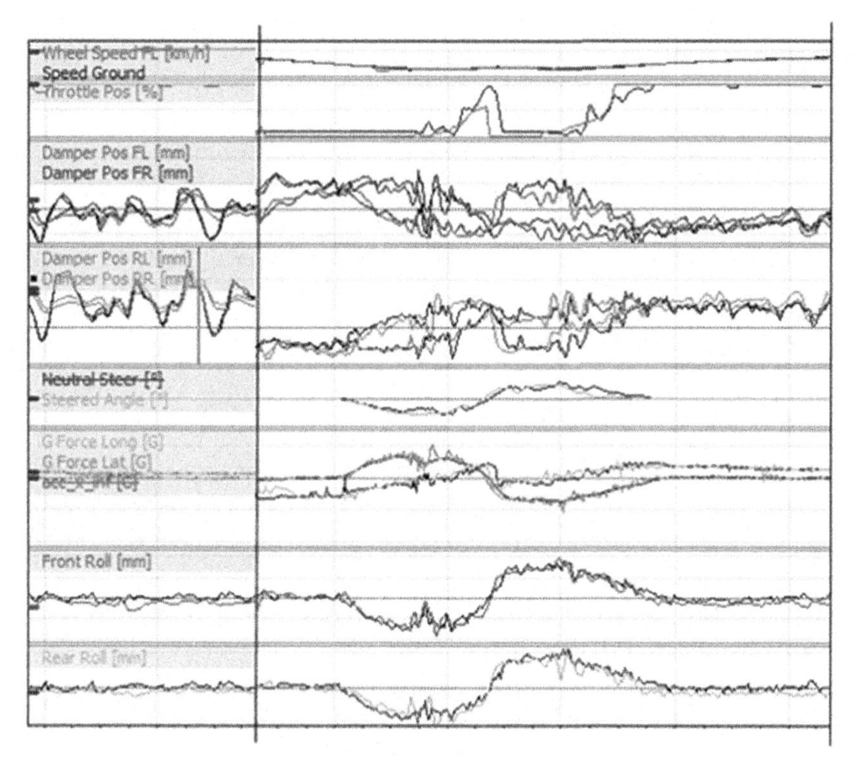

Fig. 9.12 Comparison of lap-time simulation with real data.

ChassisSim in Fig. 9.12. In the figure, we see graphs, including throttle position, the steering wheel position for a neutral car overlaid with the actual position and roll angle. Once a good level of agreement has been developed between measured data and simulated data, then one can make adjustments in the model, find changes that bring improvements and then reasonably expect that when these changes are made on the real car, the same improvements will result.

Thus, lap-time simulation can be used to determine the overall best package for a given circuit. This certainly makes it seem as if, in vehicle dynamics, only lap-time simulation is needed, but of course, this is not the case. Lap-time simulators determine *what* the lap time will be, whereas vehicle dynamics determines *why*.

And finally, to quote from the introduction at the start of this book,

> *'The issue is that until you put numbers on things you're wasting your time, you're just playing about'*

We have, at last, reached the point.

9.6 INTERNET-BASED RESEARCH AND SEARCH SUGGESTIONS

- Visit or search on mscsoftware.com and select the material about Adams Car. Research the case studies and sales support material.
- Visit YouTube and search on Adams Car.
- Conduct a review of major packages offering full-vehicle simulation and understand the differences between what's on offer.
- Search on ChassisSim and investigate all that is on offer within the package and from its vendor.

9.7 DIRECTED READING

In addition to the text listed at the end of each chapter above, a number of other texts are of general interest in the field. These are texts that provide good coverage in some area and should be used to supplement the directed reading for each chapter:

[19] Bastow, D. and Howard, G., 'Car Suspension and Handling'
[20] Beikmann R. 'Physics for Gearheads'
[21] Dukkipati R. V. 'Vehicle Dynamics'
[22] McBeath. 'Competition car Data logging'
[23] Reimpell J., Stoll H. Betzler J. W. 'The Automotive Chassis'
[24] Smith. C. 'Tune to Win'
[25] Wong J. Y. 'Theory of Ground Vehicles'

REFERENCES

[1] F.W. Lanchester, Some problems peculiar to the designs of the automobile, Proc. Inst. Automot. Eng. 2 (1908) 187–192.
[2] R. Nader, Unsafe at Any Speed, 25 reprint ed., Knightsbridge Pub Co., MA, USA, 1991. ISBN: 13: 978-1561290505.
[3] P. Haney, The Racing and High-Performance Tire, Society of Automotive Engineers Inc., Warrendale, PA, 2003. ISBN: 0-9646414-2-9
[4] W.F. Milliken, D.L. Milliken, Chassis Design Principles and Analysis Based on Previously Unpublished Work by Maurice Olley, Society of Automotive Engineers, 2002. ISBN: 1-86058-389-X.
[5] L. Segel, Theoretical prediction and experimental substantiation of the response of the automobile to steering control, Proc. Inst. Mech. Eng. Automob. Div. 10.1 (1956) 310–330.
[6] I.J. Newton, Philosphiae Naturalis Principia Mathmatica, Trin. Coll. Cantab. Soc. Mathefeos Profoffore Lucafiano & Societatis Regalis Sodali, Imprimatur – S. Pepys Reg. Soc. Praeses Julii 5, London Anno MDCLXXXVII, 1686.
[7] H.B. Pacejka, Tyre and Vehicle Dynamics, Elsevier, 2002. ISBN: 13: 980-0-7506-6918-4.
[8] J. Dixon, Tires Suspension and Handling, Society of Automotive Engineer, 1996. ISBN: 0-340-67796-1.
[9] W.F. Milliken, D.L. Milliken, Race Car Vehicle Dynamic, Society of Autmotive Engineers, USA, 1995. ISBN: 1-56091-526-9.
[10] R.N. Jazar, Vehicle Dynamics Theory and Applications, Springer Science-Business Media, New York, 2009. ISBN: 978-0-387-74243-4.
[11] T.D. Gillespie, Fundamentals of Vehicle Dynamics, Society of Automotive Engineers, 1992. ISBN: 1-56091-199-9.
[12] G. Genta, Motor Vehicle Dynamics Modeling and Simulation, World Scientific Publishing Co. Pte. Ltd., Singapore/NJ/London, 1997. ISBN: 9810229119
[13] W.E. Boyce, R.C. DiPrima, Elementary Differential Equations and Boundary Value Problems, Wiley, 2010. ISBN: 978-1-118-32361-8.
[14] V. Cossalter, Motorcycle Dynamics, ISBN: 978-1-4303-0861-4, 2016. Amazon.co.uk.
[15] J. Dixon, The roll-centre concept in vehicle handling dynamics, Proc. Inst. Mech. Eng. 201 (1) (1987).
[16] A. Staniforth, Competition Car Suspension – Design Construction, Tunin, second ed., Haynes Publishing, 1994. ISBN: 1 85960 644 X.
[17] M. Blundell, D. Harty, The Multibody Systems Approach to Vehicle Dynamics, Elsevier Buttworth-Heinemann, 2004. ISBN: 0-7506-51121.
[18] J.A. Dominy, R.G. Dominy, Aerodynamic influences on the performance of the Grand Prix racing car, Proc. Inst. Mech. Eng. Automob. Div. 198 (12) (1984).
[19] D. Bastow, G. Howard, Car Suspension and Handling, Penchworth Press and SAE, 1993. ISBN: 0-7273-0318-x.
[20] R. Beikmann, Physics for Gearheads, Benty Publishers, MA, USA, 2015. ISBN: 978-0-8376-1615-5.
[21] R.V. Dukkipati, Vehicle Dynamics, Alpha Science International Ltd., 2000. ISBN: 1-84265-014-9
[22] S. McBeath, Competition Car Data Logging, Haynes Publishing, 2002. ISBN: 1-85960-653-9.

[23] J. Reimpell, H. Stoll, J.W. Betzler, The Automotive Chassis, Reed Educational Publishing, 2001. ISBN: 0-7506-5054-0.
[24] C. Smith, Tune to Win, Areo Publishers, 1978.
[25] J.Y. Wong, Theory of Ground Vehicles, John Wiley & Sons, 2001. ISBN: 0-471-35461-9.
[26] S. Rao., Mechanical Vibrations, Addison-Wesley Publishing Company, 1995. ISBN: 0-201-59289-4.

INDEX

Note: Page numbers followed by *f* indicate figures, and *t* indicate tables.

D

M

N

O

P

Q

R

S

Printed in the United States
By Bookmasters